MW00837550

Transceiver and System Design for Digital Communications

Transceiver and System Design for Digital Communications

4th Edition

Scott R. Bullock

SciTECH
PUBLISHING
an imprint of the IET

Edison, NJ
theiet.org

Published by SciTech Publishing, an imprint of the IET.
www.scitechpub.com
www.theiet.org

Copyright © 2009, 2014 by SciTech Publishing, Edison, NJ. All rights reserved.

Third edition 2009
Fourth edition 2014

No part of this publication may be reproduced, stored in a retrieval system or transmitted in any form or by any means, electronic, mechanical, photocopying, recording, scanning or otherwise, except as permitted under Sections 107 or 108 of the 1976 United States Copyright Act, without either the prior written permission of the Publisher, or authorization through payment of the appropriate per-copy fee to the Copyright Clearance Center, 222 Rosewood Drive, Danvers, MA 01923, (978) 750-8400, fax (978) 646-8600, or on the web at copyright.com. Requests to the Publisher for permission should be addressed to The Institution of Engineering and Technology, Michael Faraday House, Six Hills Way, Stevenage, Herts, SG1 2AY, United Kingdom.

While the author and publisher believe that the information and guidance given in this work are correct, all parties must rely upon their own skill and judgement when making use of them. Neither the author nor publisher assumes any liability to anyone for any loss or damage caused by any error or omission in the work, whether such an error or omission is the result of negligence or any other cause. Any and all such liability is disclaimed.

10 9 8 7 6 5 4 3 2 1

ISBN 978-1-61353-203-4 (hardback)
ISBN 978-1-61353-204-1 (PDF)

Typeset in India by MPS Limited
Printed in the USA by Lightning Source
Printed in the UK by CPI Group (UK) Ltd, Croydon, CR0 4YY

To my loving wife, Debi;
to Crystal, Cindy, Brian, Andy, and Jenny;
and to my mother, Elaine.

Contents

Preface

This book covers basic communications theory and practical implementation of transmitters and receivers. In so doing, the focus of this book is on digital modulation, demodulation methods, probabilities, detection of digital signals, spread spectrum system design and analysis, and cognitive system processes. This book was written for those who want a good understanding of the basic principles of digital wireless communication systems, including spread spectrum techniques. It also provides an intuitive and practical approach to digital communications. Therefore, it is a valuable resource for anyone involved in wireless communications and transceiver design for digital communications. The reader will gain a broad understanding of basic communication principles for transceiver design, digital communications, spread spectrum, and cognition, along with examples of many types of commercial and military data link systems.

Chapter 1 describes transceiver design using a link budget to analyze possible trade-offs. This includes tracking of signal power and noise levels in the system, calculation of the noise figure, the gains and losses of the link, and the required signal level with respect to noise, including the link margin to provide the specified probability of error. This chapter also discusses frequency band designations along with the definitions and uses of decibels. Spread spectrum techniques and process gain, as well as coding gain, are discussed. The chapter concludes with an example of a link budget and a transceiver design that coincides with it.

Chapter 2 evaluates the basic functions of the transmitter, including antennas, transmit/receive (T/R) control, classes of power amplifiers, the upconversion process, sum and difference frequencies and the requirement to eliminate one of the conversion products, voltage standing wave ratio, and maximum power transfer principle. This chapter discusses advantages of digital versus analog communications, digital modulation techniques including phase-shift keying (PSK) and frequency-shift keying, phasor constellations and noise immunity, and error vector magnitude as a quality metric of the transmission. This chapter also addresses the advantages of continuous PSK modulation and spectral regrowth. Shaping filters for the digital waveforms are discussed, with ideal and practical solutions. Direct sequence spread spectrum systems are addressed, along with the advantages of using spread spectrum, including process gain and antijam with jamming margin. Maximal length sequence codes, including Gold codes and others, are included, along with spectral lines that are generated in the frequency domain. Several digital modulation techniques using PSK are provided, along with block diagrams and phasor diagrams to help analyze the different types of PSK systems that are used today. Variations of PSK systems and other types of spread spectrum systems are also discussed, such as frequency hopping, time hopping, and chirped frequency modulation. In addition, multiuser techniques are explained, including time, code, and frequency access systems. Finally, orthogonal techniques including orthogonal frequency division multiplexing are considered, along with power control to reduce near–far problems.

Chapter 3 covers the basic functions of the receiver, including the antenna, T/R control, image-reject filter, low-noise amplifier, downconversion process, third-order intercept, and various methods to determine dynamic range. Phase noise, mixers, spur analysis, bandwidths, and filters are also addressed. The discussion of digital processing includes principles such as group delay, sampling theorem and aliasing, anti-aliasing filters, and analog-to-digital converters (ADCs) including piecewise linear ADCs.

Chapter 4 discusses the design and analysis of automatic gain control (AGC). The main elements of a good AGC design are provided, including the amplifier curve, linearizer, detector, loop filter, threshold level, and integrator for zero steady-state error for a step response. Control theory is used to define the stability characteristics of the AGC to design the optimal AGC for the system. Chapter 4 also details the phase-locked loop (PLL), particularly for the lock condition, and compares the similarities between the AGC and the PLL. Feedback systems and oscillations, including the Barkhausen criteria, are reviewed, along with Bode diagrams to determine both gain and phase margin.

Chapter 5 describes the demodulation process portion of the receiver, which includes the different methods of correlating the incoming digital waveform, such as the matched filter, coherent sliding correlator, pulse position modulation and demodulation, code tracking loops like early-late gate analysis, and the autocorrelation function. The advantages and disadvantages of coherent versus differential demodulation techniques are discussed. Chapter 5 also contains different types of carrier recovery loops, including the squaring loop, Costas loop, and modifications of the Costas loop. The chapter concludes with a conversation about the symbol synchronizer, the eye pattern, intersymbol interference, scrambler–descrambler methods, phase-shift detection for intercept receivers, and Shannon's limit for information channel capacity.

Chapter 6 contains a basic discussion of the principles of digital communications. This includes an intuitive and analytical approach to understanding probability theory, which is used in designing and analyzing digital communications. The explanation shows the basic Gaussian distribution and how to apply it to probability of error. Quantization and sampling errors, as well as the probability of error for different types of spread spectrum systems, along with the curves and how to apply them in a design, are evaluated for the system. Also examined is the probability of false alarm and probability of detection and methods for detecting errors, including parity, checksum, and cyclic redundancy check. Error correction using forward error correction is assessed, including topics such as interleaving, block codes, convolutional codes, linear block codes, and hamming codes. The Viterbi algorithm, multi-h, low-density parity check codes, and turbo codes are also discussed. Chapter 6 also provides basic theory on pulsed systems, which includes spectral plots of the different pulse types.

Chapter 7 focuses on multipath. This chapter discusses the basic types of multipath, including specular reflection of both smooth and rough surfaces and diffuse reflections of a glistening surface as well as the Rayleigh criteria for determining if the reflections are specular or diffuse. The curvature of the earth is included for systems such as those used for satellite communications. The advantages of using leading edge tracking for radars to mitigate most of the multipath present are discussed. Several approaches to the analysis of multipath are provided, including vector analysis and power summation. Included are discussions on several different multipath mitigation techniques and information on antenna diversity.

Chapter 8 describes methods that improve the system operation against jamming signals. Discussed are burst clamps to minimize the effects of burst jammers, adaptive filtering to reject narrowband signals, and a Gram-Schmidt orthogonalizer that uses two antennas to suppress the jamming signal. An in-depth analysis is provided on the adaptive filter method using the adaptive filter configured as an adaptive line enhancer using the least mean square algorithm. A discussion of the suppression results due to amplitude and phase variations is included as well. An actual wideband system, providing simulation and hardware results of the adaptive filter, is discussed. In addition, many different types of intercept receivers for detection of signals in the spectrum are included.

Chapter 9 deals with cognitive and adaptive techniques in making a "Smart" cognitive system. This shows the various ways to monitor the environment and then to adapt to the changing environment to provide the best solution using the available capabilities. The basic cognitive techniques covered are dynamic spectrum allocation, which changes frequencies; adaptive power and gain control; techniques using modulation waveforms, spread spectrum, adaptive error correction, and adaptive filters; dynamic antenna techniques using active electronically scanned arrays such as multiple-in, multiple-out capabilities; and adapting networks using multi-hop for meshed network like mobile ad hoc networks.

Chapter 10 covers broadband communications and networking, including high-speed data, voice, and video. Many generations of mobile wireless communication products have evolved through the years and are designated as 1G through 4G. Broadband is also used in the home to connect to the outside world without having to run new wires using power line communications, phone-line networking alliance, and radio frequency such as IEEE 802.xx and Bluetooth. Along with the distribution of information, networking plays an important role in the connection and interaction of different devices in the home. Worldwide Interoperability for Microwave Access is another radio frequency wireless link based on the IEEE 802.16 standard, and LTE has emerged as a viable solution for the next generation of wireless products. The military is investigating several networking techniques to allow multiple users for communications, command, control, and weapon systems, including the Joint Tactical Radio System and Link 16. Software-defined radio; cognitive techniques; software communications architecture; five clusters development; the network challenge including gateways; and different network topologies including Star, Bus, Ring, and Mesh are all discussed in this chapter.

Chapter 11 covers satellite communications for various applications in both the commercial and military sectors. The infrastructure for distributing signals covers the widest range of methods of communications. The most remote places on the earth have the means for communication via satellite. The infrastructure, bandwidth, and availability of satellite communications, along with combining this technology with other types of communications systems, makes this method an ideal candidate for providing ubiquitous communications to everyone worldwide. This chapter discusses frequency bands, modulation such as quadrature phase-shift keying and adaptive differential pulse code modulation, geosynchronous and geostationary orbits, and different types of antennas such as primary focus, Cassegrain, and Gregorian. Noise, equivalent temperature, and gain over temperature are discussed as well as how they are used to evaluate different systems and the link budget. Multiple channels and multiple access techniques are discussed for increased capacity. Propagation delay; the cost of use depending on the types of transmission, which includes permanently assigned multiple access, demand assigned multiple access, or occasional; and the different types of satellites used for communications are addressed, including low earth orbit satellites, geosynchronous earth orbit satellites, and medium earth orbit satellites.

Chapter 12 discusses the global positioning system (GPS), which uses a direct sequence spread spectrum binary phase-shift keying data link. This chapter includes coarse acquisition code, precision code, data signal structure, receiver characteristics, errors due to the atmosphere, multipath, Doppler, and selective availability, which has been turned off. It also discusses the pros and cons of using narrow correlation, carrier smoothing of the code (integrated Doppler), differential GPS, and relative GPS. Kinematic carrier phase tracking, the double difference, and wide lane versus narrow lane techniques are also discussed. In addition, other satellite positioning systems are discussed, including the Global Navigational Satellite System from the Soviet Union commonly known as GLONASS, and the Galileo In-Orbit Validation Element in Europe.

Chapter 13 discusses direction finding and interferometer analysis using direction cosines and coordinate conversion techniques to provide the correct solution. This chapter provides information on the limitations of the standard interferometer equation and details the necessary steps to design a three-dimensional interferometer solution for the yaw, pitch, and roll of antennas.

Each of the chapters contains problems, with answers provided in the back of the book.

The appendices contain the definitions of heading, roll, pitch, and yaw for coordinate conversions; true north calculations; phase ambiguities for interferometers; elevation effects on azimuth error; and Earth's radius compensation for elevation angle calculation.

If you have any suggestions, corrections, or comments, please send them to Scott Bullock at scottrbullock@gmail.com.

Acknowledgments

Special thanks to Larry Huffman, senior scientist and system engineer, for thoroughly reviewing this text and enhancing the book with his experience and expertise.

Thanks to Don Shea, a recognized expert in the field of antennas and interferometers, for providing the technical expertise and consulting for Chapter 13.

Thanks to Brian Adams, Tom Ashburn, and Bill Brettner for providing technical input in their fields of expertise.

Thanks to James Gitre for his thorough review, providing questions and suggestions from the standpoint of an end user of the book.

About the Author

Scott R. Bullock received his BSEE degree from Brigham Young University in 1979 and his MSEE degree from the University of Utah in 1988. Mr. Bullock worked in research and development for most of his career developing a radar simulator, a spread spectrum micro-scan receiver, and a new spread spectrum receiver, for which he applied for a patent and was awarded company funds as a new idea project to develop the concept. Mr. Bullock also developed a spread spectrum environment simulator for a spread spectrum wideband countermeasures receiver using binary phase-key shifting, quadrature phase-shift keying, offset quadrature phase-shift keying, minimum-shift keying, frequency hopper, hybrids, amplitude modulation, frequency modulation, voice generator, jammers, and noise. He also designed a high-frequency adaptive filter used to reduce narrowband jammers in a wideband signal; a broadband, highly accurate frequency hop detector; an instantaneous Fourier transform receiver; a chopper modulated receiver; a Ku-band radio design for burst spread spectrum communications through a troposcatter channel; a Gram-Schmidt orthogonalizer to reduce jammers; an advanced tactical data link; radio frequency analysis of an optical receiver study; a portable wideband communications detector; and an acoustic-optic spectrum analyzer photodiode array controller.

Mr. Bullock developed the first handheld PCS spread spectrum telephone with Omnipoint in the 902–928 MHz ISM band. He also received a patent for his work on reducing spectral lines to meet the Federal Communications Commission power spectral density requirements.

He was responsible for various types of spread spectrum data links for the SCAT-1 program related to aircraft GPS landing systems. He was an active participant in the RTCA meetings held in Washington D.C. for the evaluation and selection of the D8PSK data link to be used as the standard in all SCAT-1 systems. He also worked on the concepts of the Wide Area Augmentation System, low probability of intercept data link, DS/FH air traffic control asynchronous system, JTRS, and Link-16.

Mr. Bullock developed several commercial products such as wireless jacks for telephones, PBXs, modems, wireless speakers, and other various wireless data link products.

He has designed directional volume search, tracking algorithms, and cognitive systems and networks for both the commercial and military communities. In addition, he designed, assembled, and successfully field tested a network of common data links from multiple vendors using a multibeam active electronic steerable array.

Mr. Bullock has held many high-level positions, such as vice president of engineering for Phonex Broadband, vice president of engineering for L-3 Satellite Network Division, senior director of engineering for MKS/ENI, engineering fellow for Raytheon, and engineering manager and consulting engineer for Northrop Grumman. He specializes in wireless data link design, cognitive systems, and system analysis and directs the design and development of wireless products for both commercial and military customers.

Mr. Bullock holds numerous patents in the areas of spread spectrum wireless data links, adaptive filters, frequency hop detectors, and wireless telephone and data products. He has published several articles dealing with spread spectrum modulation types, multipath, AGCs, PLLs, and adaptive filters. He is the author of this book and another book titled *Broadband Communications and Home Networking*. He is a licensed professional engineer and a member of IEEE and Eta Kappa Nu, and he holds an Extra Class Amateur Radio License, KK7LC.

He has performed data link communications work and taught in multiple seminars for Texas Instruments, L-3, BAE, Omnipoint, E-Systems, Phonex, Raytheon, Northrop Grumman, CIA, SAIC, MKS/ENI, and Thales for over 15 years. He is currently an instructor for Besser Associates, ATI courses, and K2B International. He has taught an advanced communication course at ITT, an engineering course at PIMA Community College, and was a guest lecturer on multiple access systems at PolyTechnic University, Long Island, New York. He has also taught his course for IEEE Smart Tech Metro Area Workshop in both Baltimore and Atlanta.

CHAPTER 1

Transceiver Design

A transceiver is a system that contains both a transmitter and a receiver. The transmitter from one transceiver sends a signal through space to the receiver of a second transceiver. After receiving the signal, the transmitter from the second transceiver sends a signal back to the receiver of the first transceiver, completing a two-way communications data link system, as shown in Figure 1-1.

There are many factors to consider when designing a two-way communications link. The first one is to determine what the frequency is going to be for this design. Several considerations need to be evaluated to select the frequency that is going to be used.

1.1 Frequency of Operation

In a transceiver design, we first determine the radio frequency (RF) of operation. The frequency of operation depends on the following factors:

- RF availability: This is the frequency band that is available for use by a particular system and is dependent on the communications authority for each country. For example, in the United States it is specified by the Federal Communications Commission (FCC), and in the United Kingdom it is specified by the British Approvals Board for Telecommunications (BABT). These two groups have ultimate control over frequency band allocation. Other organizations that help to establish standards are the International Telecommunications Union Standardization Sector (ITU-T), the European Conference of Postal and Telecommunications Administrations (CEPT), and the European Telecommunications Standards Institute (ETSI).
- Cost: As the frequency increases, the components in the receiver tend to be more expensive. An exception to the rule is when there is a widely used frequency band, such as the cellular radio band, where supply and demand drive down the cost of parts and where integrated circuits are designed for specific applications. These are known as application-specific integrated circuits (ASICs).
- Range and antenna size: As a general rule, decreasing the frequency will also decrease the amount of loss between the transmitter and the receiver. This loss is mainly due to free-space attenuation and is calculated using the frequency or wavelength of the

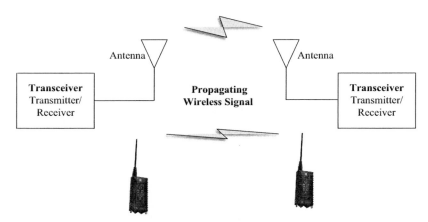

Figure 1-1   Two-way wireless transceiver block diagram.

transmission. This results in an increase in range for line-of-sight applications or a decrease in the output power requirement, which would also affect cost. However, another factor that affects range is the ability of the signal to reflect or bounce off the atmosphere, mainly the ionosphere and sometimes the troposphere. For specific frequencies, this can increase the range tremendously. Amateur radio operators use frequencies that can bounce off the atmosphere and travel around the world with less than 100 watts of power. Also, the size of the antenna increases as the frequency decreases. Therefore, the size of the antenna might be too big for practical considerations and could also be a factor in the cost of the design.

- Customer specified: Oftentimes the frequency of operation is specified by the customer. If the application is for commercial applications, the frequency selection must follow the rules currently in place for that specific application to obtain the approval of the FCC and other agencies.
- Band congestion: Ideally, the frequency band selected is an unused band or serves very little purpose, especially with no high-power users in the band. This also needs to be approved by the FCC and other agencies. Generally the less used bands are very high, which increases the cost. Many techniques available today allow more users to operate successfully in particular bands, and some of these techniques will be discussed further in the book.

The frequency of operation is then selected by taking into consideration the aforementioned criteria.

A listing of the basic frequency bands is shown in Table 1-1 with some applications specified. More detailed frequency allocations can be obtained from the FCC website or in the literature.

The frequency bands are all allocated to different users, which makes it virtually impossible to obtain a band that is not already allocated. However, some band reorganizing and renaming have occurred; for example, some of the old existing analog television bands have been reallocated to use for multiple digital wireless applications.

Table 1-1   Frequency bands

- **ELF**   Extremely low frequency   0–3 KHz
- **VLF**   Very low frequency   3–30 KHz
- **LF**   Low frequency   30–300 KHz
- **MF**   Medium frequency   300–3000 KHz – AM Radio Broadcast
- **HF**   High frequency   3–30 MHz – Shortwave Broadcast Radio
- **VHF**   Very high frequency   30–300 MHz – TV, FM Radio Broadcast, Mobile/fixed radio
- **UHF**   Ultra high frequency   300–3000 MHz – TV
- **L band**   500–1500 MHz – PCS/Cell phones
- **ISM bands**   902–928 MHz, 2.4–2.483 GHz, 5.725–5.875 GHz PCS and RFID
- **SHF**   Super high frequencies – (Microwave)   3–30.0 GHz
- **C band**   3600–7025 MHz – Satellite communications, radios, radar
- **X band**   7.25–8.4 GHz – Mostly military communications
- **Ku band**   10.7–14.5 GHz – Satellite communications, radios, radar
- **Ka band**   17.3–31.0 GHz – Satellite communications, radios, radar
- **EHF**   Extremely high frequencies – (Millimeter wave signals)   30.0–300 GHz – Satellite
- **Infrared radiation**   300–430 THz (terahertz) – infrared applications
- **Visible light**   430–750 THz
- **Ultraviolet radiation**   1.62–30 PHz (petahertz)
- **X-rays**   30–30 EHz (exahertz)
- **Gamma rays**   30–3000 EHz

Table 1-2   Frequency band designations

Legacy Designations		New Designations	
Band	**Frequency range**	**Band**	**Frequency range**
I band	to 0.2 GHz	A band	to 0.25 GHz
G band	0.2 to 0.25 GHz	B band	0.25 to 0.5 GHz
P band	0.25 to 0.5 GHz	C band	0.5 to 1.0 GHz
L band	0.5 to 1.5 GHz	D band	1 to 2 GHz
S band	2 to 4 GHz	E band	2 to 3 GHz
C band	4 to 8 GHz	F band	3 to 4 GHz
X band	8 to 12 GHz	G band	4 to 6 GHz
K_u band	12 to 18 GHz	H band	6 to 8 GHz
K band	18 to 26 GHz	I band	8 to 10 GHz
K_a band	26 to 40 GHz	J band	10 to 20 GHz
V band	40 to 75 GHz	K band	20 to 40 GHz
W band	75 to 111 GHz	L band	40 to 60 GHz
		M band	60 to 100 GHz

In addition, there are basically two accepted frequency band designations using letters of the alphabet, and these are listed in Table 1-2. This can cause some confusion when designating the band of operation. For example, both frequency band designations include the K band with different frequency ranges.

Even though a new designation of the frequency bands has been introduced, the old legacy designation is still commonly used today.

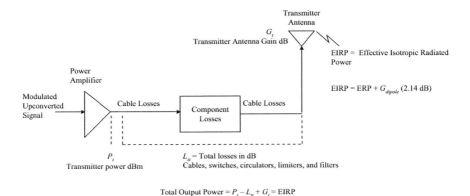

Figure 1-2   RF section of the transmitter.

1.2 Transmitter

The transmitter is the part of the transceiver that creates, modulates, and transmits the signal through space to the receiver. The transmitter is responsible for providing the power required to transmit the signal through the link to the receiver. This includes the power amplifier, the transmitter antenna, and the gains and losses associated with the process, such as cable and component losses, to provide the effective isotropic radiated power (EIRP) out of the antenna (see Figure 1-2).

1.2.1 Power from the Transmitter

The power from the transmitter (P_t) is the amount of power output of the final stage of the power amplifier. For ease in the analysis of power levels, the power is specified in dBm or converted to dBm from milliwatts (mW). The power in mW is converted to power in dBm by

$$P_{\mathrm{dbm}} = 10\log P_{\mathrm{mW}}$$

Therefore, 1 mW is equal to 0 dBm. The unit dBm is used extensively in the industry, and a good understanding of this term and other dB terms is important. The term dBm is actually a power level related to 1 mW and is not a loss or gain as is the term dB.

A decibel (dB) is a unit for expressing the ratio of two amounts of electric or acoustic signal power. The decibel is used to enable the engineer to calculate the resultant power level by simply adding or subtracting gains and losses instead of multiplying and dividing.

Gains and losses are expressed in dB. A dB is defined as a power ratio:

$$\mathrm{dB} = 10\log(P_o/P_i)$$

where

P_i = the input power (in mW)
P_o = the output power (in mW)

For example:

Given:

$$\text{Amplifier power input} = 0.15 \text{ mW} = 10\log(0.15) = -8.2 \text{ dBm}$$
$$\text{Amplifier power gain } P_o/P_i = 13 = 10\log(13) = 11.1 \text{ dB}$$

Calculate the power output:

$$\text{Power output} = 0.15 \text{ mW} \times 13 = 1.95 \text{ mW using power and multiplication}$$
$$\text{Power output (in dBm)} = -8.2 \text{ dBm} + 11.1 \text{ dB} = 2.9 \text{ dBm using dBm and dB and addition}$$
$$\text{Note: } 2.9 \text{ dBm} = 10\log(1.95)$$

Another example of using dBm and dB is as follows:

$$P_i = 1 \text{ mW or } 0 \text{ dBm}$$
$$\text{Attenuation} = 40 \text{ dB}$$
$$\text{Gain} = 20 \text{ dB}$$
$$P_o = 0 \text{ dBm} - (40 \text{ dB attenuation}) + (20 \text{ dB gain}) = -20 \text{ dBm}$$

In many applications, dB and dBm are misused, which can cause errors in the results. The unit dB is used for a change in power level, which is generally a gain or a loss. The unit dBm is used for absolute power; for example, $10\log(1\text{milliwatt}) = 0$ dBm. The unit dBw is also used for absolute power; for example, $10\log(1\text{watt}) = 0$ dBw. The terms dBm and dBw are never used for expressing a change in signal level. The following examples demonstrate this confusion.

Example 1: Suppose that there is a need to keep the output power level of a receiver at 0 dBm ± 3 dB. If ± 3 dB is mistakenly substituted by ± 3 dBm in this case, the following analysis would be made:

Given:

0 dBm = 1 mW (power level in milliwatts)
3 dBm = 2 mW (power level in milliwatts)
3 dB = 2 (or two times the power or the gain)
−3 dB = ½ (half the power or the loss)

The erroneous analysis using ± 3 dBm would be as follows:

0 dBm $\pm$ 3 dBm
0 dBm + 3 dBm = 1 mW + 2 mW = 3 mW; 4.8 dBm for the wrong answer
0 dBm − 3 dBm = 1 mW − 2 mW = −1 mW; cannot have negative power

The correct analysis using ± 3 dB would be as follows:

0 dBm $\pm$ 3 dB
0 dBm + 3 dB = 1 mW $\times$ 2 = 2 mW; equal to 3 dBm for the correct answer
0 dBm − 3 dB = 1 mW $\times$ ½ = 0.5 mW; equal to −3 dBm for the correct answer

In addition:

0 dBm $\pm$ 3 dB
0 dBm + 3 dB = 3 dBm = 2 mW; correct answer
0 dBm − 3 dB = −3 dBm = 0.5 mW; correct answer

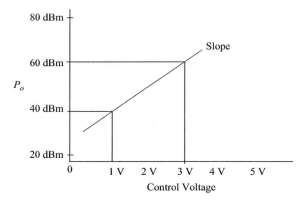

Slope = Rise/Run = (60 dBm − 40 dBm)/(3 V − 1 V) = 20/2 = 10 dB/V

Figure 1-3   Voltage-controlled amplifier curve.

Therefore, they are not interchangeable; the incorrect answers are the result of using the wrong term. Remember, dBm represents an actual power level, and dB represents a change in power level, gain, or loss.

Example 2:   Figure 1-3 shows a graph of a voltage-controlled amplifier curve, with the control voltage on the x axis and power output (in dBm) on the y axis. The slope is equal to

Slope = Rise/Run = (60 dBm − 40 dBm)/(3 V − 1 V) = 20/2 = 10 dB/V, not dBm/V

This example is showing a change in power level, which is specified in dB.

The term dB can be described both in power and voltage. For voltage, the same equation for converting losses and gains into dB is used, substituting V^2/R for P:

$$P_i = V_i^2/R_i$$
$$P_o = V_o^2/R_o$$
$$dB = 10\log\frac{(P_o)}{(P_i)} = 10\log\frac{(V_o^2/R_o)}{(V_i^2/R_i)}$$

where

P_o = power out
P_i = power in
V_o = voltage out
V_i = voltage in
R_o = output impedance
R_i = input impedance

If $R_o = R_i$, they cancel, and the resultant gain/loss equation is

$$dB = 10\log\frac{V_o^2}{V_i^2} = 10\log\left(\frac{V_o}{V_i}\right)^2 = 20\log\frac{V_o}{V_i}$$

Assuming $R_o = R_i$, dB is the same whether voltage or power is used. Therefore, if the system has 6 dB of gain, it has 6 dB of voltage gain and 6 dB of power gain. The only difference is that the ratio of voltage is increased by two and the ratio of power is increased by four.

For example, if an amplifier has 6 dB of gain, it has four times more power and two times more voltage on the output referenced to the input. Therefore, if the input to the amplifier is 5 V and the power at the input is 0.5 W, the output voltage would be 10 V (two times) and the output power would be 2 W (four times) given that $R_i = R_o$. So the gain would be

$$\text{Voltage gain} = 20\log(V_o/V_i) = 20\log(10\text{ V}/5\text{ V}) = 20\log(2) = 6\text{ dB}$$
$$\text{Power gain} = 10\log(P_o/P_i) = 10\log(2/0.5) = 10\log(4) = 6\text{ dB}$$

If $R_i \neq R_o$, then the dB gain for voltage is different from the dB gain for power. If only voltage gain is desired, dBv is used indicating dB volts. This voltage gain can be expressed in millivolts (mV), or dBmv, indicating dB millivolts (mV). For example, going from 10 mV to 20 mV would be a 6 dBmv of gain. However, the caution is that simply changing the ratio R_i/R_o can change the voltage gain without an increase in power gain. For example,

Given:

$V_i = 1$ V
$R_i = 1$ ohm
$V_o = 2$ V
$R_o = 4$ ohm
$P_i = V_i^2/R_i = 1^2/1 = 1$ W
$P_o = V_o^2/R_o = 2^2/4 = 1$ W

Then:

Power gain in dB: $10\log(P_o/P_i) = 10\log(1/1) = 10\log(1) = 0$ dB power gain
Voltage gain in dB: $20\log(V_o/V_i)$ $20\log(2/1) = 20\log(1/2) = 6$ dBv voltage gain

$R_i \neq R_o$, so voltage and power gain are different

There is no power gain in the previous example, but by changing the resistance a voltage gain of 6 dBv was realized. Since power is equal to the voltage times the current ($P = VI$), to achieve a voltage gain of two with the same power the current gain is decreased by two.

The term dB is a change in signal level, signal amplification, or signal attenuation. It is the ratio of power output to power input. It can also be the difference from a given power level, such as "so many dB below a reference power."

The following are some definitions for several log terms that are referred to and used in the industry today:

- dB is the ratio of signal levels, which indicates the gain or loss in signal level.
- dBm is a power level in milliwatts: $10\log P$, where $P = $ power (in mW).

- dBW is a power level in watts: $10\log P$, where P = power (in W).
- dBc is the difference in power between a signal that is used as the carrier frequency and another signal, generally an unwanted signal, such as a harmonic of the carrier frequency.
- dBi is the gain of an antenna with respect to the gain of an ideal isotropic radiator.

Several other dB terms, similar to the last two, refer to how many dB away a particular signal is from a reference level. For example, if a signal's power is 20 times smaller than a reference signal's power r, then the signal is -13 dBr, which means that the signal is 13 dB smaller than the reference signal.

There is one more point to consider when applying the term dB. When referring to attenuation or losses, the output power is less than the input power, so the attenuation in dB is subtracted instead of added. For example, if we have a power of $+5$ dBm and the attenuation in dB is equal to 20 dB, then the output of the signal level is equal to

$$+5 \text{ dBm} - 20 \text{ dB losses} = -15 \text{ dBm}$$

1.2.2 Transmitter Component Losses

Most transceiver systems contain RF components such as a circulator or a transmit/receive (T/R) switch that enable the transceiver to use the same antenna for both transmitting and receiving. Also, if the antenna arrays use multiple antennas, some of their components will interconnect the individual antenna elements. Since these elements have a loss associated with them, they need to be taken into account in the overall output power of the transmitter. These losses directly reduce the signal level or the power output of the transmitter. The component losses are labeled and are included in the analysis:

$$L_{tcomp} = \text{switchers, circulators, antenna connections}$$

Whichever method is used, the losses directly affect the power output on a one-for-one basis. A 1 dB loss equals a 1 dB loss in transmitted power. Therefore, the losses after the final output power amplifier (PA) of the transmitter and the first amplifier (or low-noise amplifier [LNA]) of the receiver should be kept to a minimum. Each dB of loss in this path will either reduce the minimum detectable signal (MDS) by a dB or the transmitter gain will have to transmit a dB more power.

1.2.3 Transmitter Line Losses from the Power Amplifier to the Antenna

Since most transmitters are located at a distance from the antenna, the cable or waveguide connecting the transmitter to the antenna contains losses that need to be incorporated in the total power output:

$$L_{tll} = \text{coaxial or waveguide line losses (in dB)}$$

These transmitter line losses are included in the total power output analysis; a 1 dB loss equals a 1 dB loss in power output. Using larger diameter cables or higher quality cables can reduce the loss, which is a trade-off with cost. For example, heliax cables are used for very low-loss applications. However, they are generally more expensive and larger in diameter than standard cables. The total losses between the power amplifier and the antenna are therefore equal to

$$L_{tt} = L_{tll} + L_{tcomp}$$

Another way to reduce the loss between the transmitter and the antenna is to locate the transmitter power amplifier as close to the antenna as possible. This will reduce the length of the cable, which reduces the overall loss in the transmitter.

1.2.4 Transmitter Antenna Gain

Most antennas experience gain because they tend to focus energy in specified directions compared with an ideal isotropic antenna, which radiates in all directions. Antennas do not amplify the signal power but focus the existing signal in a given direction. This is similar to a magnifying glass, which can be used to focus the sun rays in a specific direction, increasing the signal level at a single point (Figure 1-4).

A simple vertical dipole antenna experiences approximately 2.14 dBi of gain compared with an isotropic radiator because it transmits most of the signal around the antenna, with very little of the signal transmitted directly up to the sky and directly down to the ground (Figure 1-5).

A parabolic dish radiator is commonly used at high frequencies to achieve gain by focusing the signal in the direction the antenna is pointing (Figure 1-6). The gain for a parabolic antenna is

$$G_t = 10\log[n(\pi D/\lambda)^2]$$

where

G_t = gain of the antenna (in dBi)
n = efficiency factor
D = diameter of the parabolic dish
λ = wavelength

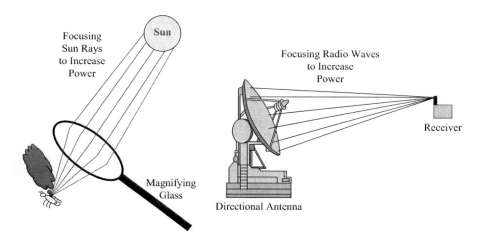

Figure 1-4   Focusing increases power, which provides gain.

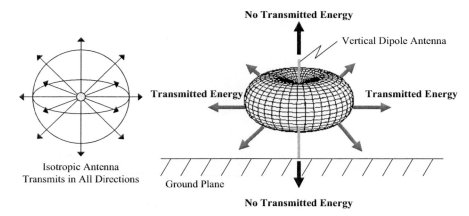

Focuses energy around the wire but does not transmit up or down
Dipole antenna (vertical wire) has ~2.14 dB gain over isotropic antenna

Figure 1-5 A simple vertical dipole antenna gain compared with an isotropic radiator.

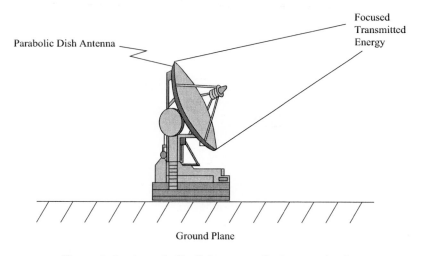

Figure 1-6 A parabolic dish antenna for improved gain.

The efficiency factor is the actual gain of the antenna compared with the theoretical gain. This can happen when a parabolic antenna is not quite parabolic, when the surface of the antenna combined with the feed is not uniform, and when other anomalies occur in the actual implementation of the antenna system. Typically this ranges from 0.5 to 0.8, depending on the design and the frequency of operation.

Notice that the antenna gain increases both with increasing diameter and higher frequency (shorter wavelength). The gain of the antenna is a direct gain where a 1 dB gain equals a 1 dB improvement in the transmitter power output. Therefore, a larger gain will increase the range of the link. In addition, the more gain the antenna can produce, the less power the power amplifier has to deliver for the same range. This is another trade-off that needs to be considered to ensure the best design and the lowest cost for a given application.

1.2.5 Transmitter Antenna Losses

Several losses are associated with the antenna. Some of the possible losses, which may or may not be present in each antenna, are as follows:

- L_{tr}, radome losses on the transmitter antenna. The radome is the covering over the antenna that protects the antenna from the outside elements. Most antennas do not require a radome.
- L_{tpol}, polarization mismatch losses. Many antennas are polarized (i.e., horizontal, vertical, or circular). This defines the spatial position or orientation of the electric and magnetic fields. A mismatch loss is due to the polarization of the transmitter antenna being spatially off with respect to the receiver antenna. The amount of loss is equal to the angle difference between them. For example, if both the receiver and transmitter antennas are vertically polarized, they would be at 90° from the earth. If one is positioned at 80° and the other is positioned at 100°, the difference is 20°. Therefore, the loss due to polarization would be

$$20\log(\cos\theta) = 20\log(\cos 20) = 0.54 \text{ dB}$$

- L_{tfoc}, focusing loss or refractive loss. This is caused by imperfections in the shape of the antenna so that the energy is focused toward the feed. This is often a factor when the antenna receives signals at low elevation angles.
- L_{tpoint}, mispointed loss. This is caused by transmitting and receiving directional antennas that are not exactly lined up and pointed toward each other. Thus, the gains of the antennas do not add up without a loss of signal power.
- L_{tcon}, conscan crossover loss. This loss is present only if the antenna is scanned in a circular search pattern, such as a conscan (conical scan) radar searching for a target. Conscan means that the antenna system is either electrically or mechanically scanned in a conical fashion or in a cone-shaped pattern. This is used in radar and other systems that desire a broader band of spatial coverage but must maintain a narrow beam width. This is also used for generating the pointing error for a tracking antenna.

The total transmitter antenna losses are

$$L_{ta} = L_{tr} + L_{tpol} + L_{tfoc} + L_{tpoint} + L_{tcon}$$

These losses are also a direct attenuation: a 1 dB loss equals a 1 dB loss in the transmitter power output.

1.2.6 Transmitted Effective Isotropic Radiated Power

An isotropic radiator is a theoretical radiator that assumes a point source radiating in all directions. Effective isotropic radiated power (EIRP) is the amount of power from a single point radiator that is required to equal the amount of power that is transmitted by the power amplifier, losses, and directivity of the antenna (antenna gain) in the direction of the receiver.

The EIRP provides a way to compare different transmitters. To analyze the output of an antenna, EIRP is used (Figure 1-2):

$$\text{EIRP} = P_t - L_{tt} + G_t - L_{to}$$

where

P_t = transmitter power in dBm

L_{tt} = total negative losses in dB; coaxial or waveguide line losses, switchers, circulators, antenna connections

G_t = transmitter antenna gain in dB referenced to a isotropic antenna

L_{ta} = total transmitter antenna losses in dB

Effective radiated power (ERP) is another term used to describe the output power of an antenna. However, instead of comparing the effective power to an isotropic radiator, the power output of the antenna is compared to a dipole antenna. The relationship between EIRP and ERP is

$$\text{EIRP} = \text{ERP} + G_{dipole},$$

where G_{dipole} is the gain of a dipole antenna, which is equal to approximately 2.14 dB (Figure 1-2). For example,

EIRP = 10 dBm

ERP = EIRP $- G_{dipole}$ = 10 dBm $-$ 2.14 dB = 7.86 dBm

1.3 Channel

The channel is the path of the RF signal that is transmitted from the transmitter antenna to the receiver antenna. This is the signal in space that is attenuated by the channel medium. The main contributor to channel loss is free-space attenuation. The other losses, such as the propagation losses and multipath losses, are fairly small compared with free-space attenuation. The losses are depicted in Figure 1-7 and are described in detail in the following paragraphs.

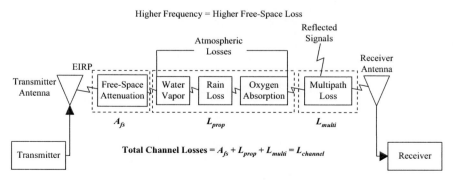

Figure 1-7 Channel link budget parameters.

1.3.1 Free-Space Attenuation

As a wave propagates through space, there is a loss associated with it. This loss is due to dispersion, the "spreading out" of the beam of radio energy as it propagates through space. This loss is consistent and relative to wavelength, which means that it increases with frequency as the wavelength becomes shorter. This is called free-space loss or path loss and is related to both the frequency and the slant range, which is the distance between the transmitter and receiver antennas. Free-space loss is given by

$$A_{fs} = 20\log[4\pi R/\lambda] = 20\log[4\pi Rf/c]$$

where

A_{fs} = free-space loss
R = slant range (same units as λ)
λ = wavelength = c/f
f = frequency of operation
c = speed of light, 300×10^6 m/sec, R is in meters

Therefore, the free-space loss increases as both the range and frequency increase. This loss is also a direct attenuation: a 1 dB loss equals a 1 dB loss in the analysis.

1.3.2 Propagation Losses

There are three main losses depending on the conditions in the atmosphere (e.g., clouds, rain, humidity):

- Cloud loss, loss due to water vapor in clouds.
- Rain loss, loss due to rain.
- Atmospheric absorption (oxygen loss), due to the atmosphere.

These values vary from day-to-day and from region to region. Each loss depends on the location, and a nominal loss, generally not the worst case, is used for calculating the power output required from the transmitter.

To get a feel for the types of attenuation that can be expected, some typical numbers are as follows:

- Rain loss = −1 dB/mile at 14 GHz with 0.5″/hr rainfall
- Rain loss = −2 dB/mile at 14 GHz with 1″/hr rainfall
- Atmospheric losses = −0.1 dB/nautical mile for a frequency range of approximately 10–20 GHz
- Atmospheric losses = −0.01 dB/nautical mile for a frequency range of approximately 1–10 GHz

A typical example gives a propagation loss of approximately −9.6 dB for the following conditions:

- Frequency = 18 GHz
- Range = 10 nm

- Rainfall rate = 12 mm/hr
- Rain temperature = 18°C
- Cloud density = 0.3 g/m^3
- Cloud height = 3–6 km

To determine the actual loss for a particular system, several excellent sources can be consulted. These propagation losses are also a direct attenuation for calculating the transmitter power output: a 1 dB loss equals a 1 dB loss in the power output.

1.3.3 Multipath Losses

Whenever a signal is sent out in space, the signal can either travel on a direct path from the transmitter antenna to the receiver antenna or take multiple indirect paths caused by reflections off objects, which is known as multipath. The most direct path the signal can take has the least amount of attenuation. The other paths (or multipath) are attenuated but can interfere with the direct path at the receiver. It is similar to a pool table, where you can hit a ball by aiming directly at the pocket or you can bank it off the table with the correct angle to the pocket. If both balls are in motion, they may interfere with each other at the input to the pocket (Figure 1-8).

The problem with multipath is that the signal takes both paths and interferes with itself at the receiving end. The reflected path has a reflection coefficient that determines the phase and the amplitude of the reflected signal, which can be different from the direct path. Also, the reflected path length is longer, which produces a signal with a different phase. If the phase of the reflected path is different, for example, 180° out of phase from the direct path, and the amplitudes are the same, then the signal is canceled out and the receiver sees very little to no signal.

Fortunately, most of the time the reflected path is attenuated, since it is generally a longer path length than the direct path. Also, the level of the multipath depends on the reflection coefficient and the type of multipath. It does not completely cancel out the signal but can alter the amplitude and phase of the direct signal. The effect of multipath can cause large variations in the received signal strength. Consequently, multipath can affect coverage and

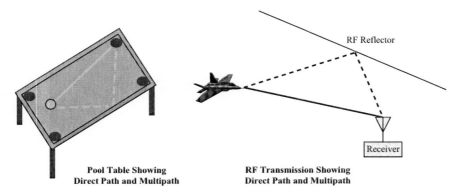

Pool Table Showing　　　　**RF Transmission Showing**
Direct Path and Multipath　　**Direct Path and Multipath**

Figure 1-8　Pool table analysis can be used to describe RF signal reflections.

accuracy where reliable amplitude or phase measurements are required. A further discussion on multipath is presented in Chapter 7. The losses are as follows:

$$L_{multi} = \text{losses due to multipath cancellation of the direct path signal (in dB)}$$

This loss is generally hard to quantize, since there are many variables and many potential paths. Thus, the analysis of the amount of loss that is used is associated with a probability number (see Chapter 7). Multipath is constantly changing, and certain conditions can adversely affect the coverage and the phase measurement accuracy. This loss is also a direct attenuation: a 1 dB loss equals a 1 dB loss in the link analysis. Careful positioning or siting of the antennas in a given environment is the most effective way to reduce the effects of multipath. Also, blanking methods to ignore signals received after the desired signal for long multipath returns are often used to help mitigate multipath. And finally, antenna diversity is another method to reduce multipath.

1.4 Receiver

The receiver accepts the signal that was sent by the transmitter through the channel via the receiver antenna (Figure 1-9). The losses from the antenna to the LNA, which is the first amplifier in the receiver, should be kept as small as possible. This includes cable losses and component losses that are between the antenna and the LNA (Figure 1-9). The distance between the receiver antenna and the LNA should also be as small as possible. Some systems actually include the LNA with the antenna system, separate from the rest of the receiver, to minimize this loss. The main job of the receiver is to receive the transmitted signal and detect the data that it carries in the most effective way without further degrading the signal.

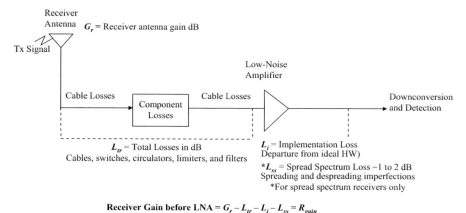

Figure 1-9 RF section of the receiver.

1.4.1 Receiver Antenna Losses

Antenna losses for the receiver are very similar to those for the transmitter, some of which are listed as follows:

- L_{rr}, radome losses on the receiver antenna. The radome is the covering over the antenna that protects the antenna from the outside elements.
- L_{rpol}, polarization loss. Many antennas are polarized (i.e., horizontal, vertical, or circular). This defines the spatial position or orientation of the electric and magnetic fields.
- L_{rfoc}, focusing loss or refractive loss. This is a loss caused by imperfections in the shape of the antenna, so that the energy is focused toward the feed. This is often a factor when the antenna receives signals at low elevation angles.
- L_{rpoint}, mispointed loss. This is caused by transmitting and receiving directional antennas that are not exactly lined up and pointed toward each other. Therefore, the gains of the antennas do not add up without a loss of signal power. Note that this loss may be combined into one number so that it is not included in both the receiver analysis and the transmitter analysis.
- L_{rcon}, conscan crossover loss. This loss is present if the antenna is scanned in a circular search pattern.

The total losses for the receiver antenna can be calculated by adding all of the losses together, assuming that their values are in dB:

$$L_{ra} = L_{rr} + L_{rpol} + L_{rfoc} + L_{rpoint} + L_{rcon}$$

This total loss, as was the case in the transmitter section, is a direct attenuation of the signal: a 1 dB loss equals a 1 dB loss in the analysis.

1.4.2 Receiver Antenna Gain

The gain of the receiver antenna (in dB) is calculated in the same way as the transmitter antenna gain:

$$G_r = 10\log[n(\pi(D)/\lambda)^2]$$

where

$n = $ efficiency factor < 1
$D = $ diameter of the parabolic dish
$\lambda = $ wavelength

The antenna receiver gain is achieved by focusing the received antenna pattern in a specific direction toward the transmitter, which reduces the amount of noise that is received. There are basically two ways to increase the signal-to-noise ratio (SNR): increasing the signal level or reducing the noise. The transmitter increases the signal level by focusing the available power in the direction of the receiver. The receiver reduces the overall noise level by focusing the antenna beam toward the transmitter.

The receiver antenna is not required to have the same antenna as the transmitter. The receiver can use an omnidirectional antenna and receive transmissions from a transmitter that uses a parabolic dish antenna, or the transmitter can be an omnidirectional antenna and the receiver can use a parabolic dish. However, if this is a direct line-of-sight system, the antennas should have the same polarization. For example, if a transmitter is using a vertically polarized antenna, the receiver should have a vertically polarized antenna. An exception to this rule is that if the system is using the ionosphere to bounce the signal for maximum range using lower frequencies, then the polarization can be reversed by the reflection off the ionosphere. For example, if a transmitter is using a horizontally polarized antenna, the optimal receiver antenna may be vertically polarized if the reflection off the ionosphere causes reversal of the polarization. The gain of the antenna is a direct gain in the communications link: a 1 dB gain equals a 1 dB improvement in the link.

1.4.3 Receiver Line Losses from the Antenna to the LNA

The cable that connects the antenna to the first amplifier, which is designated as a LNA, is included in the total losses:

$$L_{rll} = \text{coaxial or waveguide line losses (in dB)}$$

The amplifier is referred to as a LNA because it is designed to have very low noise or noise figure (NF). This is important in setting the NF of the system, since it is determined mainly by the first amplifier in the receiver. A discussion and calculation of the NF of a system are provided later in this chapter. The important thing is that the NF increases the noise level on a one-for-one dB basis. The cable loss between the antenna and the LNA, as was the case in the transmitter section, is a direct attenuation of the signal: a 1 dB loss equals a 1 dB loss in the analysis. Therefore, as mentioned earlier, the cable length should be kept as short as possible, with the option of putting the LNA with the antenna assembly.

1.4.4 Receiver Component Losses

Any components between the antenna and LNA will reduce the SNR of the system. For example, often a limiter is placed in the line between the antenna and the LNA to protect the LNA from damage by high-power signals. This can be included in the NF of the receiver or viewed as a loss of the signal. Both methods are used in the industry, with the same end results, since the first method increases the noise and the second method decreases the signal level, producing the same SNR results for the receiver.

However, since this loss does not add noise to the system but only attenuates the signal, a more straightforward approach would be to treat it as a loss in signal level and calculate the NF separately. The noise before and after the lossy devices is the same, since the temperature before and after the device is the same. Only the signal level is attenuated. The noise on the front end of the receiver before the LNA is equal to kTB, where

k = Boltzmann constant (1.38×10^{-23} J/°K)
T = nominal temperature (290°K)
B = bandwidth

For a 1 Hz bandwidth,

$$kTB = 1.38 \times 10^{-23} \text{ J/°K} * 290°K * 1 \text{ Hz} = 4.002 \times 10^{-21} \text{ W} = 4.002 \times 10^{-18} \text{ mW}$$

Converting to dBm,

$$10\log(kTB) = 10\log(4.002 \times 10^{-18}) = -174 \text{ dBm}$$

A convenient way to calculate the *kTB* noise is to use the aforementioned 1 Hz bandwidth number and simply take $10\log(B)$ and add to it. For example, for a 1 MHz bandwidth,

$$-174 \text{ dBm} + 10\log(1 \text{ MHz}) \text{ dB} = -174 \text{ dBm} + 10\log(10^6 \text{ Hz}) \text{ dB}$$
$$= -174 \text{ dBm} + 60 \text{ dB} = -114 \text{ dBm}$$

The LNA amplifies the input signal and noise, and during this amplification process additional noise is present at the output, mainly due to the active transistors in the amplifier. This additional noise is referred to as the NF of the LNA and increases the overall noise floor, which reduces the SNR. Since this is a change in noise level, NF is in dB. The resultant noise floor is equal to *kTBF*, where *F* is the increase in the noise floor due to the LNA. The noise factor F is the increase in noise due to the amplifier, and the noise figure NF is the increase in noise converted to dB. The rest of the components of the receiver increase the *kTBF* noise floor, but this contribution is relatively small. The amount of increased noise is dependent on how much loss versus gain there is after the LNA and if the noise floor starts approaching the *kTB* noise. Also, if the bandwidth becomes larger than the initial *kTB* bandwidth at the LNA, then the NF can degrade. Calculation of the actual receiver noise is discussed later. The important concept is that the LNA is the main contributor in establishing the noise for the receiver and must be carefully designed to maintain the best SNR for detection of the desired signal with the lowest NF. The *kTB* noise is a constant noise floor unless the temperature or bandwidth is changed. Therefore, the receiver component losses before the LNA are applied to the signal, thus reducing the SNR:

$$L_{rcomp} = \text{switches, circulators, limiters, and filters}$$

This loss, as was the case in the transmitter section, is a direct attenuation of the signal: a 1 dB loss equals a 1 dB loss in the link analysis.

The total losses between the receiver antenna and the LNA are therefore equal to

$$L_{tr} = L_{rll} + L_{rcomp}$$

1.4.5 Received Signal Power at the Output to the LNA

The received signal level P_s (in dBm) at the output of the LNA is calculated as follows:

$$P_{s \text{ dBm}} = \text{EIRP}_{\text{dBm}} - A_{fs \text{ dB}} - L_{p \text{ dB}} - L_{multi \text{ dB}} - L_{ra \text{ dB}} + G_{r \text{ dB}} - L_{rll \text{ dB}} - L_{rcomp \text{ dB}} + G_{\text{LNA dB}}$$

where

EIRP_{dBm} = effective radiated isotropic power
$A_{fs \text{ dB}}$ = free-space attenuation
$L_{p \text{ dB}}$ = propagation loss
$L_{multi \text{ dB}}$ = multipath losses

$L_{ra\ \text{dB}}$ = total receiver antenna losses
$G_{r\ \text{dB}}$ = receiver antenna gain
$L_{rll\ \text{dB}}$ = coaxial or waveguide line losses
$L_{rcomp\ \text{dB}}$ = switches, circulators, limiters, and filters
$G_{\text{LNA dB}}$ = gain of the LNA

Most often dBm is used as the standard method of specifying power.

The noise out of the LNA (in dBm) is equal to

$$N_{\text{LNA dBm}} = kTB_{\text{dBm}} + NF_{\text{dB}} + G_{\text{LNA dB}}$$

Thus, a preliminary SNR can be calculated to obtain an estimate of the performance of a receiver by using the SNR at the output of the LNA. Since the SNR is a ratio of the signal and noise, and since both the signal power and the noise power are in dBm, a simple subtraction of the signal minus the noise will produce a SNR in dB as follows:

$$SNR_{\text{dB}} = P_{s\ \text{dbm}} - N_{\text{LNA dbm}}$$

where

SNR = signal-to-noise ratio (in dB)
P_s = power out of the LNA (in dBm)
N_{LNA} = noise power out of the LNA (in dBm)

One point to note is that the gain of the LNA (G_{LNA}) is applied to both the signal level and the noise level. Therefore, this gain can be eliminated for the purposes of calculating the SNR at the output of the LNA.

$$SNR_{\text{dB}} = (P_{s\ \text{dbm}} - G_{\text{LNA dB}}) - (N_{\text{LNA dbm}} + G_{\text{LNA dB}}) = P_{s\ \text{dbm}} - N_{\text{LNA dbm}}$$

1.4.6 Receiver Implementation Loss

When implementing a receiver in hardware, there are several components that do not behave ideally. Losses associated with these devices degrade the SNR. One of the contributors to this loss is the phase noise or jitter of all of the oscillator and potential noise sources in the system, including the upconversion and downconversion oscillators, code and carrier tracking oscillators, phase-locked loops (PLLs), match filters, and analog-to-digital converters (ADCs). Another source of implementation loss is the detector process, including nonideal components and quantization errors. These losses directly affect the receiver's performance. Implementation losses (L_i) are included to account for the departure from the ideal design due to hardware implementation.

A ballpark figure for implementation losses for a typical receiver is about −3 dB. With more stable oscillators and close attention during receiver design, this can be reduced. This loss will vary from system to system and should be analyzed for each receiver system. This loss is a direct attenuation of the signal: a 1 dB loss equals a 1 dB loss in the link analysis.

1.4.7 Received Power for Establishing the Signal-to-Noise Ratio of a System

The detected power P_d or S (in dBm) that is used to calculate the final SNR for the analysis is

$$P_{d \text{ dBm}} = P_{s \text{ dBm}} - L_{i \text{ dB}} + G_{receiver \text{ dB}} = S_{\text{dBm}}$$

where

$$P_{s \text{ dBm}} = \text{power to the LNA}$$
$$L_{i \text{ dB}} = \text{implementation losses}$$
$$G_{receiver \text{ dB}} = \text{receiver gain}$$
$$P_{d \text{ dBm}} = S_{\text{dBm}} \text{ detected power or signal level}$$

Carrying the receiver gain through the rest of the analysis is optional for SNR calculations, as mentioned previously.

1.4.8 Received Noise Power

The noise power (N) of the receiver is compared to the signal power (S) of the receiver to determine a power SNR. The noise power is initially specified using temperature and bandwidth. System-induced noise is added to this basic noise to establish the noise power for the system.

1.4.9 Noise Figure

The standard noise equation for calculating the noise out of the receiver is

$$N = kT_oBF$$

where

$$T_o = \text{nominal temperature } (290°\text{K})$$
$$F = \text{noise factor}$$
$$k = \text{Boltzmann constant } (1.38 \times 10^{-23} \text{ J/°K})$$
$$B = \text{bandwidth in Hz}$$

The noise factor is the increase in noise level that is caused by an active device generating noise that is greater than the kTB noise. For example, if the noise at the output of an active device increases the kTB noise by 100, then the noise factor would be 100. The noise figure (NF) is the noise factor (F) in dB. For example,

$$\text{Noise factor} = F = 100$$
$$NF = F \text{ dB} = 10\log(100) = 20 \text{ dB}$$

If the NF is used, then the noise out of the receiver is in dBm and the equation is

$$N_{\text{dBm}} = 10\log(kTB) + NF \text{ dB}$$

Therefore, when solving for the noise of the receiver using the SNR, the equations become

$$\text{Noise factor } F = SNR_{\text{in}}/SNR_{\text{out}}$$
$$NF_{\text{dB}} = 10\log(SNR_{\text{in}}) - 10\log(SNR_{\text{out}}) = SNR_{\text{in dB}} - SNR_{\text{out dB}}$$

The noise out of the receiver will be kT_oBF plus any effects due to the difference in temperature between the sky temperature (at the antenna) and the nominal temperature (at the LNA), with F being the receiver noise factor:

$$N = kT_oBF + (kT_sB - kT_oB) = kT_oBF + kB(T_s - T_o) = kT_oBF_t$$

Solving for F_t

$$F_t = F + (T_s - T_o)/T_o$$

where

T_s = sky temperature
T_o = nominal temperature ($290°K$)

If $T_s = T_o$, then the noise is $N = kT_oBF$. Since generally T_s is less than T_o, the noise will be less than the standard kT_oBF and the noise factor is reduced by $(T_s - T_o)/T_o$.

The noise factor for the entire receiver is

$$F_t = F_1 + [(F_2 \times \text{Losses}) - 1]/G_1 + [(F_2 \times \text{Losses}) - 1]/G_1G_2 \ldots$$

As more factors are added, they contribute less and less to the total noise factor due to the additional gains. The receiver noise factor is approximated by the LNA noise factor F_1, assuming that the losses are not too great between stages. For the LNA, noise factor of F_1 and a gain of G_1, and one additional amplifier stage, noise factor F_2 with losses in between them, the total noise factor is calculated as follows (Figure 1-10, Method 1).

$$F_t = F_1 + [(F_2 \times \text{Losses}) - 1]/G_1$$

Another method (Figure 1-10, Method 2) to utilize the Friis noise equation is to treat the losses as individual elements in the chain analysis as shown.

$$F_{total} = F = F_1 + [F_2 - 1]/G_1 + [F_3 - 1]G_1G_2$$

Both of these methods produce the same results, as shown in Figure 1-10. The first method is the preferred one and will be used in the context of this publication.

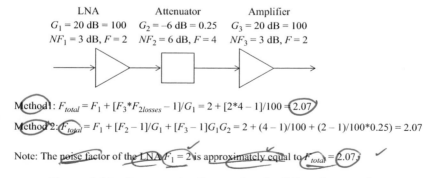

LNA
$G_1 = 20 \text{ dB} = 100$
$NF_1 = 3 \text{ dB}, F = 2$

Attenuator
$G_2 = -6 \text{ dB} = 0.25$
$NF_2 = 6 \text{ dB}, F = 4$

Amplifier
$G_3 = 20 \text{ dB} = 100$
$NF_3 = 3 \text{ dB}, F = 2$

Method 1: $F_{total} = F_1 + [F_3 * F_{2losses} - 1]/G_1 = 2 + [2*4 - 1]/100 = 2.07$

Method 2: $F_{total} = F_1 + [F_2 - 1]/G_1 + [F_3 - 1]G_1G_2 = 2 + (4 - 1)/100 + (2 - 1)/100*0.25) = 2.07$

Note: The noise factor of the LNA $F_1 = 2$ is approximately equal to $F_{total} = 2.07$.

Figure 1-10 System noise figure using the Friis noise equation.

Note that the noise factor of the LNA, $F_1 = 2$, is approximately equal to the total noise factor for the receiver, $F_{total} = 2.07$. Therefore, the NF of the LNA (F_1) is the main contributor of the NF of the receiver. The exceptions would be if the losses are much greater in the chain or the bandwidth changes.

The losses before the LNA are not included in the NF calculation since they do not add noise to the system and are treated as losses to the signal. The reason for this is that the noise temperature is the same before and after the losses, so the kT noise is the same but the signal is attenuated. Therefore, the SNR is reduced on the output. There is no additional noise added to the system, but the NF has been defined as $SNR_{in} - SNR_{out}$ (in dB). This is true only if the bandwidths of the input and the output are the same. One major assumption is that the bandwidth is smaller as the signal progresses through the receiver. If the bandwidth increases, then the NF increases. This NF increase could be substantial and may need to be included in the analysis.

1.4.10 Received Noise Power at the Detector

The noise power at the detector is computed by converting the parameters of the noise factors into dB to simplify the calculation. The gain of the receiver, $G_{receiver}$, is included in this example:

$$N_{dBm} = kT_{o\ dBm} + B_{dB} + NF_{LNA\ dB} + G_{receiver\ dB}$$

Note that this assumes $T_s = T_o$ and that the line losses are included in the link losses and not in the NF.

Eliminating $G_{receiver}$ for analysis purposes is optional since the gain is common to both the noise power and the signal power. However, it is often included to show the actual signal and noise levels through the receiver.

1.4.11 Receiver Bandwidth

The SNR is used for analog signals where the signal (S) is the signal power of the analog signal, and noise (N) is the amount of noise power in the required bandwidth for sending the analog signal. For digital modulation, square pulses or bits are used to send information from the transmitter to the receiver. The amount of energy in a single bit is denoted as E_b. The bit rate or the number of bits per second is denoted as R_b. The signal power for a digital modulation system is equal to the energy in a single bit times the bit rate:

$$S = E_b * R_b$$

The noise power (N) is equal to the noise power spectral density (N_o, which is the amount of noise per Hz, or in other words, the amount of noise in a 1 Hz bandwidth), times the bandwidth (B) of the digital system:

$$N = N_o * B$$

Therefore, the SNR for the digital system is equal to

$$S = E_b * R_b$$
$$N = N_o * B$$
$$SNR = E_b/N_o * R_b/B$$

The bandwidth for a basic digital system is equal to the bit rate. In this case, the SNR is equal to

$$SNR = E_b/N_o$$

For more complex digital communications, the bandwidth may not equal the bit rate, so the entire equation needs to be carried out.

For example, one method of digital modulation is to change the phase of a frequency in accordance with the digital signal, a digital "0" is 0° phase and a digital "1" is 180° phase. This is known as binary phase-shift keying (BPSK). (The types of digital modulation are discussed later in this book). For BPSK, the bandwidth is equal to the bit rate:

$$SNR = E_b/N_o * R_b/B$$

$$SNR = E_b/N_o$$

However, if a more complex digital modulation waveform is used that contains four phase states—0°, 90°, 180°, −90°—such as quadrature phase-shift keying (QPSK), then the bit rate is twice as fast as the bandwidth required because there are two bits of information for one phase shift. Further explanation is provided in Chapter 2. Consequently, the SNR is equal to

$$SNR = E_b/N_o * 2R_b/B$$

$$SNR = 2E_b/N_o$$

The bandwidth used in the BPSK example uses a bit rate equal to the bandwidth. This provides an approximation so that the SNR approximates E_b/N_o. E_b/N_o is used to find the probability of error from the curves shown in Chapter 6 (see Figure 6-6). The curves are generated using the mathematical error functions, $erf(x)$.

For example, using BPSK modulation, the probability of error, P_e, is equal to

$$P_e = 1 \times 10^{-8} \text{ for } E_b/N_o = 12 \text{ dB}$$

This equation gives the probability of bit error for an ideal system. If this is too low, then the E_b/N_o is increased. This can be accomplished by increasing the power out of the transmitter. This can also be accomplished by changing any of the link parameters discussed previously, such as by reducing the line loss. Either increasing E_b or decreasing N_o will increase the ratio.

For QPSK, since it has four phase states, it contains 2 bits of information for every phase state or symbol:

$$2^{\#bits} = \text{states}, 2^2 = 4$$

$$\text{Symbol rate of QPSK} = \text{bit rate of BPSK}$$

Although it sends twice as many bits in the same bandwidth, it requires a higher SNR:

$$SNR = E_b/N_o * 2BR/BW$$

If the symbol rate is reduced by 1/2 so that the bit rate is equal for both QPSK and BPSK, then both the bandwidth and noise are reduced by 1/2 and the probability of error is approximately the same.

1.4.12 Received E_b/N_o at the Detector

The received E_b/N_o (in dB) is equal to the received SNR (in dB) for BPSK. For higher order modulations, the bandwidth will be larger than the bit rate, so the SNR (in dB) $= XE_b/N_o$, where X relates to the order of modulation.

E_b/N_o is the critical parameter for the measurement of digital system performance. Once this has been determined for a digital receiver, it is compared to the required E_b/N_o to determine if it is adequate for proper system operation for a given probability of error.

1.4.13 Receiver Coding Gain

Another way to improve the performance of a digital communication system is to use codes that have the ability to correct bits that are in error. This is called error correction. The most common error correction is known as forward error correction (FEC). This type of correction uses code to correct bit errors without any feedback to the transmitter. Coding will be discussed in more detail in subsequent chapters. With FEC, the coding gain is the gain achieved by implementing certain codes to correct errors, which is used to improve the BER, thus requiring a smaller E_b/N_o for a given probability of error. The coding gain, G_c, depends on the coding scheme used. For example, a Reed-Solomon code rate of 7/8 gives a coding gain of approximately 4.5 dB at a BER of 10^{-8}. Note that coding requires a trade-off, since adding coding either increases the bandwidth by 8/7 or the bit rate is decreased by 7/8.

1.4.14 Required E_b/N_o

The required E_b/N_o for the communications link is

$$E_b/N_o(\text{req}) = E_b/N_o(\text{uncoded}) - G_c$$

This is the minimum required E_b/N_o to enable the transceiver to operate correctly. The parameters of the communications link need to be set to ensure that this minimum E_b/N_o is present at the detector.

1.5 The Link Budget

The link budget a method used to determine the necessary parameters for successful transmission of a signal from a transmitter to a receiver. The term link refers to linking or connecting the transmitter to the receiver, which is done by sending out RF waves through space (Figure 1-11). The term budget refers to the allocation of RF power, gains, and losses and tracks both the signal and the noise levels throughout the entire system, including the link between the transmitter and the receiver. The main items that are included in the budget are the required power output level from the transmitter power amplifier, the gains and losses throughout the system and link, and the SNR for reliable detection; the E_b/N_o to produce the desired bit error rate (BER); or the probability of detection and probability of false alarm at the receiver. Therefore, when certain parameters are known or selected, the link budget allows the system designer to calculate unknown parameters.

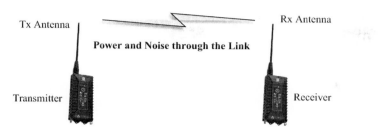

Figure 1-11   Link connects the transmitter to the receiver for link budget analysis.

Several of the link budget parameters are given or chosen during the process and the rest of the parameters are calculated. There are many variables and trade-offs in the design of a transceiver, and each one needs to be evaluated for each system design. For example, there are trade-offs between the power output required from the power amplifier and the size of the antenna. The larger the antenna (producing more gain), the less power is required from the power amplifier. However, the cost and size of the antenna may be too great for the given application. On the other hand, the cost and size of the power amplifier increase as the power output increases, which may be the limiting factor. If the power output requirement is large enough, a solid-state amplifier may not be adequate, and therefore a traveling-wave tube amplifier (TWTA) may be needed. The TWTA requires a special high-voltage power supply, which generally increases size and cost. So by making these kinds of trade-off studies, an optimum data link solution can be designed for a specific application.

Before starting the link budget, all fixed or specified information concerning the transceiver needs to be examined to determine which parameters to calculate in the link budget. These concessions need to be evaluated before the link budget is performed and then must be reevaluated to ensure that the right decisions have been made. The parameters for a link budget are described previously in this chapter.

Proper transceiver design is critical in the cost and performance of a data link. To provide the optimal design for the transceiver, a link budget is used to allocate the gains and losses in the link and to perform trade-offs of various parts of the system. The link budget also uses the required SNR or the ratio of bit energy to noise spectral density (E_b/N_o) for a given probability of error. These required levels are derived by using probability of error curves given a certain type of modulation. Probability of error curves are discussed in Chapter 6. Generally, since there are both known and unknown variances in the link budget, a link budget will provide an additional SNR or E_b/N_o, which is referred to as the link margin. The link margin is equal to

$$\text{Link margin for analog systems} = \text{SNR (calculated)} - \text{SNR (required)}$$
$$\text{Link margin for digital systems} = E_b/N_o \text{ (calculated)} - E_b/N_o \text{ (required)}$$

This link margin is used to provide a margin of error in the analysis, hardware implementation, and other factors that can affect the desired performance.

1.5.1 Spread Spectrum Systems

Spread spectrum is a technique for using more bandwidth than is required to send digital data. This is generally accomplished by using a faster code sequence and combining it with the digital data sequence before modulation. Spread spectrum is used mainly to mitigate jamming signals. To complete the link analysis for a spread spectrum system, spreading losses need to be included in the link budget.

If spread spectrum is used, then the signal level is modified by the spreading losses associated with nonideal spreading and despreading techniques, which reduce the received signal power (sometimes referred to as match filter loss). This is an additional loss separate from the implementation loss, since it deals with the spreading and despreading of a spread spectrum system. Also, since many digital systems are not spread spectrum systems, the loss is kept separate for clarity. The spread spectrum loss is generally around 1 to 2 dB and varies from system to system. Note that this is not included in a nonspread spectrum system. The losses are specified as follows:

$$L_{ss} = \text{spread spectum loss (1 to 2 dB)}$$

The value of the spread spectrum loss is dependent on the type of spread spectrum used and the receiver design for spreading and despreading. This loss is a direct attenuation of the signal: a 1 dB loss equals a 1 dB loss in the link analysis.

1.5.2 Process Gain

One of the main reasons for using spread spectrum is the ability of this type of system to reject other signals or jammers. This ability is called process gain (G_p). The process gain minus spreading losses provides the ideal jamming margin for the receiver. The jamming margin is the amount of extra power, referenced to the desired signal, which the jammer must transmit to jam the receiver:

$$J_m = G_p - L_{ss}$$

1.5.3 Received Power for Establishing the Signal-to-Noise Ratio for a Spread Spectrum System

The detected power, P_d or S (in dBm), that is used to calculate the final SNR for the analysis of spread spectrum systems includes the spreading loss, L_{ss}:

$$P_{d\ \text{dBm}} = P_{s\ \text{dBm}} - L_{i\ \text{dB}} - L_{ss\ \text{dB}} + G_{receiver\ \text{dB}} = S_{\text{dBm}}$$

where

$$P_{s\ \text{dBm}} = \text{power to the LNA}$$
$$L_{ss\ \text{dB}} = \text{spreading losses}$$
$$G_{receiver\ \text{dB}} = \text{the receiver gain}$$

The receiver gain is applied to both the signal and the noise and is canceled out during the SNR calculation. Again, carrying the receiver gain through the rest of the analysis is optional.

1.5.4 Link Budget Example

A typical link budget spreadsheet used for calculating the link budget is shown in Table 1-3. This link budget tracks the power level and the noise level side by side so that the SNR can be calculated anywhere in the receiver. The altitude of an aircraft or satellite and its angle incident to the earth are used to compute the slant range, and the atmospheric loss is analyzed at the incident angle. The bit rate plus any additional bits for FEC determines the final noise

Table 1-3   Typical link budget analysis

Link Budget Analysis:									
	Slant Rng(km)	**Freq.(GHz)**	**Power(W)**	**Conversions:**					
Enter Constants...	92.65	3	1	**Slant Rng:**	Enter Range	Kilometers			
					2 mi	3.218			
Enter Inputs.........	**Inputs**	**Power Levels**			50 nmi	92.65			
Transmitter	Gain/Loss (dB)	Sig. (dBm)	Noise (dBm)		5280 ft	1.61			
Tran. Pwr (dBm) =		30							
Trans. line loss =	−1	29		**Trans. Pwr:**	Enter Power	Pwr (watts)			
Other (switches)	−2	27			50 dBm	100			
Trans Ant Gain =	10	37		**Trans. Ant Gain:**	Enter	Gain (dB)	Parabolic Dish		
Ant. Losses =*	−2	35		Diameter (inches) =	24	22.63			
ERP		35		1/2 pwr beamwidth =	11	23.01	Bm Meth		
Channel				n = (efficiency) =	0.5				
Free Space Loss =	−141.32	−106.32							
Rain Loss =	−2.00	−108.32							
Cloud Loss =	−1.00	−109.32		**Rx Ant. Gain:**	Enter	Gain (dB)	Parabolic Dish		
Atm loss (etc) =	−0.50	−109.82		Diameter (inches) =	24	22.63			
Multipath Loss =	−2.00	−111.82		1/2 pwr beamwidth =	11	23.01	Bm Meth		
Receiver				n = (efficiency) =	0.5				
RF BW (MHz)**	100.00		−94.00						
Ant. losses =*	−2.00	−113.82		**Calculated Constants:**		**Prob. of Err**	E_b/N_o(dB)	E_b/N_o	P_e
Rx Ant Gain =	10.00	−103.82		Lambda =	0.1 meters	Coh.PSK	10.53	11.29796	1.00E-06
Ant. ohmic loss =	−2.00	−105.82		Lambda =	3.94 inches	Coh.FSK	13.54	22.59436	1.00E-06
Other (switches)	−1.00	−106.82		Free Space Loss	−141.32 dB	Non.(DPSK)	11.18	13.122	1.00E-06
Rec. Line loss =	−2.00	−108.82		Boltzman's =	1.4E-23	Non.FSK	14.19	26.24219	1.00E-06
LNA Noise Fig. =	3.00		−91.00			Coh.QPSK/MSK	10.53	11.29796	1.00E-06
LNA Gain =	25.00	−83.82	−66						
NF Deviation =	0.10		−65.90	Receiver Noise Figure:	8.00 dB		6.31	Factor	
LNA levels =		−83.82	−65.90	Losses, Post Preamp:	10.00 dB		10.00		
Receiver Gain =	60.00	−23.82	−5.90	LNA Gain:	25.00 dB		316.23		
Imp. Loss	−4.00	−27.82		LNA NF:	3.00 dB		2.00		
Despread BW**	10.00		−15.9	NF Final:	3.10 dB		2.04		
PG 10log# of bits	30.00	2.18		**NF Deviation:**	0.10				
Spreading Loss =	−2.00	0.18							
Match Filt. loss				* Ant. losses		Transmit	Receive		
Res. Doppler				Lpointing (misalign)	0.5	0.5			
Det. Peak offset				Lpolarization =	0.5	0.5			
Det. Levels =		0.18	−15.9	Lradome loss =	0	0			
S/N =		**16.08**		Lconscan (x over) =	0	0			
Req. E_b/N_o				Lfoc (refract, L-angle) =	1	1			
Req. E_b/N_o		12	$P_e = 10 \exp^{-8}$						
Coding Gain =	4.00	8.00		** Enter the actual bandwidths in MHz in the column even though they are not					
E_b/N_o **Margin =**		8.08		gains and losses. The noise power is adjusted accordingly.					

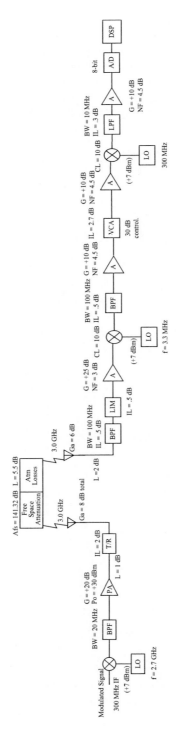

Figure 1-12 Data link system for link budget analysis.

bandwidth. A block diagram of a typical transmitter and receiver is shown in Figure 1-12. This block diagram refers to the link analysis that was performed in accordance with the typical link budget analysis (Table 1-3).

1.6 Summary

The transceiver design is applicable to all wireless communication systems. The link budget provides a means to design a transceiver and to perform the necessary analysis to ensure that the design is optimal for the given set of requirements. The link budget provides a way to easily make system trade-offs and to ensure that the transceiver operates properly over the required range. Once known or specified parameters are entered, the link budget is used to solve for the unknown parameters. If there is more than one unknown parameter, then the trade-offs need to be considered. The link budget is continually reworked, including the trade-offs, to obtain the best system design for the link. Generally, a spreadsheet is used for ease of changing the values of the parameters and monitoring the effects of the change (Table 1-3).

References

Crane, R. K. *Electromagnetic Wave Propagation through Rain.* New York: Wiley, 1996.

Gandhi, O. P. *Microwave Engineering and Applications.* New York: Pergamon Press, 1981.

Haykin, S. *Communication Systems,* 5th ed. New York: Wiley, 2009.

Holmes, J. K. *Coherent Spread Spectrum Systems.* New York: Wiley & Sons, 1982, pp. 251–267.

Problems

1. Show that a specification of 0 dBm ± 2 dBm = ??? is an impossible statement and write a correct statement for it.
2. Solve for the total signal level through the receiver in Figure 1-P2 using milliwatts to obtain the answer, then solve for the output in dBm for the above receiver. Compare your answers. Note: Try the dB method without using a calculator.

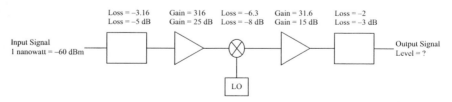

Figure 1-P2   A receiver diagram for calculating power output.

3. An oscillator generates a center frequency of 1 GHz at 1 mW. The largest spurious response or unwanted signal is at a power level of 0.01 mW. How would you specify the spurious response in dBc? What would the power be in milliwatts for a spur at −40 dBc?

4. A system has been operational for the past 3 years. A need arose to place a limiter in the path between the antenna and the LNA to avoid overload from a newly built high-power transmitter. The limiter has 1.5 dB of loss. How will this affect the ability to receive low-level signals? What can you do to overcome the loss?

5. What is the diameter of a parabolic antenna operating at 5 GHz, at an efficiency of 0.5 and a gain of 30 dBi?

6. What is the free-space attenuation for the system in Problem 5 with a range of 10 nautical miles?

7. What is the noise level out of the LNA given that the bandwidth is 10 MHz and the LNA NF is 3 dB with $T_o = T_s$ at room temperature?

8. What is the NF of the receiver, given that the LNA has an NF of 3 dB, a gain of 20 dB, a second amplifier after the LNA with a gain of 20 dB and an NF of 6 dB, and there is a loss of 5 dB between the amplifiers?

CHAPTER 2

The Transmitter

The transmitter is responsible for formatting, encoding, modulating, and upconverting the data communicated over the link using the required power output according to the link budget analysis. The transmitter section is also responsible for spreading the signal using various spread spectrum techniques. Several digital modulation waveforms are discussed in this chapter. The primary types of digital modulation using direct sequence methods to phase modulate a carrier are detailed, including diagrams and possible design solutions. A block diagram showing the basic components of a typical transmitter is shown in Figure 2-1.

2.1 Basic Functions of the Transmitter

The basic functions of the transmitter are the following:

- Transmitter antenna: Provides an antenna to deliver power into space toward the receiver.
- Transmit/receive (T/R) device: T/R switch, circulator, diplexer, or duplexer allows the same antenna to be used for both the transmitter and the receiver.
- Radio frequency (RF) power amplifier: Provides an RF power amplifier, which amplifies the signal power to an antenna.
- Upconverter: Utilizes upconversion to select a carrier for the modulated signal that provides a frequency of operation to use a practical, smaller antenna. This upconverter could contain multiple upconverters as necessary to reach the final RF for transmission.
- Modulator: Provides data modulation, which puts data on the carrier frequency. This can be accomplished using various types of modulation techniques, including spread spectrum and coding.

2.1.1 Transmit Antenna

The antenna receives the RF signal from the power amplifier through the T/R device and translates this RF power into electromagnetic waves so that the signal can be propagated through space. The antenna is dependent on the RF frequency selected by the upconverter according to the specified operational requirements. The proper design of the antenna ensures that the maximum signal power is sent out in the direction of the receiver or in the

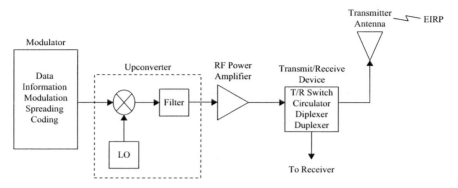

Figure 2-1   Block diagram of a basic transmitter.

required service volume. The design of the antenna is a critical parameter in the link budget and must be optimized to produce the best performance for a given size and cost.

The gain of the antenna improves the link budget on a one-for-one basis; that is, for every dB of gain the antenna exhibits, the link is improved by a dB. Therefore, careful design of the antenna can reduce the power output required from the power amplifier, which reduces the cost and thermal dissipation of the system. The frequency, amount of gain required, and size are factors in determining the type of antenna to use. Parabolic dishes are frequently used at microwave frequencies. The gain of the parabolic dish was calculated in Chapter 1. The antenna provides gain in the direction of the beam to reduce power requirements or increase range. The gain of the antenna is usually expressed in dBi, which is the gain in dB referenced to what an ideal isotropic radiator antenna would produce. In other words, this is the amount of amplifier gain that would be delivered to an isotropic radiator antenna to transmit the same amount of power in the same direction as a directional antenna. Even a vertical dipole antenna provides gain in the direction of the receiver, which is typically 2.14 dBi.

2.1.2 Transmit/Receive Device

To reduce the number of antennas for a transceiver system, which ultimately reduces size and cost, the same antenna is generally used for both the transmitter and the receiver at each location. If the same antenna is going to be used for the system, a T/R device is required to prevent the transmitter from interfering with or damaging the receiver. This device can be a duplexer or diplexer, T/R switch, circulator, or some combination of these devices. The method chosen must provide the necessary isolation between the transmitter circuitry and the receiver circuitry to prevent damage to the receiver during transmission. The transmitted signal is passed through this device and out to the antenna with adequate isolation from the receiver. The received signal is passed through this device from the antenna to the receiver with adequate isolation from the transmitter.

Duplexers and diplexers are commonly used to provide the isolation between the transmit signal and the receiver, but they are often misunderstood. A duplexer is a device that provides a passband response to the transmitted signal and a notch filter at the receiver frequency. Therefore, the transmitted output is passed through the duplexer to the antenna, but the transmitted signal at the receiver frequency is attenuated by the notch filter. Generally, the receiver frequency and the transmitter frequency are in the same band but separated far enough in frequency so that the bandpass filter and the notch filter can be separated. A diplexer is basically two bandpass filters, one tuned to the transmitter frequency band and the other tuned to the receiver frequency band. Diplexers are usually used to split multiple frequencies, such as two RF bands in the receiver, so that the same antenna can be used to receive two different frequency bands. For example, in the amateur radio bands, a single antenna can be used to receive the 2 m band and the 70 cm band because a diplexer can separate these bands using two bandpass filters in the receiver.

2.1.3 RF Power Amplifier

The RF power amplifier (PA) is used to provide the necessary power output from the transmitter to satisfy the link budget for the receiver. The general classifications for power amplifiers are as follows:

- Class A: Power is transmitted all of the time; used for linear operation.
- Class A/B: Power is transmitted somewhere in between Class A and Class B. This is a trade-off between the linear operation of a Class A amplifier and the efficiency of a Class B amplifier.
- Class B: The conduction angle for Class B is $180°$ or half the time. A classic implementation of Class B amplifiers for a linear system is push–pull, that is, one stage is on and the other stage is off. The only drawback of this configuration is crossover distortion when transitioning from one device to the other. Class B amplifiers are more efficient than Class A and A/B amplifiers, but they are more nonlinear.
- Class C: Power is transmitted less than half of the time. This amplifier class is more efficient than the previous classes, but it is more nonlinear. Class C amplifiers may present a potential problem for some types of digital modulation techniques.

PAs can be classified in many other ways depending on their design, function, and application, but the aforementioned list outlines the basic power amplifiers used for communications.

To generate higher power outputs, power amplifiers can be combined (Figure 2-2). The input is divided into multiple signal paths that are amplified by separate power amplifiers

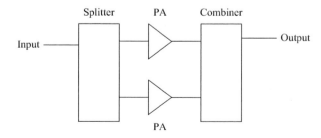

Figure 2-2   Combining power amplifiers for high power output.

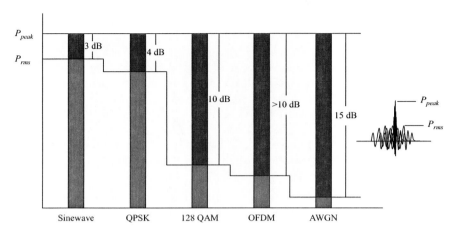

Figure 2-3   Crest factors for different signals.

and then coherently combined to generate higher power outputs. The phase/delay of each of the paths should be matched properly before they are combined, and a quadrature 90° hybrid can be used to reduce the reflected power and provide a better match. In addition, feedback techniques have been used to provide better matching characteristics for the combining.

An objective for most transmitters is to obtain the highest power out of an amplifier without distortion due to saturation. To accomplish this, the waveform needs to be evaluated with respect to its crest factor (CF). The CF is equal to the required peak-to-average ratio (PAR) or peak-to-average power ratio (PAPR) for each of the waveforms (Figure 2-3). This ratio is equal to the peak power divided by the average, or RMS power:

$$CF = PAPR = P_{peak}/P_{rms}$$
$$CF = PAPR_{dB} = P_{peak\ dBm} - P_{rms\ dBm}$$

The requirement specifies that the transmitter needs to transmit the peak power without saturation. This causes the average power to be less so that the system requires a larger power amplifier for the same performance. For example, a sinewave has a crest factor of 3 dB compared with an orthogonal frequency-division multiplexing (OFDM) waveform that has a crest factor of >10 dB. That means that to transmit an OFDM waveform the transmitter would have to have an additional 7 dB or greater power output requirement.

Note that the crest factor for noise is high since some large noise spikes are well above average noise power output.

2.1.4 Upconverter

The information to be transmitted is made up of low frequencies, such as voice or a digital data stream. The low frequencies are mixed or upconverted to higher frequencies before the signal is radiated out of the antenna. One reason for using higher frequencies for radiation is the size of the antenna. The lower the frequency, the longer the antenna is required.

For example, many antenna designs require antenna lengths of $\lambda/2$. The antenna length is calculated as follows:

$$\lambda = c/f$$

where

λ = wavelength
c = speed of light = 3×10^8 m/sec
f = frequency

Given

$f = 10$ MHz
$\lambda/2 = (c/f)/2 = (3 \times 10^8/10 \text{ MHz})/2 = 15$ m = length of the antenna

This is much too large to use for practical applications. If the frequency is upconverted to a higher frequency, then the antenna can be small as shown:

$f = 2.4$ GHz
$\lambda/2 = (c/f)/2 = (3 \times 10^8/2.4 \text{ GHz})/2 = 0.0625$ m = 6.25 in = length of the antenna

This length of antenna is much more practical for multiple applications.

Another reason for using higher frequencies is band congestion. The lower usable frequency bands are overcrowded. Operating frequencies continue to increase in hopes that these bands will be less crowded. Regardless of the reason, the main function of the transmitter is to upconvert the signal to a usable radiating frequency.

A local oscillator (LO) is used to translate the lower frequency band to a higher frequency band for transmission. For example, a single lower frequency is mixed or multiplied by a higher frequency, which results in the sum and difference of the two frequencies as shown:

$$\cos(\omega_c t + \varphi_c)\cos(\omega_l t + \varphi_l) = \tfrac{1}{2}\cos[(\omega_c - \omega_l)t + \varphi_c - \varphi_l] + \tfrac{1}{2}\cos[(\omega_c + \omega_l)t + \varphi_c + \varphi_l]$$

where

ω_c = carrier or higher frequency
φ_c = phase of the carrier
ω_l = the lower frequency containing the data
φ_l = phase of the lower frequency
t = time

A simpler form showing the concept of upconversion to a higher frequency is as follows:

$$f_c \times f_m = (f_c + f_m) + (f_c - f_m) = 2.4 \text{ GHz} \times 10 \text{ MHz}$$
$$= (2.4 \text{ GHz} + 10 \text{ MHz}) + (2.4 \text{ GHz} - 10 \text{ MHz}) = 2.41 \text{ GHz} + 2.39 \text{ GHz}$$

where

f_c = high frequency carrier = 2.4 GHz
f_m = low modulating frequency = 10 MHz or .010 GHz

A diagram of the results of this mixing process is shown in Figure 2-4.

This is an ideal case that ignores any of the harmonics of the two input frequencies and only uses one low frequency (usually the data include a band of frequencies). A bandpass

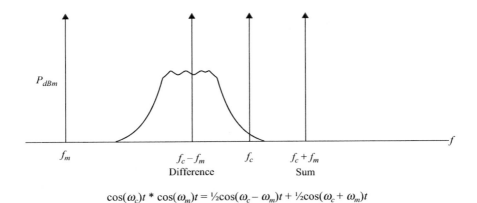

$$\cos(\omega_c)t * \cos(\omega_m)t = \tfrac{1}{2}\cos(\omega_c - \omega_m)t + \tfrac{1}{2}\cos(\omega_c + \omega_m)t$$

Figure 2-4   Ideal frequency spectrum of a CW modulating signal mixed with a CW carrier.

filter is used at the output to select either the sum term or the difference term. The difference term is selected in this example. A bandpass filter is centered at $\cos[(\omega_c - \omega_l)t + \varphi_c - \varphi_l]$ with a roll-off sharp enough to virtually eliminate the sum term $\cos[(\omega_c + \omega_l)t + \varphi_c + \varphi_l]$, with the following result:

$$\tfrac{1}{2}\cos[(\omega_c - \omega_l)t + \varphi_d] \text{ or in simple form } (f_c - f_m)$$

where

$$\varphi_d = \text{difference of the phases}$$
$$\omega_c - \omega_l = \text{difference of the frequencies}$$
$$f_c = \text{high frequency carrier}$$
$$f_m = \text{low modulating frequency}$$

2.1.5 Sum and Difference Frequencies Generated in the Upconversion Process

An LO is used to upconvert lower frequencies to higher frequencies for transmitting out the antenna. During this upconversion process, two frequency bands are produced in the ideal case: the sum of the LO frequency and the input frequencies; and the difference of the LO frequency and the input frequencies. One of these bands of frequencies must be filtered out to recover the signal at the receiver. In addition, since the same data are contained in both frequency bands, more power is required to transmit both sum and difference channels, and there is more chance of generating interference to other users by transmitting both bands. If both bands are allowed to be transmitted, there is a possibility that the phases will align at the receiver, which will cancel the desired signal. For example, suppose that the signal to be transmitted is defined as $A\cos(\omega_s)t$ and the carrier frequency is defined as $B\cos(\omega_c)t$. The output of the upconversion process is found by multiplying the two signals:

$$A\cos(\omega_s)t \times B\cos(\omega_c)t = AB/2\cos(\omega_s + \omega_c)t + AB/2\cos(\omega_s - \omega_c)t$$

Generally, one of these terms is filtered out. If both of these terms are allowed to be transmitted, a worst-case phase situation at the receiver would be when the LO or carrier frequency ω_c is 90° out of phase $[C\cos(\omega_c t + 90) = C\sin(\omega_c)t]$ with the incoming signal:

$$
\begin{aligned}
&[AB/2\cos(\omega_s + \omega_c)t + AB/2\cos(\omega_s - \omega_c)t][C\sin(\omega_c)t] \\
&= ABC/4[\sin(\omega_s + \omega_c + \omega_c)t + \sin(\omega_s - \omega_c + \omega_c)t] \\
&\quad - ABC/4[\sin(-\omega_s + \omega_c - \omega_c)t + \sin(\omega_s - \omega_c - \omega_c)t] \\
&= ABC/4[\sin(\omega_s + 2\omega_c)t + \sin(\omega_s)t] - ABC/4[\sin(\omega_s)t + \sin(\omega_s - 2\omega_c)t] \\
&= ABC/4[\sin(\omega_s + 2\omega_c)t - \sin(\omega_s - 2\omega_c)t]
\end{aligned}
$$

The results show that only the sum of $\omega_s + 2\omega_c$ and the difference of $\omega_s - 2\omega_c$ frequencies were detected by the receiver. The frequency ω_s was not retrieved, and the output was degraded substantially. This demonstrates the requirement to filter out one of the unwanted sidebands during transmission.

If the sum term is filtered out at the transmitter, and using the worst-case phase situation at the receiver as before, yields

$$
\begin{aligned}
&[AB/2\cos(\omega_s - \omega_c)t][C\sin(\omega_c)t] \\
&= ABC/4[\sin(\omega_s - \omega_c + \omega_c)t] - ABC/4[\sin(\omega_s - \omega_c - \omega_c)t] \\
&= ABC/4[\sin(\omega_s)t - ABC/4[\sin(\omega_s - 2\omega_c)t]
\end{aligned}
$$

The results of filtering the sum band with this worst-case scenario shows that the signal waveform ω_s is retrieved along with the difference of $\omega_s - 2\omega_c$. A low-pass filter will eliminate the second term, since the carrier frequency is generally much higher than the signal frequency, giving only the desired signal output of $ABC/4[\sin(\omega_s)t]$.

Another way to eliminate the unwanted band in the transmitter is to use an image-reject mixer. The image-reject mixer quadrature upconverts the signal into two paths: an in-phase path called the I-channel, and a path that is in quadrature or −90° out of phase called the Q-channel:

I-channel output $= \cos\omega_s t^* \cos\omega_c t = \frac{1}{2}\cos(\omega_s - \omega_c)t + \frac{1}{2}\cos(\omega_s + \omega_c)t$

Q-channel output $= \cos\omega_s t^* \cos(\omega_c t - 90) = \frac{1}{2}\cos[(\omega_s - \omega_c)t + 90] + \frac{1}{2}\cos[(\omega_s + \omega_c)t - 90]$

The I-channel output is then phase shifted by −90°, and the I and Q channels are then summed together, eliminating the difference band as follows:

$$
\begin{aligned}
\text{I-channel phase shifted} &= \frac{1}{2}\cos[(\omega_s - \omega_c)t - 90] + \frac{1}{2}\cos[(\omega_s + \omega_c)t - 90] \\
I + Q &= \frac{1}{2}\cos[(\omega_s - \omega_c)t - 90) + \frac{1}{2}\cos[(\omega_s + \omega_c)t - 90] \\
&\quad + \frac{1}{2}\cos[(\omega_s - \omega_c)t + 90] + \frac{1}{2}\cos[(\omega_s + \omega_c)t - 90] \\
&= \cos[(\omega_s + \omega_c)t - 90]
\end{aligned}
$$

Therefore, only the sum of the frequency band is transmitted, which eliminates the problem on the receiver side.

2.1.6 Modulator

A modulator is used to transmit data information by attaching it to a higher frequency carrier to transmit it to the receiver using a small, practical antenna. For a typical analog system, the analog signal modulates the carrier directly by using either amplitude modulation (AM) or

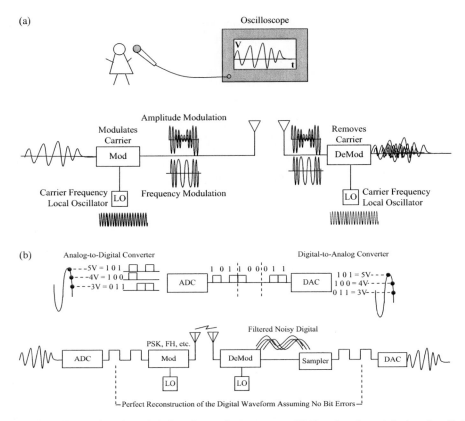

Figure 2-5 (a) Baseband modulation for analog systems. (b) Baseband modulation for digital systems.

frequency modulation (FM) (Figure 2-5a). The output of the modulator is a higher frequency signal that can be easily transmitted using a practical small antenna. For a typical digital modulation system, the analog signal (e.g., voice) is digitized using a device known as an analog-to-digital converter (ADC). The ADC takes samples of the analog input and creates a digital pattern of "1"s and "0"s representing the value of the analog sample. This digital data stream is used to modulate the carrier frequency by shifting either the phase or frequency by the value of the digital signal (Figure 2-5b). In addition, the modulator is used to generate spread spectrum and coding to enhance the digital signal.

2.2 Voltage Standing Wave Ratio

The voltage standing wave ratio (VSWR) is caused by a mismatch in the impedance between devices. It determines how much power is reflected back from the load that is not matched to the source. Typically, this is specified using a 50 Ω system; the VSWR is the measurement used to indicate the amount of impedance mismatch. The amount of power that is reflected to the source is a loss in the signal level, which is a 1:1 dB loss in the link analysis.

In the link analysis, the power amplifier is chosen along with the gain of the antenna to meet the slant range of the data link. The power amplifier is generally a high-cost item in the transceiver, and therefore careful transceiver design to minimize losses in the link analysis will reduce the power requirement of the power amplifier and also the overall cost of the transceiver. Therefore, designing the power amplifier and the antenna to obtain the best impedance match or the lowest VSWR will help minimize the loss.

The VSWR defines a standing wave that is generated on the cable between the power amplifier and the antenna. The ratio of the minimum and maximum voltages measured along this line is the VSWR. For example, a VSWR of 2:1 defines a mismatch where the maximum voltage is twice the minimum voltage on the line. The first number defines the maximum voltage, and the second is the normalized minimum. The minimum is defined as one, and the maximum is the ratio of the two. The match between the power amplifier and the antenna is often around 2:1, depending on the bandwidth of the antenna.

The standing wave is caused by reflections of the signal, due to a mismatch in the load, that are returned along the transmission line. The reflections are added by superposition, causing destructive and constructive interference with the incidence wave depending on the phase relationship between the incidence wave and the reflected wave. This results in voltages that are larger and smaller on points along the line. The main problem with mismatch and a high VSWR is that some of the power is lost when it is reflected to the source and thus is not delivered to the load or antenna. The standing wave is minimized by making the impedances equal so that there are virtually no reflections, which produces a VSWR of 1:1.

2.2.1 Maximum Power Transfer Principle

The maximum power transfer principle states that for maximum power transfer to a load, the load impedance should be equal to the source impedance. For example, if the source impedance is equal to 50 Ω, then the load impedance should be equal to 50 Ω. This produces a VSWR of 1:1, which means that no power is reflected to the source; it is all delivered to the load (Figure 2-6).

However, the reverse is not true. If the impedance of the load is given, the source impedance is not chosen as equal to the load but is selected to minimal, or zero. Therefore,

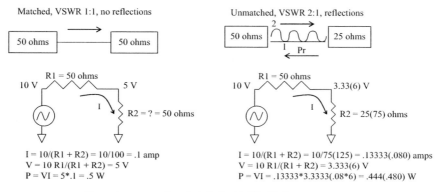

Figure 2-6   Maximum power transfer to the load with a given source impedance.

R1 = ? = 50 ohms

10 V   5 V

I   R2 = 50 ohms

I = 10/(R1 + R2) = 10/100 = .1 amp
V = 10 R1/(R1 + R2) = 5 V
P = VI = 5*.1 = .5 W

R1 = 0 ohms

10 V   10 V

I   R2 = 50 ohms

I = 10/(R1 + R2) = 10/50 = .2 amp
V = 10 R1/(R1 + R2) = 10 V
P = VI = 10*.2 = 2 W

Maximum power transfer to the load is when R1 = 0 ohms

Figure 2-7   Maximum power transfer to the load with a given load impedance.

if the load impedance is given as 50 Ω, the source impedance is not selected as 50 Ω for maximum power transfer to the load but is chosen as 0 Ω. If the source is zero, then there is no loss in the source resistance (Figure 2-7). This applies especially for operational amplifiers (op amps), whose source impedance is low, around 10 Ω.

2.3 Digital Communications

Digital communications is becoming the standard for nearly all communications systems. The advantages of digital communications are as follows:

• Perfect reconstruction of the transmitted digital waveform assuming no bit errors. After digitizing the analog signal, the digital bit stream can be reconstructed at the receiver, and if there are no bit errors it is an exact replica of what was sent by the transmitter. There is no gradual build-up as there is in analog systems, so the quality of the signal is excellent in a digital system until bit errors occur in the digital waveform (Figure 2-8).

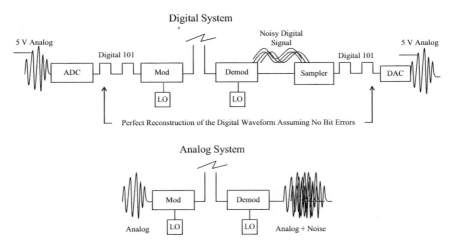

Figure 2-8   Digital versus analog communications.

- Smaller size due to custom application-specific integrated circuits (ASICs) and programmable chips such as digital signal processors (DSPs) and field-programmable gate arrays (FPGAs).
- Software-defined radios (SDRs) utilize the programmability of these DSPs and FPGAs to change several parameters including modulation to provide highly diverse radios that can be used for multiple applications and networking. Basically it is like having multiple radios in one unit that is controlled by software.
- Cognitive radios (CRs) can be realized using digital communications and can be adaptive to the changing environment. The CRs can change functionality on the fly including frequency, modulation, bandwidths, and many other parameters made possible by the ability to program the digital circuitry to accommodate changes in the environment.

2.3.1 Digital versus Analog Communications

As mentioned earlier, digital transmissions can be perfectly reconstructed at the receiver, assuming there are no bit errors. However, when analog signals, such as voice, are used in a digital communication system, the analog signal needs to be digitized. This becomes a source of error, since this is only a digital approximation of the analog signal. With over-sampling, this error becomes very small.

For an analog communication system, the actual analog signal modulates the carrier, and the analog signal is sent to the receiver. However, as noise and external signals distort the analog signal, it slowly degrades until it is undetectable or until it reaches the minimum detectable signal (MDS) (Figure 2-8).

The advantage of a digital signal is that as it degrades the quality of the signal is not affected, assuming the receiver can still detect the bits that were sent from the transmitter with no bit errors. Therefore, it does not matter how noisy the signal is coming into the receiver: if the bits can be detected without bit errors or the errors can be corrected, then the quality of the signal does not change or sound noisy.

An example of this is the comparison between analog cell phones and digital cell phones. Analog cell phones continue to get noisier until it is difficult to discern the voice because of so much noise, whereas digital cell phones are perfectly clear until they start getting bit errors, which results in dropouts of the voice or missing parts of the conversation.

2.3.2 Software Programmable Radios and Cognitive Radios

Software programmable radios are often referred to as SDRs. A big advantage of digital communications is that the modulation and demodulation can be accomplished using software. This makes the design much more flexible since this technique is used to change to different modulation schemes. The principle is that the software is programmed in both the transmitter and the receiver to provide multiple modulation schemes that can be selected on a real-time basis. If there are multiple transmitters with different modulation types, the receiver can be programmed to accept these different modulation types by changing the software. An example of this is using a data link that is designed and built to transmit and receive amplitude modulation minimum shift keying (AM-MSK). By simply changing the software or changing the software load of either a DSP or FPGA or both, this modulation can

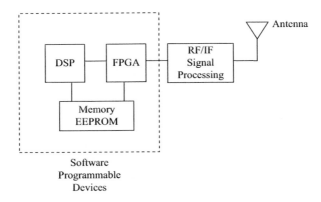

Figure 2-9 Basic software-defined radio.

change to differential 8-level phase-shift keying (D8PSK). This was accomplished in setting the standards for special category 1 (SCAT-1) landing systems. The software can be changed at the factory, or for a real-time system the software loads can reside in memory on the data link and the desired software can be loaded on command.

The hardware uses DSPs, FPGAs, or both. FPGAs are generally used for high-speed processing, while DSPs are generally used for low-speed processing. Many data link systems today use both FPGAs and DSPs and partition the functionality between these parts. This technique can be used to optimize the function of the data link. A basic software-defined radio is shown in Figure 2-9.

Cognitive radios are an extension of SDRs. They use smart radios that can sense the present environment for existing radios with given modulations and data links and program the transmitter, receiver, or both for interoperability. In addition, cognitive radios can automatically determine which bands or parts of bands have the minimum amount of noise, jammers, or other users and incorporate this information to determine the best band for operation of the cognitive data link. This maximizes the number of users by optimizing the data link with respect to frequency, modulation, time, and space and provides the best noise-free and jammer-free operation.

2.4 Digital Modulation

The most common form of digital modulation is changing the phase of the carrier frequency in accordance with the digital signal that is being sent. This is known as phase-shift keying (PSK) or direct sequence digital modulation. This digital sequence can be either the digitized data or a combination of digitized data and a spread spectrum sequence. There are many different levels and types of PSK; only basic modulation methods are discussed here. However, the principles can be extended to higher order PSK modulations. The basic form of digital communication is shown in Figure 2-10. This shows the carrier frequency and the digital data being fed into a modulator. The modulator is binary phase-shift keying (BPSK) and changes the phase between 0° and 180° according to the digital data being sent.

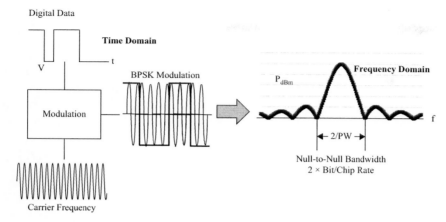

Figure 2-10   Shifts the phase of the carrier frequency according to the data.

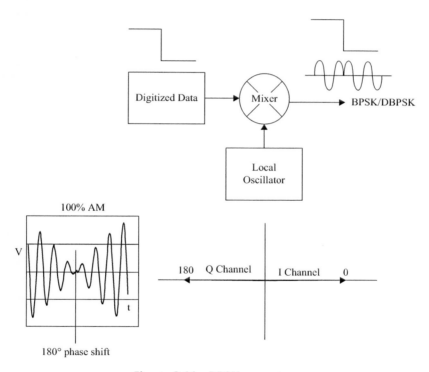

Figure 2-11   BPSK generator.

2.4.1 Binary Phase-Shift Keying

The BPSK modulator and phasor diagram is shown in Figure 2-11. BPSK is defined as shifting the carrier $0°$ or $180°$ in phase, depending on the digital waveform. For example, a binary "0" gives $0°$ phase of the carrier, and a binary "1" shifts the carrier by $180°$.

To produce the digital waveform, the data or information signal is digitized, encoded, and sent out in a serial bit stream (if not already). The end result is a serial modulating digital waveform representing the data to be transmitted. This digital output contains 0 and 1 and often needs to be changed to ±1 to directly modulate the carrier frequency in a typical mixer application. However, certain forms of hardware can bypass this step and modulate the mixer directly. The output of the mixer is a BPSK modulated carrier signal that is transmitted and sent over to the receiver for demodulation and detection. As the carrier phase changes 180°, the hardware does not allow for an instantaneous change in phase, so the amplitude goes to zero, which produces 100% amplitude modulation (Figure 2-11). This is called the zero crossover point and is an unwanted characteristic of BPSK and can cause degradation to the waveform and performance.

A high-speed pseudo-noise (PN) code and modulo-2 adder/exclusive-or functions are added to the basic BPSK modulator to produce a spread spectrum waveform (Figure 2-12). The high-speed code generates a wider spectrum than the data spectrum needed for communications, which is known as a spread spectrum.

Applying minus voltage to an actual RF mixer reverses the current through the balun of the mixer and causes the current to flow in the opposite direction, creating a net 180° phase shift of the carrier. Thus, the carrier is phase shifted between 0° and 180° depending on the input waveform. A simple way of generating BPSK spread spectrum is shown in Figure 2-13. The LO is modulated by this digital sequence producing a 0° or 180° phase shift on the output of the mixer. Other devices such as phase modulators and phase shifters can create the same waveform as long as one digital level compared with the other digital level creates a 180° phase difference in the carrier output.

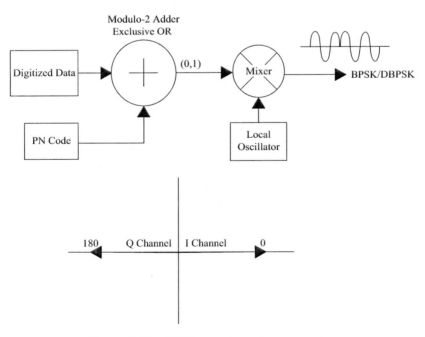

Figure 2-12   BPSK spread spectrum generator.

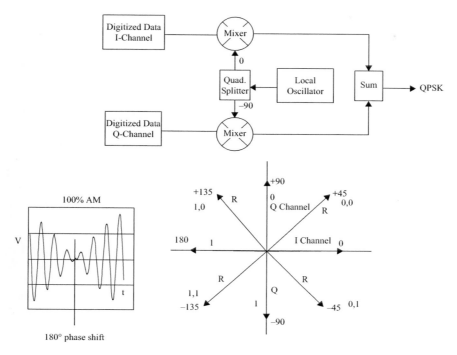

Figure 2-13   QPSK generator.

2.4.2 Differential Phase-Shift Keying

The BPSK waveform above can be sent out as absolute phase (i.e., a 0° phase shift is a "1" and a 180° phase shift is a "0"). This type of system is known as a coherent system, or coherent BPSK. Another way to perform this function is to use differential phase-shift keying (DPSK), which generates and detects a change of phase. A change of phase (0° to 180° or 180° to 0°) represents a "1", and no change (0° to 0° or 180° to 180°) represents a "0." This scheme is easier to detect because the absolute phase does not need to be determined; just the change of phase is monitored. The differential mode can be applied to various phase shifting schemes and higher order phase shift schemes. DPSK results in a degradation of the signal, compared with using the absolute phase or coherent BPSK. However, it is much easier to demodulate and detect the data and generally is less costly due to the demands of keeping the phase constant over a longer period of time. Note that differential techniques can be applied in higher order PSK systems such as differential quadrature phase-shift keying (DQPSK) and differential 8-level phase-shift keying (D8PSK).

2.4.3 Quadrature Phase-Shift Keying

Quadrature phase-shift keying (QPSK) is generated by quadrature phase shifting the LO so that four possible phase states are produced at the output. One method of producing these phasors is by using two channels: one channel containing an LO that is in phase and the output of the mixer is BPSK modulated to produce 0° or 180°; and the other channel containing the same LO that is shifted by 90° so that the output of the mixer is BPSK modulated to produce 90° or 270°. These two channels are then combined to produce the four phase states (Figure 2-13). The two BPSK systems are summed together, which gives four possible resultant phasors—45°, 135°, −135°, and −45°—which are all in quadrature, as shown in Figure 2-13. Since the digital transitions occur at the same time, changes between any four resultant phasors can occur.

Depending on the input of both bit streams, the phase of the resultant can be at any of the four possible phases. For example, if both bit streams are 0, then the phasor would be 45°. If both bit streams changed to 1, then the phasor would be at −135°, giving a change of carrier phase of 180°. If only the first channel changes to 1, then the phasor would be at −45°, giving a change of 90°. Therefore, the four possible phase states are 45°, 135°, −135°, or −45°. The phasor diagram can be rotated, since it is continually rotating in time and only a snapshot is shown, and then the phasors would be at 0°, 90°, −90°, and 180° for QPSK generation. QPSK has the capability of 180° phase shifts depending on the code and the previous phase state, so it often goes through the zero crossover point, which produces 100% amplitude modulation during the phase transition as shown in Figure 2-13. This causes unwanted characteristic of QPSK and can cause degradation to the waveform and performance.

Usually, the LO contains the 90° phase shift that is used to provide the two quadrature channels of the BPSK phasors instead of trying to provide the 90° phase shift of the actual binary input. Either way it would provide the 90° quadrature phase rotation that is required. However, quadrature rotation of the binary input requires phase shifting all of the frequency components by 90°. This requires a more sophisticated and expensive phase shifter that is broadband. Shifting only the carrier to produce the quadrature channels requires a phase shift at one frequency, which is much simpler to build. In fact, in the latter case, a longer piece of cable cut to the right length can provide this 90° quadrature phase shift.

2.4.4 Offset QPSK

Another type of QPSK is referred to as offset QPSK (OQPSK) or staggered QPSK (SQPSK). This configuration is identical to QPSK except that one of the digital sequences is delayed by a half-cycle so that the phase shift occurs on only one mixer at a time. Consequently, the summation of the phasors can result in a maximum phase shift of only ±90° and can change only to adjacent phasors (no 180° phase shift is possible). The phasor diagram in Figure 2-14 is identical to Figure 2-13, except no 180° phase shift is possible. This prevents the zero-amplitude crossover point and provides smoother transitions with less chance of error, reducing the AM effects common to both BPSK and QPSK (Figure 2-14). OQPSK changes only 90° so that the phasor AM is much less (approximately 29%). Figure 2-14 also shows a simple way of generating OQPSK. The only difference between this and the QPSK generator is the ½-bit delay in one of the bit streams. This prevents the 180° phase shift present in the QPSK waveform.

2.4.5 Higher Order PSK

The previous analysis can be extended for higher order PSK systems. The same principles apply, but they are extended for additional phase states and phase shifts.

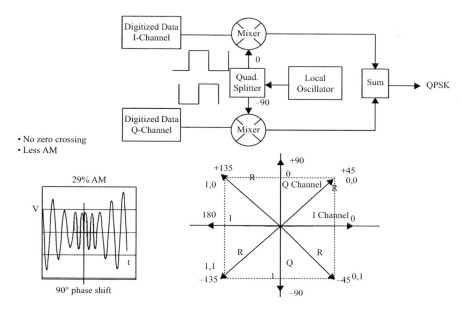

Figure 2-14   OQPSK generator.

Direct sequence waveforms are used extensively for digital modulation systems. BPSK is the simplest form of PSK, shifting the carrier 0° or 180°. QPSK uses two BPSK systems, one in quadrature, to achieve four phase shifts: 0°, 90°, −90°, and 180°. OQPSK is used to minimize AM and is identical to QPSK with the phase transitions occurring at ½-bit intervals and at different times for each of the BPSK channels. This provides phase shifts of 0°, 90°, and −90° degrees and eliminates the 180° phase shift. Higher order PSK systems can be analyzed much the same way, only with more phase states and phase transitions. However, the more phase shift possibilities, the harder it is to detect and resolve the different phase states. Therefore, there is a practical limit on how many phase states can be sent for good detection. This limit seems to grow with better detection technology, but caution must be observed in how many phase states can be sent out reliably for practical implementation.

2.4.6 $\pi/4$ Differential QPSK

The $\pi/4$ DQPSK is a modulating scheme that encodes 2 bits of data as ±45° and ±135° phase shifts that are in quadrature. The next phase shift will be from that resultant phasor. Therefore, if the first phase shift is +135° and the second phase shift is +135°, the resultant phasor's absolute phase would be at 270°. Because it is differential, once the phase shift occurs, the next phase shift starts at the last phase state. Then, according to the data bits, it shifts the phase either ±45° or ±135°. Another example: if a +45° shift occurs referenced to the last phasor sent, then the bits that are sent equal 0,0. Proceeding with the other phase shifts, a −45° shift is 0,1; a +135° shift is 1,0; and a −135° shift is 1,1 (Figure 2-15). The shifts for the next bits sent start from the last phasor measured. Over time this will eventually fill up the constellation and will look similar to the D8PSK diagram. Note that none of the

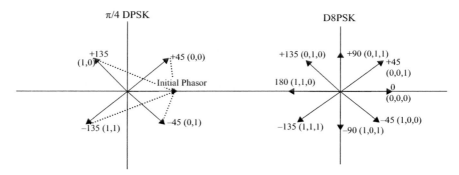

Figure 2-15   Phasor diagrams for a $\pi/4$ DPSK and D8PSK modulation waveforms.

phase shifts go through zero amplitude, so there is less AM for this type of modulation compared with DQPSK, which does go through zero amplitude.

2.4.7 Differential 8-Level PSK

The D8PSK type of modulation is the same as $\pi/4$ DQPSK but includes the phase-shift possibilities of $0°$, $90°$, $-90°$, and $180°$, thus providing $0°$, $45°$, $-45°$, $90°$, $-90°$, $135°$, $-135°$, and $180°$ phase shifts. This provides eight possible phase shifts, or 3 bits (2^3) of information, as shown in Figure 2-15. Since this is a differential system, these phase shifts are referenced to the previous bit, not the absolute phase. Therefore, the previous bit is mapped to the reference phasor with zero degrees for every bit received, and the next bit is shown to have one of the eight possible phase shifts referenced to $0°$. If this was not a differential system, then the phasors would be absolute, not referenced to the previous bit.

Thus, for the same bandwidth as BPSK, the D8PSK can send three times as many bits (i.e., the bit rate is three times the bit rate of BPSK); however, it requires a higher E_b/N_o. The actual phase shifting occurs at the same rate as BPSK. The rate of the phase shifts is called the symbol rate. The symbol is the rate at which the phase is shifted, but the actual bit rate in this case is three times the symbol rate since there are 3 bits of information in each phase shift or symbol. The symbol rate is important because it describes the spectral waveform of the signal in space. For example, if the symbol rate is 3 ksps (1000 samples per second), then the null-to-null bandwidth would be 6 kHz wide (2 times the symbol rate), which would decode into 9 kbps. Note that this type of modulation allows a $180°$ phase shift, which experiences AM.

2.4.8 16-Offset Quadrature Amplitude Modulation

Another common modulation scheme is 16-offset quadrature amplitude modulation (16-OQAM), which is very similar to OQPSK. In 16-OQAM, each of the phasors has two amplitude states before summation. The resultant phasor in one quadrant has four possible states (R1, R2, R3, and R4), as shown in the phasor diagram in Figure 2-16. Offset is used here for the same reasons as in OQPSK—to prevent transitions through zero amplitude and to reduce the AM on the output.

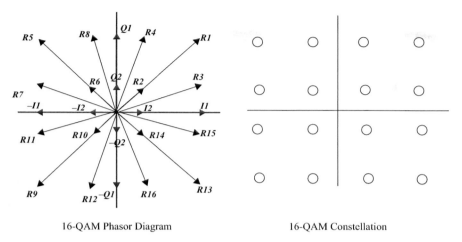

16-QAM Phasor Diagram   16-QAM Constellation

Figure 2-16   Amplitude/phasor diagram for 16 OQAM.

Since there are 4 possible amplitude/phase positions in one quadrant and a total of four quadrants, there are 16 possible amplitude/phase positions in this modulation scheme (Figure 2-16). These 16 states equates to 4 bits of information for each change in state or symbol, which provides four times the data rate compared with BPSK. However, a higher E_b/N_o is required.

Since this is 16-offset QAM, the "offset" means that the phasors in the defined quadrant can change only to the adjacent quadrants. For example, the phasors shown in quadrant 1 can change only to quadrants 2 and 4. These phasors cannot change from quadrant 1 to quadrant 3. In other words, only the I or Q channel can change in phase at one time, not both at the same time (Figure 2-16).

2.4.9 Phasor Constellations and Noise Immunity

A phasor constellation is a technique used to determine the quality of phase modulated waveforms and to distinguish between phase states. A phasor constellation shows the magnitude and phase of the resultant phasor, including noise. A vector analyzer is used to display the constellations. The constellation diagram for a 16-OQAM signal is shown in Figure 2-16. The 16 phasors appear as points in a two-dimensional display representing both phase and amplitude and with no noise in the system. These constellation points show the amount of separation between one phasor state point and another. The data rate of the 16-state quadrature amplitude modulation (16-QAM) is four times higher than BPSK, as shown in Figure 2-17.

To understand the noise immunity of different modulation types, Figure 2-17b shows two types of phase modulations, including noise, which varies the phasor's end points for BPSK and 16-OQAM modulated signals. For BPSK, the constellations are close enough together that it is very easy to detect which value of phase was sent, 0° or 180°. Therefore, BPSK is much more resistant to noise and variations in phasor values. Although the 16-OQAM signal contains more bits per phasor value or symbol, which provides a faster data rate for a given

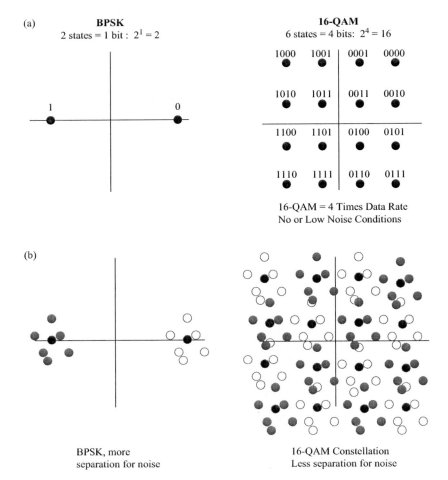

Figure 2-17   (a) Higher order modulation results in higher data rates. (b) Constellation and noise immunity between 16-QAM and BPSK.

bandwidth, the actual phasor value is more susceptible to noise, and it is much more difficult to separate the noisy phasor values, which results in more bit errors. This is a trade-off between noise resistance and higher data rates.

The quality of the phasor values can also be measured by determining the error vector magnitude (EVM). An error vector is a vector in the I–Q plane between the ideal vector or constellation point and the received vector or constellation point (Figure 2-18). The constellation points are the ends of the vectors and appear as points on a vector analyzer. These points represent the magnitude and phase of the vectors. The average length or magnitude of the error vector, defined as the distance between these two points, is the EVM. The ratio of the EVM (the distance between constellation points) and the ideal vector magnitude multiplied by 100 provides the EVM as a percentage. For example, if the ideal phasor magnitude is equal to 10 and the EVM is equal to 3, then the EVM in percent would be

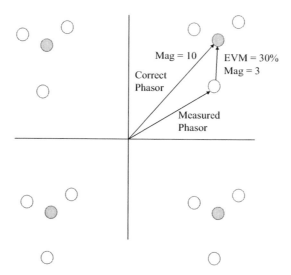

Figure 2-18   Error vector magnitude (EVM).

$3/10 \times 100 = 30\%$ EVM (Figure 2-18). The smaller the percentage EVM, the less noise and the higher the probability of detecting the correct signal phase and amplitude.

2.4.10 BPSK versus QPSK Constellation Comparison

BPSK uses two quadrants or two phase states to transmit the data. QPSK, on the other hand, uses four quadrants or four phase states, each state representing two bits of data, referred to as symbol rate. Therefore, QPSK can send twice as many data bits as BPSK. However, using four quadrants increases the bit error rate since the symbols do not have enough separation as shown in Figure 2-19.

This can be overcome by either increasing the signal power or reducing the noise by decreasing the bandwidth as shown in Figure 2-20. Reducing the bandwidth requires reducing the symbol rate by ½ so the bit rate for both BPSK and QPSK are the same.

BPSK: $P_e = 1/2\,\text{erfc}((E_b/N_o)^{1/2})$
QPSK: $P_e = 1/2\,\text{erfc}((E_b/N_o)^{1/2}) - 1/4\,\text{erfc2}((E_b/N_o)^{1/2})*$

Increasing the signal power allows QPSK to send twice the data as BPSK in the same bandwidth.

2.4.11 Variations in PSK Schemes

Many other PSK configurations are in use today, and there will be other types of modulating schemes in the future. One of the main concerns with using PSK modulators is the AM that is

---

* the second term can be eliminated if $E_b/N_o \ggg 1$.

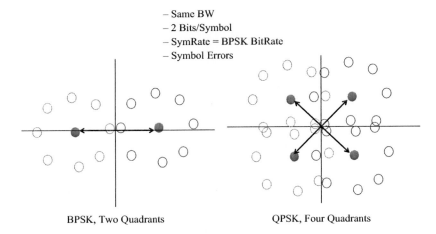

Figure 2-19   Constellations of BPSK and QPSK with the same signal power.

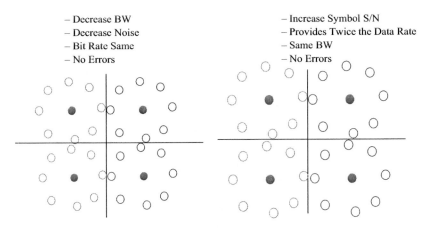

Figure 2-20.   Improvements by decreasing bandwidth or increasing S/N for QPSK.

inherent when phase states are changed. For example, BPSK switches from 0° to 180°. Since this is not instantaneous, due to bandwidth constraints, the phasor passes through zero amplitude in the process. The more band-limiting, the more AM there is in the resultant waveform. Various schemes have been developed to reduce this AM problem. For example, OQPSK allows only a maximum phase shift of 90° so that the AM problem is significantly reduced. Also, minimum shift keying (MSK) is used to smooth out the phase transitions, which further reduces the AM problem and makes the transition continuous. Continuous phase PSK (CP-PSK), also referred to as constant envelop modulation, is another modulation scheme that prevents AM, where the phase shifts are continuous around the phasor diagram with no change in amplitude.

• Sinusoidally transitions from one phase state to another
• Remains at a phase state for a period of time
• Used for packet radio and other burst type systems

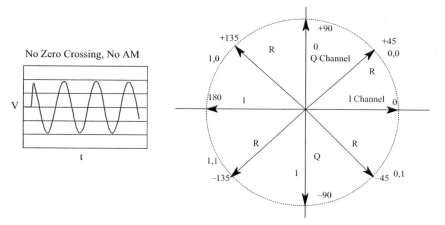

Figure 2-21   Continuous phase-phase shift keying (CP-PSK).

2.4.12 Continuous Phase PSK

Continuous phase modulation is being used extensively in data links due to spectral characteristics, burst-type modulation, and packet radios, where data are sent in bursts or packets. The modulation causes the phase of the carrier to shift to another state, where it can remain in that given state for a period of time. Continuous phase modulation uses sinusoidal transitions from one phase state to another. This does not allow zero amplitude crossovers between phase shifts, so this type of modulation does not exhibit AM (Figure 2-21). This prevents spectral regrowth or sidelobe regeneration due to nonlinearities in the hardware.

2.4.13 Spectral Regrowth

Digital modulation causes sidelobes to be generated by the digital signal. These sidelobes extend continuously in the frequency domain. To prevent interference with other out-of-band signals and to comply with FCC regulations, these sidelobes need to be filtered. However, when the digital modulation is not continuous, these sidelobes are regenerated when introduced to a nonlinear function in the hardware. For example, after the sidelobes are filtered, when they are amplified by the power amplifier—which most of the time is nonlinear—the sidelobes reappear (Figure 2-22). This can be a major concern when complying with regulations and preventing out-of-band interference. Continuous phase modulation schemes minimize this spectral regrowth and keep the sidelobes down even when they are exposed to a nonlinear device such as CP-PSK.

2.4.14 Minimum Shift Keying

If the OQPSK signal is smoothed by sinusoidally weighting the BPSK signals before summation, then an MSK signal is produced (Figure 2-23).

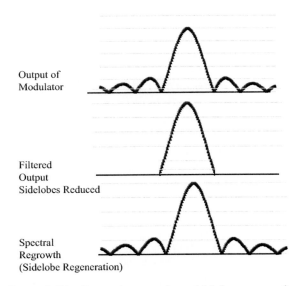

Output of
Modulator

Filtered
Output
Sidelobes Reduced

Spectral
Regrowth
(Sidelobe Regeneration)

Figure 2-22 Spectral regrowth or sidelobe regeneration.

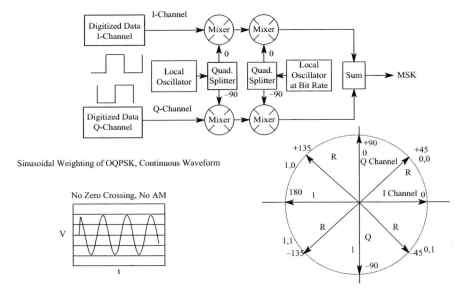

Figure 2-23 MSK generator.

The sinusoidal weighting frequency is equal to half the bit rate, so that it eases the transition of a digital data bit by one cycle of the sine wave weighting signal. This provides a smoothing of the 180° phase shifts that occur in both the channels. Then the phase transitions of the two channels are slowed, which reduces the high-frequency content of the spectrum and results in attenuation of the sidelobes containing the high frequencies. Therefore, the spectrum is said to be efficient compared with standard PSK systems (e.g., BPSK, QPSK) since more power is contained in the main lobe and less in the sidelobes. The main lobe for MSK is 1.5 times larger than for OQPSK for the same number of bits sent due to the sinusoidal weighting.

Since the modulation is half the bit rate, the spectrum is widened by ½ bit width on both sides, which equals a 1 bit width bigger spectrum. Therefore, the total width would be three times the bit rate, which is 1.5(2 × bit rate). MSK is a continuous constant envelop modulation scheme, so it contains minimal AM and prevents sidelobe regeneration after filtering.

MSK can be generated in two ways. The first is sinusoidally weighting the two BPSK channels in the OQPSK design before the summation takes place. This reduces the sidelobes and increases the main lobe. The other way to generate MSK is to use a frequency-shift keying (FSK) system, ensuring that the frequency separation is minimal.

2.4.15 Frequency-Shift Keying

Minimum shift keying can be generated using FSK by setting the frequency separation between the two frequencies equal to ½ the frequency shift rate or data rate. Therefore, MSK is a continuous phase FSK with a modulation index of 0.5. This is where MSK got its name, since it is the minimum spacing between the two frequencies that can be accomplished and still recover the signal with a given shift rate. It is rather remarkable that the same waveform can be produced via two different generation methods, each of which provides a different way of understanding, designing, and analyzing MSK systems. A simple way of generating MSK is shown in Figure 2-24.

A two-frequency synthesizer is frequency shifted by the binary bit stream according to the digital data. If the rate is set to the minimum bit rate, then MSK is the resultant output.

The frequency spacing of the FSK needs to be equal to ½ the bit rate to generate MSK. The frequency shift provides one bit of information, as does the BPSK waveform, but the null-to-null bandwidth is only 1.5 times the bit rate compared with 2 times the bit rate in the BPSK waveform. A simulation shows different types of FSK with different spacings with respect to the bit rate. If the frequency spacing is closer than ½ the bit rate, then the information cannot be recovered. If the spacing is too far apart, the information can be retrieved using FSK demodulation techniques, but MSK is not generated. As the frequency spacing approaches ½ the bit rate, then the resultant spectrum is MSK (Figure 2-25).

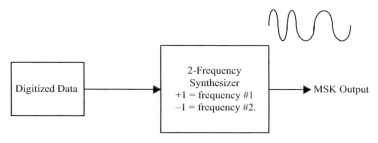

Figure 2-24 MSK generator using FSK.

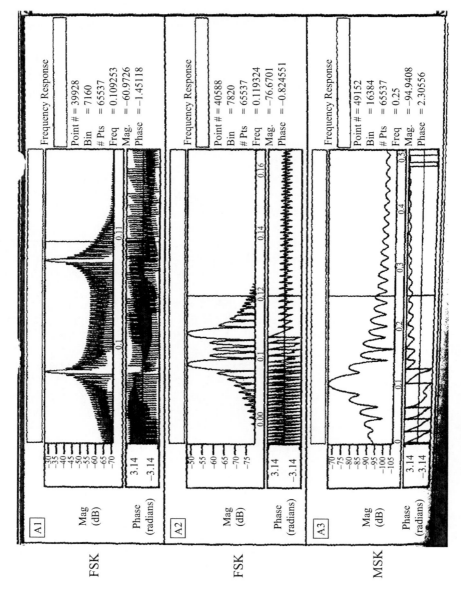

Figure 2-25 FSK spectrums with spacing minimized to generate MSK.

2.4.16 Sidelobe Reduction Methods

One of the problems with PSK systems is that the sidelobes can become fairly large and cause a problem with adjacent channel operation. The sidelobes continue out theoretically to infinity. The main concern is usually the first or second sidelobes, which are larger in magnitude. To confine the bandwidth for a particular waveform, a filter is required. The main problem with filtering a PSK signal is that this causes the waveform to be dispersed or spread out in time. This can cause distortion in the main signal and also more intersymbol interference (ISI), which is interference between the digital pulses.

2.4.17 Ideal Shaping Filter

For no ISI, the sampling time of the pulse needs to occur when the magnitude of all other pulses in the digital signal are zero amplitude. A square wave impulse response in the time domain would be ideal, since there is no overlap of digital pulses and thus sampling anywhere on the pulse would not have ISI. However, the bandwidth in the frequency domain would be very wide, since a square wave in the time domain produces a $\sin(x)/x$ in the frequency domain.

Therefore, the best shaping filter to use for digital communications is a $\sin(x)/x$ impulse response in the time domain, which produces a square wave in the frequency domain. This provides ideal rejection in the frequency domain and also a sampling point in the center of the $\sin(x)/x$, where all the other digital pulses have an amplitude of zero. This produces the minimum ISI between symbols (Figure 2-26). However, a $\sin(x)/x$ impulse response is not possible since the sidelobes extend out to infinity. Thus practical filter implementations are used to approximate this ideal filter. Some of the practical filters used in communications are Gaussian, raised cosine, raised cosine squared, and root raised cosine, which are all approximations of the $\sin(x)/x$ impulse response. The $\sin(x)/x$ impulse response approximates a square wave in the frequency domain, which reduces the noise bandwidth, reduces out-of-band transmissions, and is designed to minimize ISI (Figure 2-27).

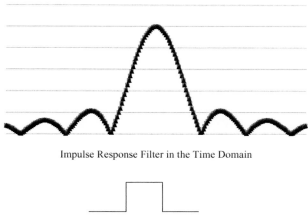

Impulse Response Filter in the Time Domain

Ideal Frequency Response of a Time Domain Impulse Response Sin$x$/x Filter

Figure 2-26   Ideal sin$x$/x impulse response in the time domain.

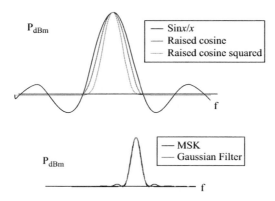

Figure 2-27   Practical digital waveform impulse responses.

Root raised cosine filters are used extensively in communications and data links. They are employed in both the transmitter and the receiver, which basically splits the raised cosine filter, with half of the filter in the transmitter and half of the filter in the receiver. Therefore, the net result is a combination of the square root of the raised cosine filters, which leads to a raised cosine filter for the total data link system. The root raised cosine filter has slightly faster transitions in the time domain for the transmitter pulse, and since a matched filter is used on the receiver side splitting the raised cosine filter response provides a slight improvement in the performance.

Another type of filtering scheme uses a Gaussian-shaped pulse. A Gaussian-shaped curve is a standard bell-shaped curve showing a Gaussian distribution, which is used extensively in probability theory (see Chapter 6). The Gaussian-shaped impulse response also provides a good approximation to the ideal $\sin(x)/x$ impulse response (Figure 2-27). This provides effective use of the band and allows multiple users to coexist in the same band with minimum interference. One type of modulation that includes this Gaussian impulse response is Gaussian MSK (GMSK), a continuous phase modulation scheme that reduces the sidelobe energy of the transmitted spectrum. The main lobe is similar to MSK and is approximately 1.5 times wider than QPSK.

2.5 Direct Sequence Spread Spectrum

Many systems today use spread spectrum techniques to transport data from the transmitter to the receiver. One of the most common forms of spread spectrum uses PSK and is referred to as direct sequence spread spectrum (DSSS). The data are usually exclusive-or'd or modulo-2 added with a pseudo-random or pseudo-noise (PN) code that has a much higher bit rate than the data bit rate to produce the spread spectrum waveform. The bits in this high-speed PN code are called chips, to distinguish them from data bits. This produces a much wider occupied spectrum in the frequency domain (spread spectrum) (Figure 2-28).

Direct sequence spread spectrum systems use phase-shift generators (PSGs) to transfer data from the transmitter to the receiver by phase shifting a carrier frequency. This is done for both spread spectrum systems and nonspread spectrum systems. There are several ways to build a PSG depending on the waveform selected.

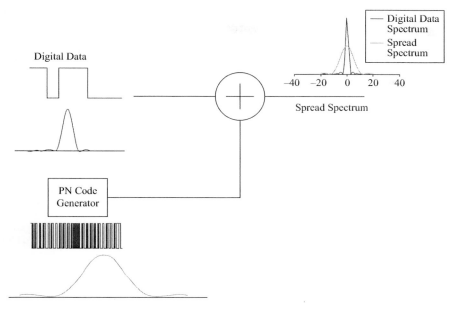

Figure 2-28   Direct sequence spread spectrum.

Fundamentally, spread spectrum uses more bandwidth than is required to send the signal. It utilizes techniques such as a faster pseudo-random code to spread the bandwidth. A higher code rate requires more bandwidth to send and receive the code.

2.5.1 Frequency-Hopping Spread Spectrum

Most frequency-hopping spread spectrum transmitters use a direct frequency-hopping synthesizer for speed. Direct frequency-hopping synthesizers use a reference oscillator and directly multiply or translate the reference frequency to higher frequencies for use in the hopping sequence. An indirect frequency synthesizer uses a phase-locked loop (PLL) to generate the higher frequencies, which slows the process due to the loop bandwidth of the PLL. A PN code is used to determine which of the frequency locations to hop to. The frequency hop pattern is therefore pseudo-random to provide the spread spectrum and the processing gain required for the system (Figure 2-29).

The process gain assumes that the jammer does not jam multiple frequency cells. Usually the best jammer is a follower jammer. A follower jammer detects the frequency cell that the frequency hopper is currently in and immediately jams that frequency. The delay is associated with the time to detect the frequency and then to transmit the jammer on that frequency. For slow frequency-hopping systems, this method works quite well because most of the time the signal will be jammed. The faster the frequency hopping, the less effective the follower jammer is on the system. Also, many times a combination of frequency hopping and direct sequence is used, slowing the detection process and reducing the effectiveness of the follower jammer.

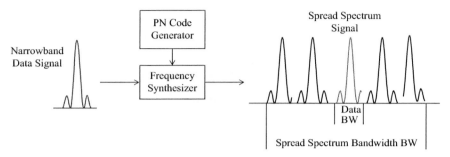

Figure 2-29 Frequency-hopped spread spectrum.

2.5.2 Spread Spectrum

The main reasons for using spread spectrum are as follows:

- It provides jammer resistance, reducing the effects of jamming.
- It allows multiple users to operate in a given band.
- It reduces interference with existing narrowband systems.
- It minimizes multipath effects; multipath is narrowband and dependent on frequency.
- It provides a low probability of intercept/low probability of detect (LPI/LPD); the transmitted signals are hard to detect or intercept and are spread out so that they look like noise.

Spread spectrum was initially invented and patented in 1942 by an actress named Hedy Lamarr and a composer named George Antheil. It was used by the military during World War II for covert operations to transmit signals undetected. However, it was not fully utilized until many years later. Today spread spectrum is commonplace and is used extensively in communications and in the development and design of data links.

2.5.3 Jammer Resistance

One of the main advantages of spread spectrum systems is their ability to operate in a jamming environment, both friendly interfering signals and hostile jammers where there is an attempt to jam communications. The more spreading that is used, the more jammer resistance the system has. An analysis that is used to determine a system's ability to resist jamming is called process gain (G_p). Process gain is essentially the spread bandwidth over the required bandwidth or how much band spreading the system is using.

The process gain for DSSS systems or PSK systems is

$$G_p = 10\log(\text{RF bandwidth/detected bandwidth})$$

The process gain for frequency-hopping systems is equal to the occupied bandwidth over the signal bandwidth and is

$$G_p = 10\log(\text{number of frequency channels})$$

The number of frequency channels assumes that they are independent and adjacent.

When calculating G_p for a frequency-hopping system, the separation of the frequencies used can alter the actual jamming margin. For example, if the frequencies are too close together, the jammer may jam multiple frequencies at a time. Also, the frequencies used need to be adjacent, with no missing frequencies, or the previous calculation becomes invalid.

The jamming margin (J_m), or the level of a jamming signal that will interfere with the desired signal, is equal to the process gain (G_p) minus the spreading losses (L_{ss}):

$$J_m = G_p - L_{ss}$$

L_{ss} assumes that the signal-to-jammer ratio (SJR) is zero.

In addition, since most systems need a certain amount of signal level above the jamming level, the required SJR is specified. The required SJR will vary depending on the modulation used and the bit error rate (BER) required. This needs to be included to calculate the actual jamming margin.

What follows is an example of calculating the jamming margin of a direct sequence system with a required SJR:

$$G_p = 10\log(1000 \text{ MHz}/10 \text{ MHz}) = 20 \text{ dB}$$
$$J_m = 20 \text{ dB} - 3 \text{ dB} = 17 \text{ dB}$$
$$J_m \text{ (actual)} = 17 \text{ dB} - \text{SJR (required)} = 17 \text{ dB} - 9.5 \text{ dB} = 7.5 \text{ dB}$$

A signal can be received at the specified BER with the jammer 7.5 dB higher than the desired signal.

Two methods to obtain the process gain are shown in Figure 2-30. They include a frequency-hopping system showing that during frequency dehopping process gain is realized. The direct sequence method uses a sliding correlator to strip off the wide-bandwidth high-frequency code, thus leaving the low-frequency narrower-bandwidth data signal.

One frequently asked question is, "Does spreading the signal increase the signal-to-noise ratio (SNR) of the system so that it requires less power to transmit?" In general, the spreading process does not improve the SNR for a system. Sometimes this can be a bit

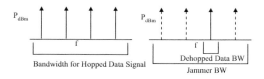

Frequency Dehopping

S/J Is Improved by Reducing the Bandwidth

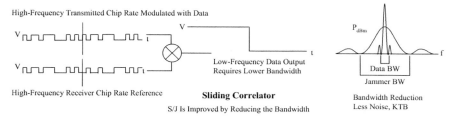

Figure 2-30 Methods for process gain.

confusing when only looking at the receiver. The receiver by itself improves the SNR related to the input signal at the antenna. This improvement is included in the link budget because the bandwidth required for the spread signal is much larger on the input to the receiver than is needed for a nonspread system. Since the bandwidth is large on the input to the receiver, it is either reduced to lower the noise power, which produces a better SNR, or the signals are summed together, producing a larger amplitude and thus a better SNR.

However, if the entire spread spectrum system from the transmitter to the receiver is examined, the narrowband signal is not spread or despread and produces the same ideal SNR as the spread spectrum system. Therefore, in theory, the SNR required at the receiver for a nonspread system is the same as that required for a spread system.

In actual spread spectrum systems, however, the SNR is degraded due to the spreading losses associated with the spreading and despreading processes. Therefore, without jammers present and not considering possible multipath mitigation, it takes more power to send a signal using spread spectrum than it does to send the same signal using a nonspread spectrum system.

2.5.4 Despreading to Realize Process Gain in the Spread Spectrum System

One of the ways to achieve the process gain in a spread spectrum system is to reduce the bandwidth, thereby lowering the noise. For example, a continuous BPSK signal uses a sliding correlator for dispreading, which reduces the bandwidth. The bandwidth is spread due to the phase shift rate (chip rate) in the transmitter, and the receiver uses the same code and slides or lines up the code using a sliding correlator that multiplies the incoming signal with the chip code. This process strips off the high-speed code and leaves the slow data rate bit stream, which requires less bandwidth to process. The narrow bandwidth reduces the noise level, but the signal level is not increased. The jamming signal is spread out by the sliding correlator so that the amplitude of the jammer is lower than the desired signal in the narrow bandwidth, providing the process gain (Figure 2-31).

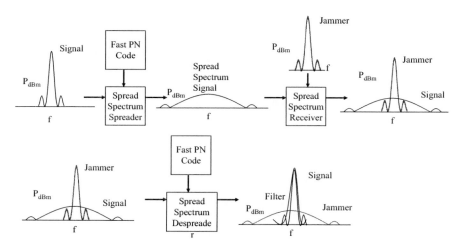

Figure 2-31   Direct sequence spread spectrum signal flow.

Another example of reducing the bandwidth to achieve process gain is using a frequency-hopping signal that has constant amplitude but changes frequency over a large bandwidth at the transmitter. The dehopper in the receiver follows the hopped signal, and the resultant bandwidth is greatly reduced so that the noise is reduced by the same amount. However, the signal level is not increased, but the bandwidth is reduced.

The narrowband jammer is spread out by the dehopping process so that the average amplitude of the jammer is lower than the desired signal (Figure 2-32). The desired signal is jammed only for a short period of time as the jammer hops into the narrow bandwidth where the desired signal resides. The rest of the time the frequency hopper is hopping to different frequencies outside the desired signal bandwidth, which gives the process gain against jammers. However, although the average power of the jammer is reduced because the energy is spread out over the bandwidth, for an instant in time the signal will be jammed while the jammer is in band. This may be critical in some systems where each hop frequency contains data that are lost for that short period of time. However, if the hop rate is higher than the data rate, then the jammer power can be looked at as a decrease in average power.

Another way to realize process gain is by summing the power in each bit of energy to increase the signal strength. This is accomplished in the analog domain by using a match filter similar to an acoustic charge transport (ACT) device or a surface acoustic wave (SAW) device. This can also be done digitally by using a finite impulse response (FIR) filter with the tap (single delay in a shift register) spacing equal to the chip time, with the weights either $+1$ or -1 depending on the code pattern. These concepts will be discussed later. This digital system produces pulses when the code is lined up and data can be extracted as the pulses are received or by using the time slot when the pulses occur for pulse position modulation (PPM). Although the bandwidth is not affected—and therefore the noise is not decreased—it does integrate the signal up to a larger level. The process gain is 10log (number of bits combined). The jammer is not coherent to the matched filter, so it produces sidelobes only at

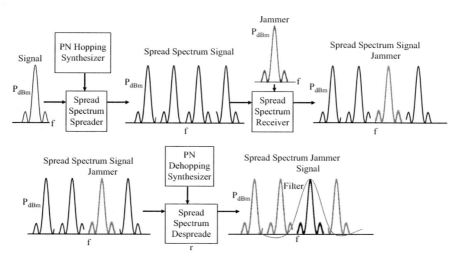

Figure 2-32 Frequency-hopped spread spectrum.

the time of the system pulses and in between the system pulses. These sidelobe levels are reduced due to the length of the matched filter.

All of these approaches are used for low probability of intercept/low probability of detect (LPI/LPD) systems since the signal is spread out and hard to detect unless a matched receiver with the known spread spectrum characteristics are known.

The main reasons for using spread spectrum are to reduce the effects of jamming and interfering signals, to reduce the ability for detection by an unwanted receiver, and to facilitate multipath mitigation. However, remember that it takes more power to send a spread spectrum signal due to spreading losses.

2.5.5 Maximal Length Sequence Codes

Maximal length sequence codes, referred to as m-sequence codes, are used extensively in digital communication systems. The reason is that they have very good properties for minimizing interference between codes by reducing the cross-correlation products or the amount of correlation with the wrong code.

Some of the code properties that they contain are the following:

- Runs (sections of code) are defined: half of the runs are of length 1, one-quarter of the runs are of length 2, one-eighth of the runs are of length 3. The length is the amount of consecutive "1"s or "0"s in a code.
- Contains one more "1" as compared to the number of "0"s.
- Their autocorrelation function is 1 for zero delay, $-1/(2^m - 1)$ for any delay (m is the number of shift registers and $2^m - 1$ is the length of the code), positive or negative, greater than one bit with a triangular shape between ± 1 bit.
- Generated by a linear feedback shift register (LFSR).
- The taps of the LFSR are defined depending on the polynomial.
- Tap numbers will all be relatively prime.

These codes, or a modification of these codes, are used in nearly all digital communication systems. In the case of the global positioning system (GPS), the different m-sequence codes for the different satellites are generated by using an m-sequence code and by multiplying this code with different delays of the same code. The details are provided later in this book.

2.5.6 Maximal Length PN Code Generator

A method of building a PN generator consists of a shift register and a modulo-2 adder. The taps (digital delays) used from the shift register to the modulo-2 adder are specified for each of the shift register lengths to generate a maximal length sequence (MLS) code. If a MLS code can be chosen so that only two taps are required, a simple two-input exclusive-or gate can be used. A "1" is loaded into the shift registers on power-up to get the code started (Figure 2-33).

A counter is provided to ensure that the "all-zeros case" is detected (which would stop the process) and a "1" is automatically loaded into the PN generator. The basic building block for generating direct sequence PSK systems is the MLS generator. There are other methods of generating MLS codes. One process is to store the code in memory and then recall the code serially to generate the MLS code stream. This provides the flexibility to alter the code, for example, to make a perfectly balanced code of the same number of "0"s as

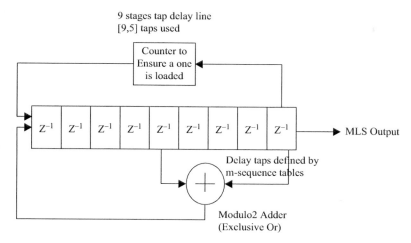

Figure 2-33 Maximal length sequence generator.

"1"s (orthogonal code), which is a modification to the MLS generator. This provides a code with no direct current (DC) offset. The DC offset in the system causes the carrier to be less suppressed because it biases the mixer on an average of zero. In addition, if there is not a balanced input, then there is an added bias that causes the carrier to bleed through the mixer to the output. Also, other types of codes can be generated, such as the Gold codes used to provide orthogonal codes that reduce the cross-correlation between different code sets.

2.5.7 Maximal Length PN Code Taps

To generate a maximal length code using the code generator previously discussed, the taps of the shift register are defined by a maximal length tap table. As the length of the shift register increases, the number of possible tap configurations increases. A list of tap values is shown in Figure 2-34.

2.5.8 Gold Codes

Gold codes are a special case of maximal length codes. Simply put, they are generated by linearly combining two MLSs and then selecting only certain pairs, known as preferred pairs. These codes provide minimum cross-correlation properties. An extension of the Gold codes is making the code orthogonal. This is accomplished by adding a zero to the Gold code.

2.5.9 Other Codes

Many other codes are used in digital communication, some of which are as follows:

- Kasami sequences. These codes are used and also have low cross-correlation properties.
- Orthogonal codes. The Hadamard transform generates these codes, which, if they are ideal, have no cross-correlation. However, if they are nonideal, any offset causes problems with cross-correlation.

```
3 stages: [3, 2]
4 stages: [4, 3]
5 stages: [5, 3][5, 4, 3, 2][5, 4, 3, 1]
6 stages:[6, 5][6, 5, 4, 1][6, 5, 3, 2]
7 stages:[7, 6][7, 4][7, 6, 5, 4][7, 6, 5, 2][7, 6, 4, 2][7, 6, 4,
1][7, 5, 4, 3][7, 6, 5, 4, 3, 2][7, 6, 5, 4, 2, 1]
8 stages:[8, 7, 6, 1][8, 7, 5, 3][8, 7, 3, 2][8, 6, 5, 4][8, 6, 5,
3][8, 6, 5, 2][8, 7, 6, 5, 4, 2][8, 7, 6, 5, 2, 1]
9 stages:[9, 5][9, 8, 7, 2][9, 8, 6, 5][9, 8, 5, 4][9, 8, 5, 1][9, 8,
4, 2][9, 7, 6, 4][9, 7, 5, 2][9, 6, 5, 3][9, 8, 7, 6, 5, 3][9, 8, 7, 6,
5, 1][9, 8, 7, 6, 4, 3][9, 8, 7, 6, 4, 2][9, 8, 7, 6, 3, 2][9, 8, 7, 6,
3, 1][9, 8, 7, 6, 2, 1][9, 8, 7, 5, 4, 3][9, 8, 7, 5, 4, 2][9, 8, 6, 5,
4, 1][9, 8, 6, 5, 3, 2][9, 8, 6, 5, 3, 1][9, 7, 6, 5, 4, 3][9, 7, 6, 5,
4, 2][9, 8, 7, 6, 5, 4, 3, 1]
10 stages:[10, 7][10, 9, 8, 5][10, 9, 7, 6][10, 9, 7, 3][10, 9, 6,
1][10, 9, 5, 2][10, 9, 4, 2][10, 8, 7, 5][10, 8, 7, 2][10, 8, 5, 4][10,
8, 4, 3][10, 9, 8, 7, 5, 4][10, 9, 8, 7, 4, 1][10, 9, 8, 7, 3, 2][10,
9, 8, 6, 5, 1][10, 9, 8, 6, 4, 3][10, 9, 8, 6, 4, 2][10, 9, 8, 6, 3,
2][10, 9, 8, 6, 2, 1][10, 9, 8, 5, 4, 3][10, 9, 8, 4, 3, 2][10, 9, 7,
6, 4, 1][10, 9, 7, 5, 4, 2][10, 9, 6, 5, 4, 3][10, 8, 7, 6, 5, 2][10,
9, 8, 7, 6, 5, 4, 3][10, 9, 8, 7, 6, 5, 4, 1][10, 9, 8, 7, 6, 4, 3,
1][10, 9, 8, 6, 5, 4, 3, 2][10, 9, 7, 6, 5, 4, 3, 2]
```

Figure 2-34 Maximally length PN code taps.

- Walsh codes. These codes are generated from an orthogonal set of codes defined using a Hadamard matrix of order N. The generator block outputs a row of the Hadamard matrix specified by the Walsh code index.

In this simple example of a Walsh orthogonal code, first of all the number of bits equals the number of users. Therefore, two bits = two users. Since the codes are orthogonal:

$$Code1 = (1,-1)\ Code2 = (1,1)\ Orthogonal$$

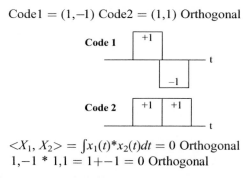

$$<X_1, X_2> = \int x_1(t)*x_2(t)dt = 0\ Orthogonal$$
$$1,-1 * 1,1 = 1+-1 = 0\ Orthogonal$$

Incorporating data with the codes we have:

Code1 = (1,−1) Data1 = **1011** (0 = −1)
Code2 = (1,1) Data2 = **0011** (0 = −1)
Bit Stream 1 = (1,−1)*(1,−1,1,1) = (1, −1, −1, 1, 1, −1, 1, −1)
Bit Stream 2 = (1,1)*(−1,−1,1,1) = (−1,−1,−1,−1,1,1,1,1)
(1,−1,−1,1,1,−1,1,−1) + (−1,−1,−1,−1,1,1,1,1) = (0,−2,−2,0,2,0,2,0) Orthogonal
Decode Data 1: (1,−1)* (0,−2,−2,0,2,0,2,0) = (1,−1)*[(0,−2)(−2,0)(2,0)2,0)]
  = (0+2)(−2+0)(2+0)(2+0) = (2,−2,2,2) = **1,0,1,1** = Data 1
Decode Data 2: (1,1)* (0,−2,−2,0,2,0,2,0) = (1,1)*[(0,−2)(−2,0)(2,0)2,0)]
  = (0−2)(−2+0)(2+0)(2+0) = (−2,−2,2,2) = **0,0,1,1** = Data 2

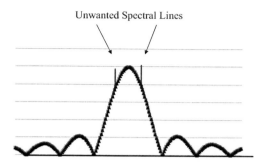

Figure 2-35   Spectral lines caused by repetitions in the code.

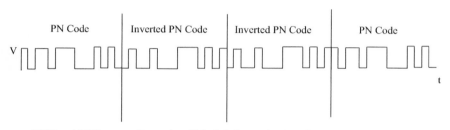

Figure 2-36   Invert the short code using a pseudo-random sequence to reduce spectral lines.

2.5.10 Spectral Lines in the Frequency Domain

Another criterion to be considered when selecting codes for communications is the characteristics of the spectral lines (unwanted frequencies) generated in the frequency domain. This is especially important for short codes, which will produce spectral lines caused by the code repetition frequency and sections of the code that repeat (Figure 2-35). Basically, anything that repeats in the time domain will show up as a spectral frequency in the frequency domain. This can pose a problem with requirements for power spectral density (frequency content) in the specified band. One method of reducing these spectral lines in the frequency domain is to invert the existing PN-code according to an additional pseudo-noise code. This additional PN-code determines when to invert or not invert the existing PN-code (Figure 2-36). This technique has been patented and has been proven successful in reducing the spectral lines to an acceptable level.

2.6 Other Forms of Spread Spectrum Transmissions

Other forms of spread spectrum include time hopping and chirped-FM. The time-hopping system uses the time of transmission of the data and ignores the rest of the time, which reduces the overall effect of the jammer that has to be on for the entire time. The chirped system generates more bandwidth by sweeping over a broad bandwidth. This system reduces the bandwidth in the detection process by reversing the process, thus providing process gain.

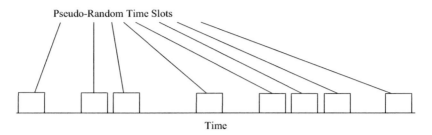

Pseudo-Random Time Slots

Time

Figure 2-37   Time-hopping spread spectrum signal.

2.6.1 Time Hopping

Time hopping entails transmitting the signal only at specified times; that is, the transmissions are periodic, and the times to transmit are pseudo-random using a pseudo-random code (Figure 2-37). The process gain is equal to 1/duty cycle. This means that if the duty cycle is short—on for a short period of time and off for a long period of time—the process gain is high.

The receiver demodulates this by looking only at the times that the signal was sent, knowing the pseudo-random sequence of time slots. This increases the level of the signal during the time slots and decreases the average power of the jamming signal by ignoring the other time slots since the jamming signal is spread over all the pulses.

2.6.2 Chirped-FM

Chirped-FM signals are generated by sliding the frequency in a continuously changing manner, similar to sweeping the frequency in one direction. The reason that these signals are called chirped is that when spectrally shifted to audible frequencies they sound like a bird chirping. There are up-chirps and down-chirps, depending on whether the frequencies are swept up or down. Chirp signals can be generated by a sweeping generator with a control signal that resets the generator to the starting frequency at the end of the chirp. Chirp signals can also be generated using SAW devices excited by an impulse response (which theoretically contains all frequencies). The delay devices or fingers of the SAW device propagate each of the frequencies at different delays, thus producing a swept frequency output. This reduces the size of chirping hardware tremendously.

For chirp-FM systems, the instantaneous frequency is calculated by

$$F = F_h - t(W/T)$$

where

$F =$ instantaneous frequency
$F_h =$ highest frequency in the chirped bandwidth
$t =$ change in time (Δt)
$W =$ bandwidth of the dispersive delay line (DDL)
$T =$ dispersive delay time

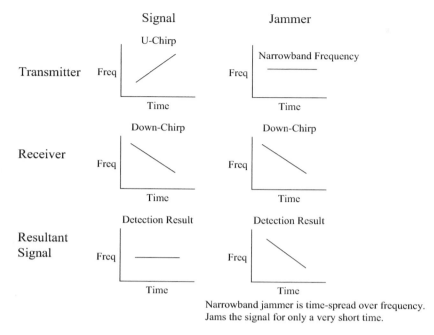

Figure 2-38   Chirped FM spread spectrum.

For chirp systems, the high frequencies have more attenuation for a given delay. For a flat amplitude response for all frequencies, the low frequencies need to be attenuated to match the high frequencies. Therefore, the total process contains more insertion loss. SAW devices are used to provide frequency-dependent delays, since it is easier to control the delay with lower frequency sound waves than electromagnetic waves. These are called acoustic delay lines or delay fingers.

The up-chirp is used by the transmitter; the down-chirp is used by the receiver, matched to the transmitter to eliminate the chirp (Figure 2-38). The up-chirp spreads the frequencies over a broad band, and the down-chirp despreads the signal. In addition, the jammer, when in the down-chirp, since it is not spread by the up-chirp, is spread across a broad band, thus reducing the impact that the jamming signal has on the system.

2.7 Multiple Users

To allow multiple users in the same geographical area, a scheme needs to be allocated to prevent interference between systems. Some of the schemes for reducing interference are the following:

- Time division multiple access (TDMA)
- Code division multiple access (CDMA)
- Frequency division multiple access (FDMA)

These techniques are shown in Figure 2-39. Some control is required in all of these systems to ensure that each system in an operating area has different assignments.

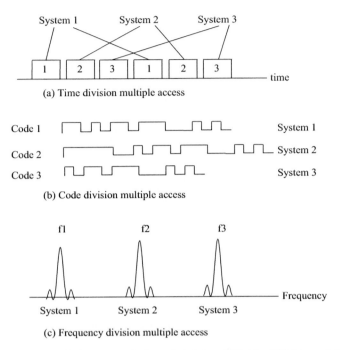

(a) Time division multiple access

(b) Code division multiple access

(c) Frequency division multiple access

Figure 2-39   Multiple user techniques applying TDMA, CDMA, and FDMA.

TDMA provides interference reduction by having the systems communicate at different times or time slots using the same frequencies and codes (Figure 2-39). If the systems have predetermined time slots, then the system is considered to be time division multiplexing (TDM). If a system accesses the time slot, say, on a first come, first served basis, then the system is a TDMA system. This definition applies to other types of systems, such as CDMA and code division multiplexing (CDM) and FDMA and frequency division multiplexing (FDM).

CDMA provides interference reduction by having the systems communicate on different codes, preferably maximal length sequence codes, orthogonal codes, or Gold codes, at the same frequencies and times, which provide minimum cross-correlation between the codes, resulting in minimum interference between the systems (Figure 2-39). A lot of research has been done to find the best code sets for these criteria. Generally, the shorter the code, the more cross-correlation interference is present and fewer optimal codes can be obtained.

FDMA provides interference reduction by having the systems communicate on different frequencies, possibly using the same times and codes (Figure 2-39). This provides very good user separation since filters with very steep roll-offs can be used. Each user has a different frequency of operation and can communicate continuously on that frequency.

Each of the previously discussed multiuser scenarios reduces interference and increases the communications capability in the same geographical area.

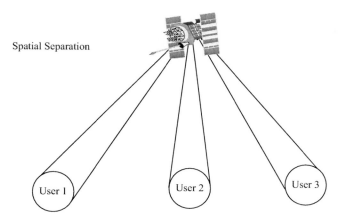

Spatial Separation

Figure 2-40 Directional beam antennas provide spatial separation for multiple users.

2.7.1 Other Methods for Multiuser Techniques

Other methods to allow multiple users in the same band of frequencies include frequency hopping, spatial separation, orthogonal techniques, and wideband–narrowband signal separation. Frequency hopping is where each user has a different hop pattern. Spatial separation is where each user is in a different area, and they are separated in space or direction (Figure 2-40). Pointing directional antennas to different sectors allows more users to operate in the same band. Orthogonal techniques are used for separating users where each user is orthogonal to the others in the band. Finally, narrowband and wideband users can be separated using correlation techniques. Wideband signals do not have autocorrelation with a chip/bit delay, whereas narrowband signals have autocorrelation with the same delay. This technique will be utilized in the adaptive filter discussion to mitigate narrowband jammers in a wideband signal environment.

2.7.2 Orthogonal Signals

Several types of orthogonal signals can be used to separate multiple communication links in the same band. This is valuable for multiple access applications and for separation of channels to reduce adjacent channel interference. Some of the techniques used to separate signals by orthogonal means include the following:

- Phase separation: separating I/Q or cosine and sine signals, for example.
- Orthogonal frequency separation: orthogonal frequency division multiplexing, which uses orthogonal frequencies.
- Antenna polarization: separation due to vertical versus horizontal, left-hand circular polarized versus right-hand circular polarized (LHCP/RHCP).
- Jammer/signal separation: Graham-Schmidt orthogonalization forces the jammer to be orthogonal to the signal using error feedback.

2.7.3 Quadrature Phase Detection of Two Signals

Quadrature phase detection is the ability to separate signals due to orthogonal phases. For example, if one signal can be put on an in-phase carrier, $\cos(\omega t)$, and another signal put on an orthogonal signal, 90° out of phase of the first signal, $\sin(\omega t)$, then these signals can be separated by the receiver (Figure 2-41).

A simple math calculation shows that the two signals will be on each of the paths, as shown as follows:

- Path 1: Low-pass filter included to eliminate the sum term.
    f1 signal: $f1(t)\cos(\omega t) * \cos(\omega t) = f1(t)[\frac{1}{2}\cos(\omega t - \omega t) + \frac{1}{2}\cos(\omega t + \omega t) = \frac{1}{2}f1(t)$
    $[\cos(0)] = \frac{1}{2}f1(t)$; signal f1 is detected.
    f2 signal: $f2(t)\sin(\omega t) * \cos(\omega t) = f2(t)[\frac{1}{2}\sin(\omega t - \omega t) + \frac{1}{2}\sin(\omega t + \omega t) = \frac{1}{2}f2(t)$
    $[\sin(0)] = 0$; signal f2 is not detected.
- Path 2: Low-pass filter included to eliminate the sum term.
    f1 signal: $f1(t)\cos(\omega t) * \sin(\omega t) = f1(t)[\frac{1}{2}\sin(\omega t - \omega t) + \frac{1}{2}\sin(\omega t + \omega t) = \frac{1}{2}f1(t)$
    $[\sin(0)] = 0$; signal f1 is not detected.
    f2 signal: $f2(t)\sin(\omega t) * \sin(\omega t) = f2(t)[\frac{1}{2}\cos(\omega t - \omega t) - \frac{1}{2}\cos(\omega t + \omega t) = \frac{1}{2}f2(t)$
    $[\cos(0)] = \frac{1}{2}f2(t)$; signal f2 is detected.

Therefore, both the signals will be retrieved, with signal f1 appearing on path 1 and signal f2 appearing on path 2. A phasor diagram shows the separation of the two signals (Figure 2-42).

A practical detector called a Costas loop uses feedback to phase lock on a carrier frequency to provide two phases at the output of the detection process (Figure 2-43). Costas loops will be discussed in detail later in this book.

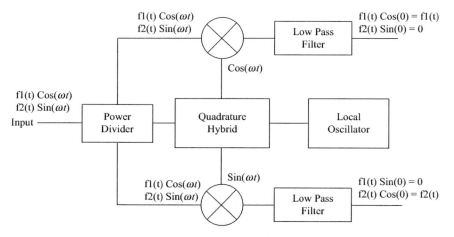

Figure 2-41   Quadrature detection using orthogonal phase separation.

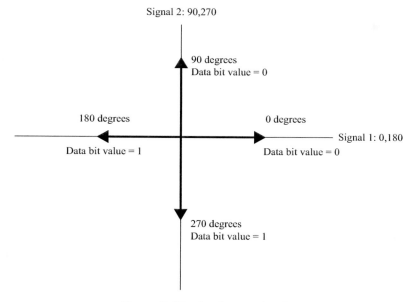

Figure 2-42   Quadrature signals.

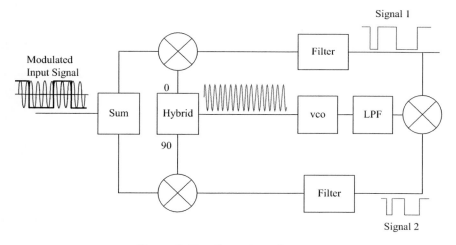

Figure 2-43   Costas loop detector.

2.7.4 Orthogonal Frequency Division Multiplexing

An extension of FDM uses the principle that if signals are orthogonal they do not interfere with each other. A technique being used in many communications applications is known as orthogonal frequency division multiplexing. This technique allows closer spacing of multiple users in a given bandwidth or allows for multiple parallel channels to be used for higher data rates by combining parallel channels. It is used for broadband communications in both powerline communications (PLC) and RF systems with different modulation schemes, including BPSK, QPSK, and QAM. The basic concept is to use frequencies that are orthogonal over a specified time period.

The basic orthogonality concept is that the inner product of two orthogonal frequency channels are integrated to zero over a specified integration time and the inner product of the same frequency is equal to one:

$$\langle Xn(t),\ Xm(t) \rangle = \int_0^t Xn(t)Xm(t)dt = 1$$

where

$$Xn(t) = Xm(t)$$

and

$$\langle Xn(t),\ Xm(t) \rangle = \int_0^t Xn(t)Xm(t)dt = 0$$

where

$Xn(t)$, $Xm(t)$ are orthogonal

Consequently, if the frequencies are orthogonal, their inner products are equal to zero. So taking the inner product of all the frequencies with the desired frequency, only the incoming desired frequency will be present.

One way to obtain orthogonal frequencies is to ensure that the frequencies are harmonics of the fundamental frequency. Then, by choosing the integration time, the inner product will be zero. An example of orthogonal frequencies containing the second and third harmonic is shown in Figure 2-44.

When any combination of these frequencies is multiplied and integrated with the limits shown in Figure 2-44, the results are equal to zero. If the inner product of any one of these

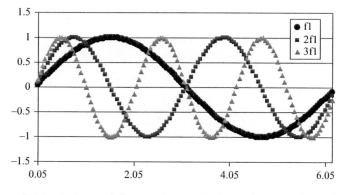

Figure 2-44   Orthogonal frequencies producing an inner product of zero.

frequencies with itself is used, then the inner product is not zero and the specified signal is detected.

A way to visualize this is to take the previous example, multiply every point on the curves, and sum the values. They will be equal to zero for the orthogonal frequencies and equal to a one for the same frequencies. A simple example was shown earlier using $\cos(\omega t)$ and $\sin(\omega t)$ as the orthogonal signals.

For OFDM, there are multiple users for each of the frequency slots, and the adjacent slots are made to be orthogonal to each other so that they can overlap and not interfere with each other. This allows more users to operate in the same bandwidth. Thus, using orthogonal frequencies reduces adjacent channel interference and allows for more channels to be used is a smaller bandwidth (Figure 2-45).

Another use for OFDM is to allow high-speed data to be transmitted in a parallel system. This is accomplished by taking the serial data stream from the transmitter and converting to parallel streams, which are multiplexed using different frequency channels in the OFDM system to send the parallel stream of data to the receiver. The receiver then performs a parallel-to-serial conversion to recover the data. In addition, the system can be made adaptive to only select the channels that are clear for transmission.

One of the drawbacks to OFDM is the high crest factor (Figure 2-3). The crest factor or PAPR is greater than 10 dB. This requires a linear power amplification, 10 dB from saturation, and a high dynamic range of A/D and D/A. Due to several waveforms with high crest factors, there have been several PAPR reduction techniques to help mitigate these problems as follows:

1. Use of Golay or Reed/Muler codes help reduce the PAPR.
2. Several crest factor reduction techniques have been implemented for multiple signals used in LTE applications.
3. Use of digital predistortion can help reduce the crest factor.

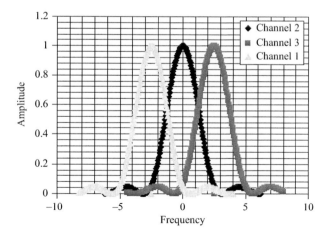

Figure 2-45 OFDM allows frequency channels to overlap with minimal adjacent channel interference.

2.7.5 Other OFDM Techniques

To improve the performance of a standard OFDM system, additional modulation techniques have been added. For example, two types of OFDM systems that have aided other techniques in performance improvement are as follows:

- Coded OFDM (COFDM): implements coding schemes in addition to orthogonal techniques to improve the performance against multipath.
- Vector OFDM: uses time division multiplexing (TDM) and packet communications to improve wireless broadband networks, including local multipoint distribution service (LMDS) systems.

2.8 Power Control

Power control is a technique to control the amount of power of multiple transmitters into a single receiver. This is especially important for systems using CDMA and other multiplexing schemes. Power control helps reduce the effects of near/far problems where one user is close to the transceiver and another user is far away. The user close to the receiver will jam the user that is far away from the receiver. If power control is used, the idea is that the power will be the same level at the receiver for both the close in and far away users.

A multiuser system or network is based on the premise that the users are all at the same power level and the separation between users is accomplished by multiplexing schemes such as CDMA. However, most users are at different locations and distances from the receiver so that the power into the receiver varies tremendously, up to 100 dB. Since most CDMA

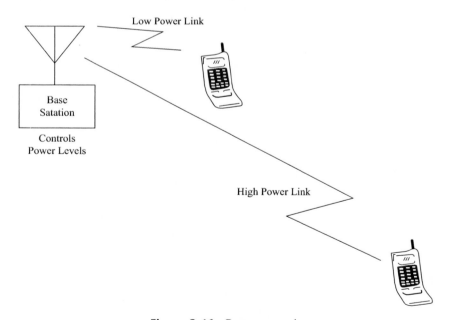

Figure 2-46   Power control.

systems can handle variations in power or jamming margin that are only much less than this, there is a problem with interference between users.

Power control helps mitigate this interference problem by having the user transmit only the amount of power required for the distance from the receiver. Therefore, for a user close to the receiver the power is reduced, and for a user at a far distance from the receiver the power is increased. The main objective is to have the power, regardless of the distance, at the same level when it reaches the receiver (Figure 2-46).

Many digital cell phone networks use power control to allow more users and less interference. The base station controls the power of all of the telephones that are communicating with that particular base station so that the power is nearly the same regardless of the distance from the base station. With the limited amount of process gain and jamming margin, a typical CDMA system is required to have power control to operate correctly. Power control is used where process gain is inadequate for good separation of users.

2.9 Summary

The transmitter is a key element in the design of the transceiver. The transmitter provides a means of sending out the information, over the channel, with the power necessary to provide coverage to the intended receiver. Several types of modulation methods were discussed with their advantages and disadvantages. Many types of spread spectrum transmitters provide process gain to reduce the effects of jammers and to allow more efficient use of the spectrum for multiple users. Digital systems have many advantages over analog systems, and different techniques were discussed for optimizing the digital data link.

References

Bullock, S. R. "Phase-Shift Keying Serves Direct-Sequence Applications." *Microwaves and RF*, December 1993.

Bullock, S. R., and R. C. Dixon, *Spread Spectrum Spectral Density Techniques*, Patents #5,436,941, #5,604,767.

Haykin, S. *Communication Systems*, 5th ed. Wiley, 2009.

Holmes, J. K. *Coherent Spread Spectrum Systems*. Wiley & Sons, 1982, pp. 251–267.

Torrieri, D. *Principles of Spread-Spectrum Communication Systems*, 2d ed. Springer, 2011.

Problems

1. Show that not filtering the output of a mixing product could result in a cancellation of the desired signal using a carrier of 1 MHz and a signal carrier of 0.1 MHz.
2. If the unwanted sideband in problem 1 was filtered, show that the desired signal can be retrieved.
3. Show by using a phasor diagram that two QPSK modulators can be used to generate 8-PSK (eight phase states). Find the phase relationship of the two QPSK modulators.

4. Determine an 8-PSK modulator that eliminates the 180° phase shift. Why is the 180° shift a problem?

5. Why is it required to load a "1" in a simple PN generator?

6. Find the process gain for a time-hop signal with a duty cycle of 20%?

7. Find the process gain of a frequency hopper using 20 frequencies with a jammer that has the ability to eliminate two frequencies at time?

8. What is the basic difference between OQPSK and 16-OQAM?

9. What is the basic difference between $\pi/4$ DQPSK and D8PSK?

10. What is the basic concept of process gain with respect to bandwidths?

11. What two factors are needed to calculate the jamming margin from process gain?

12. Why is power control needed for a CDMA data link?

13. What is the main advantage of a digital data link over an analog data link?

14. What are the advantages and disadvantages of differential systems vs coherent systems?

15. What is the advantage of an MSK or CP-PSK over standard BPSK and QPSK systems?

16. What two approaches can be used to generate MSK?

17. What are the advantages and disadvantages of BPSK and 16-QAM?

18. What types of filters are used to approach an ideal match filter? What is the result in the frequency domain of an ideal match filter in the time domain?

19. What are the reasons to use spread spectrum? What would be the reason not to use spread spectrum?

20. What movie star helped to invent spread spectrum?

21. What technique does direct sequence use to generate spread spectrum?

22. What devices can be used to generate spread spectrum codes?

23. What are the three main techniques that can be incorporated to allow multiple users?

CHAPTER 3

The Receiver

The receiver is responsible for downconverting, demodulating, decoding, and unformatting the data received over the link with the required sensitivity and bit error rate (BER) according to the link budget analysis of Chapter 1. The receiver is responsible for providing the dynamic range (DR) to cover the expected range and power variations and to prevent saturation from larger power inputs and provide the sensitivity for low-level signals. The receiver provides detection and synchronization of the incoming signals to retrieve the data sent by the transmitter. The receiver section is also responsible for despreading the signal when spread spectrum signals are used.

The main purpose of the receiver is to take the smallest input signal, the minimum detectable signal (MDS), at the input of the receiver and amplify that signal to the smallest detection level at the analog-to-digital converter (ADC) while maintaining a maximum possible signal-to-noise ratio (SNR). A typical block diagram of a receiver is shown in Figure 3-1. Each of the blocks will be discussed in more detail.

3.1 Superheterodyne Receiver

Heterodyne means to mix, and superheterodyne means to mix twice. Most receivers are superheterodyne receivers, which means that they use a common intermediate frequency (IF) and two stages or double conversion to convert the signal to baseband, as shown in Figure 3-1. The reasons for using this type of receiver include the following:

- A common IF can be used to reduce the cost of parts and simplify design for different radio frequency (RF) operating ranges.
- It is easier to filter the image frequency since the image frequency is far away at two times the IF.
- It is easier to filter intermodulation products (intermods) and spurious responses.
- It has the ability to change or hop the RF local oscillator (LO) while the IF filters and circuitry remain the same.

Multiple-stage downconverters with more than one IF are sometimes used to aid in filtering and commonality. Occasionally a single downconverter is used, and it generally contains an image-reject mixer to reduce the effects of the image frequency.

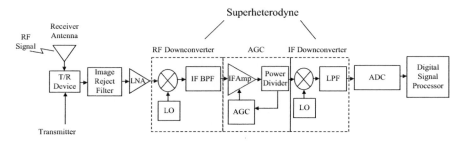

Figure 3-1   Typical superheterodyne receiver.

3.2 Basic Functions of the Receiver

The basic functions of the receiver are shown in Figure 3-1 and are listed as follows:

- Receiver antenna: Used to receive the RF signal from the transmitter.
- Transmit/Receive device: Uses a T/R switch, circulator, diplexer or duplexer to allow the same antenna for both the receiver and transmitter.
- Image reject filter: Reduces the image frequencies and frequencies in other bands from interfering with the incoming signal. It also reduces bandwidth and noise into the receiver.
- Low-noise amplifier (LNA): Helps to establish the noise figure (NF) of the receiver and provide gain.
- RF downconverter: Converts RF signals to an IF using a LO for IF processing and filtering.
- Automatic gain control (AGC): Uses feedback techniques to automatically adjust the gain with respect to the incoming signal to keep a constant signal level into the rest of the receiver.
- IF downconverter: Converts the IF to a lower frequency or baseband for detection and signal processing.
- Analog-to-digital converter: Converts the incoming lower frequency or baseband to digital signals for digital signal processing.
- Digital signal processor: Provides data detection, despreading, and demodulation of the incoming digital waveform and retrieves the desired data information.

3.3 Receiver Antenna

The receiver antenna gain is computed in the same way as for the transmitter with the calculation performed using the link budget in Chapter 1. Factors to consider in determining the type of antenna to use are frequency, the amount of gain required, and the size. These factors are similar to the ones used to determine the transmitter antenna. Parabolic dishes are frequently used at microwave frequencies. In many communications systems, the receiver uses the same antenna as the transmitter, which reduces the cost and size of the system. The antenna provides gain in the direction of the beam to reduce power requirements, reduce the noise into the antenna, and increases the sensitivity of the receiver, and also to reduce the amount of interference received or selective jamming reduction. The gain of the antenna is usually expressed in dBi, which is the gain in dB reference to what an ideal isotropic radiator antenna would produce.

3.4 Transmit/Receive Device

As was described in the transmitter section, the signal is received by the antenna and is passed through a device—a duplexer or diplexer, T/R switch, or circulator—to allow the same antenna to be used by both the transmitter and receiver. The chosen device must provide the necessary isolation between the transmitter circuitry and the receiver circuitry to prevent damage to the receiver during transmission.

3.5 Image Reject Filter

In nearly all receivers, an image reject filter/band reject filter is placed in the front end so that the image frequencies along with other unwanted signals are filtered out, with a decrease in noise. These unwanted signals are capable of producing intermodulation products (intermods). Intermods are caused by the nonlinearities of the LNA and the mixer.

The image frequencies are the other band of frequencies that, when mixed with the LO, will fall in the passband of the IF band. For example,

LO = 1000 MHz
Desired input signal = 950 MHz
IF = 1000 MHz − 950 MHz = 50 MHz

Thus,

Image frequency = 1050 MHz
1050 MHz − 1000 MHz = 50 MHz = same IF

Therefore, if the receiver did not have an image reject filter in the front end, then an interference signal at 1050 MHz would fall in the IF passband and jam the desired signal (Figure 3-2). In addition, the noise power in the image bandwidth would increase the receiver NF by 3 dB.

Image reject mixers are sometimes used to reduce the image frequency without filtering. This is useful in a receiver system that contains a wide bandwidth on the RF front end and a narrow bandwidth IF after the mixer so that the image frequency cannot be filtered. This eliminates the need for double downconversion in some types of receivers. However, double downconversion rejects the image frequency much better. This is generally a trade-off on the image frequency rejection and cost. Also, the performance and the versatility of the superheterodyne receiver are usually preferred.

3.6 Low-Noise Amplifier

An LNA is used in a receiver to help establish the NF, providing that the losses between gain elements are not too large and the bandwidth does not increase. Using the NF, the equation for calculating the receiver noise is

$$\text{Noise factor } F_s = F_1 + (F_2 - 1)/G_1 + (F_3 - 1)/G_1 G_2 \ldots$$

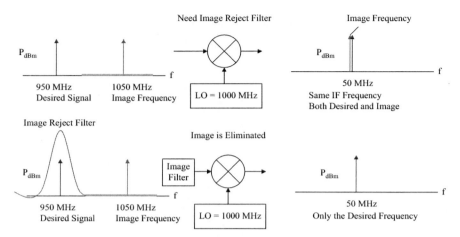

Figure 3-2 Image reject filter required to eliminate image frequency to prevent distortion.

Therefore, the NF would be

$$NF = 10\log F_s \approx 10\log F_1, \text{if } G_1 \text{ is large}$$

For example, given the following parameters:

LNA amplifier:

$NF_1 = 6$ dB, $F_1 = 4$
$G_1 = 25$ dB, $G_1 = 316$

Mixer:

$L = 8$ dB (conversion loss), $L = 6.3$

IF amplifier:

$NF_2 = 6$ dB, $F_2 = 4$
$G_2 = 20$ dB, $G_2 = 100$

Therefore, the calculation for the NF of this receiver chain would be

$$F_t = F_1 + [(F_2 \times \text{Losses}) - 1]/G_1 = 4 + (4 * 6.3 - 1)/316 = 4.08$$
$$NF = 10\log(4.08) = 6.10 \text{ dB}$$

An LNA also provides isolation between the LO used for downconversion and the antenna. This prevents LO bleed-through from appearing at the antenna port. This can be a problem for two reasons. If the system is designed for a low probability of intercept, then this LO signal is being transmitted from the antenna and can be detected. The other reason is this signal might interfere with other users.

Another benefit of using an LNA in the receiver is that less power is required from the transmitter if the NF is low. For a given desired SNR or E_b/N_o, the NF alters these according to the link budget, as specified in Chapter 1. In general, an LNA amplifies the desired signal with minimum added noise and establishes the receiver noise level.

Some reasons for not using an LNA in a system are as follows:

- Low-noise amplifiers reduce the dynamic range DR of a receiver, which is the operating range from the minimal signal to the maximum signal that can be detected and processed. For example, if the 1 dB compression point is +10 dBm, the maximum input power to an LNA with a gain of 25 dB is −15 dBm. Without a preamp, the 1 dB compression for a low-level mixer is +3 dBm. Also, the LO drive can be increased to provide an even higher DR without the LNA.
- The interference signals, clutter, and bleed-through from the transmitter may be more significant than the noise floor level, and therefore establishing a lower NF with an LNA may not be required.
- Cost, space, DC power consumption, and weight are less without a preamplifier.
- Some airborne radar systems use mixer front ends (no preamp or LNA) to reduce weight.

The results show that an LNA gives a better NF, which results in better sensitivity at the cost of reduced DR, cost, and space. An LNA also provides isolation between the LO and the antenna. Also, the low NF established by the LNA decreases the power requirement for the transmitter. In general, an LNA in the receiver provides the best overall design and should be included and carefully designed for the best NF.

3.7 RF Downconverter

After the LNA, the received RF signal is downconverted to a lower frequency or IF for processing. This is accomplished by using an LO, a mixer, and a bandpass filter (BPF) as shown in Figure 3-1. The IF BPF is used to reduce the harmonics, spurious, and intermod products that are produced in the downconverter process. The IF is usually a lower frequency than the incoming frequency; however, in some applications the IF could be higher, which would require an RF upconverter. Generally, most receivers use an RF downconverter.

3.8 Mixers

The downconversion process requires a mixer to translate the frequency carrier to an intermediate frequency and eventually to a baseband frequency. Many types of mixers and their characteristics should be examined when developing an optimal downconversion system.

3.8.1 High-Level or Low-Level Mixers

High-level mixers use higher voltages and require more power to operate than do the standard low-level mixers. The advantages of using high-level mixers include the following:

- High third-order intercept point [less cross modulation, more suppression of two-tone third-order response, $(2f_1 - f_2) \pm$ LO].

- Larger DR.
- Lower conversion loss, better NF.
- Best suppression above the bottom two rows of the mixer spurious chart (Figure 3-3).

Standard low-level mixers are used in most systems. The advantages of the low-level mixers are the following:

- Less complex, easier to balance, giving better isolation, lower DC offset, and less mixer-induced phase shift.
- More covert or low probability of intercept, less LO bleed-through.
- Less system power and less expensive.
- Best suppression of the bottom two rows of the mixer chart (Figure 3-4).

Depending on the receiver application, these advantages need to be reviewed to make the best mixer selection.

Harmonics of f_R	0	1	2	3	4	5	6	7
3	>90	>90	>90	>90	>90	>90	>90	89
2	73	83	75	79	80	80	77	82
1	24	0	34	11	42	18	49	37
0		18	10	23	14	19	17	21

Harmonics of f_L

Figure 3-3 Typical spurious chart for a high-level mixer.

Harmonics of f_R	0	1	2	3	4	5	6	7
5	80	79	78	72	82	71	x	x
4	88	80	90	80	90	82	x	x
3	58	55	65	54	66	54	x	x
2	69	64	71	64	73	61	x	x
1	16	0	25	12	33	19	x	x
0		24	30	37	41	34	X	X

Harmonics of f_L

Figure 3-4 Typical spurious chart for a low-level mixer.

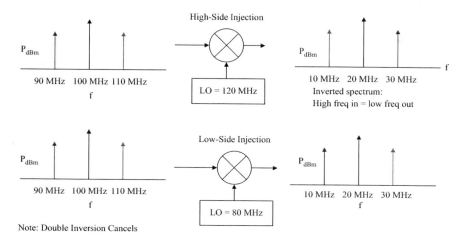

Figure 3-5 High-side versus low-side injection reverses bands and changes image frequency.

3.8.2 High-Side or Low-Side Injection

Another consideration when selecting a mixer is whether to use high-side versus low-side injection. High side refers to the LO being at a higher frequency than the input signal frequency; low side refers to the LO being at a lower frequency than the input signal frequency. Some of the main points to consider in the selection are as follows:

- Performing a spurious signal analysis assists in determining the best method to use.
- High-side injection inverts the signal; lower sideband becomes upper sideband, which may cause some problems.
- Low-side injection provides a lower frequency for the LO, which may be easier to get and less expensive.
- The image frequency is different and analysis needs to be done to determine which of the image frequencies will affect the receiver the most (Figure 3-5).

3.8.3 Mixer Spur Analysis—Level of Spurious Responses and Intermods

A mixer spur analysis is done for each mixer output to

- determine which mixer products fall in the passband of the output;
- determine the frequencies to be used to ensure that mixer spurs do not fall in the passband.

Spurious signals are generated when mixing signals up and down in frequency. They are the mixer products and are designated as $n \times m$-order spurs, where n is the harmonics of the LO, and m is the harmonics of the RF or IF (the input of the mixer could be an RF signal or an IF signal depending on the frequency translation). These mixer spurs can cause problems if they fall in the passband since they cannot be filtered.

A spurious analysis should be done to determine where the spurs are located—if they fall in the passband—and the power of the spur with respect to the desired signal. Several software programs have been written to assist in determining where the spurs are for a given system. These programs take two input frequencies and multiply them together, depending on the order specified, to generate the possibilities. For example, given this sixth-order system:

Sixth-order analysis (sums and differences need to be evaluated)

$1 \times 0 = $ LO	$0 \times 1 = $ input carrier
$2 \times 0 = $	$0 \times 2 = $
$3 \times 0 = $	$0 \times 3 = $
$4 \times 0 = $	$0 \times 4 = $
$5 \times 0 = $	$0 \times 5 = $
$6 \times 0 = $	$0 \times 6 = $
$1 \times 1 = $ desired	
$2 \times 1 = $	$1 \times 2 = $
$3 \times 1 = $	$1 \times 3 = $
$4 \times 1 = $	$1 \times 4 = $
$5 \times 1 = $	$1 \times 5 = $
$3 \times 2 = $	$2 \times 3 = $
$4 \times 2 = $	$2 \times 4 = $
$2 \times 2 = $	$3 \times 3 = $

The first number multiplies the LO and the second number multiplies the input, and then the resultants are added and subtracted to determine the frequency of the spurs. Note that the 1×1 contains the desired signal and a spur, depending whether or not the wanted signal is the sum or the difference. Usually this is obvious and the unwanted signal is filtered out.

Selecting the right LO or specifying the operational frequencies for a selected level of spur rejection is done by comparing all the possible spur locations with respect to a given amplitude threshold. This is dependent on high-side, low-side, sum, or difference. Frequency selection depends on mixer spur analysis.

3.8.4 Sixth-Order Analysis

A sixth-order analysis is standard for spur analysis. This ensures that the spurious signals are about 60 dB down from the desired output for most mixers. The analysis below is for a sixth-order spur analysis with the desired signal being the difference frequency. The LO is higher in frequency than the highest desired frequency, which eliminates the spurs 2×1, 3×1, 4×1, 5×1, 3×2, and 4×2.

The worst-case highest in-band frequency for the 1×0 spur would be $LO - f_i$. Since it is high-side injection, the 1×0 (LO) would always be greater than $LO - f_i$:

$$1 \times 0; \quad LO > LO - f_i.$$

This also applies for the following spurs: 2×0, 3×0, 4×0, 5×0, 6×0.

The worst-case lowest in-band frequency for the 1×0 spur would be $LO - f_h$. Since it is high-side injection, the 1×0 (LO) would always be less than $LO - f_h$:

$$1 \times 0; \ LO < LO - f_h.$$

This also applies for the following spurs: 2×0, 3×0, 4×0, 5×0, 6×0.

Note, the bandwidth is from $LO - f_h$ to $LO - f_l$, which none of these spurs fall into.

The analysis continues for all possible spur products to see if they fall into the operating band. If they do, then the spur chart is examined to determine how big the spurious signal will be. If the design is not to allow any sixth-order products in the band, then each of the possible products must be calculated to ensure that none of them fall into the desired bandwidth. The same type of mixer analysis can be done for each mixer configuration to ensure that the spurious signals do not fall in the desired band and can be filtered out of the system.

Another point to make about mixers is that, in general, the voltage standing wave ratio (VSWR) for a mixer is not very good, about 2:1. If the source is 2:1, then, for electrical separation of greater than ¼ wavelength, the VSWR can be as much as 4:1. This is equivalent to having a mismatched load four times larger (or ¼ as large) than the nominal impedance.

3.9 Automatic Gain Control

Automatic gain control is shown in the IF section of the receiver (Figure 3-1). However, the AGC can be placed in the RF, IF, baseband, digital processor, or multiple places in the receiver where it is important to keep the output levels constant, to increase the dynamic range, and to automatically adjust the gain with respect to the incoming signal. The AGC uses feedback to detect the level of the incoming signal, and automatically controls the receiver gain to the desired level of output signal. This keeps the signal level constant for detection and signal processing and increases the range of signals that the receiver can handle. A detailed discussion on AGC can be found in Chapter 4.

3.10 IF Downconverter

The IF downconverter is used to convert the IF to a lower frequency or baseband. This reduces the sample time of the ADC and digital processing and is composed of an LO, mixer, and low-pass filter (LPF), the latter of which limits the high-frequency components to allow the ADC to sample and process the signal. This second downconverter creates a super-heterodyne (double conversion) receiver, which is ideal to prevent and filter out unwanted harmonics, spurious, and intemod products to eliminate interference in the digital signal processer.

The double downconversion process relaxes constraints on filters and allows for common circuitry for different systems. For example, a receiver operating at 1 GHz and another operating at 2 GHz could both downconvert the signal to a common IF of 50 MHz, and the rest of the receiver could be identical.

Some of the newer receivers digitize the IF band directly and do a digital baseband downconversion before processing. This is especially true for quadrature downconversion using an in-phase (I) channel and a quadrature-phase (Q) channel. This relaxes the constraints of quadrature balance between the quadrature channels in the IF/analog portion of the receiver, since it is easier to maintain quadrature balance in the digital portion of the receiver and reduces hardware and cost.

Quadrature downconversion is a technique that is used to eliminate unwanted bands without filtering. For example, during a conversion process, both the sum and difference terms are created as shown:

$$A\cos(\omega_s)t * B\cos(\omega_c)t = AB/2\cos(\omega_s + \omega_c)t + AB/2\cos(\omega_s - \omega_c)t$$

The receiver processes the incoming signal using a quadrature downconverter so that when the worst-case situation occurs with the $\sin(\omega_c)$ multiplying, the quadrature channel will be multiplying with a $\cos(\omega_c)$ so that the signal is recovered in either (or both) the I and Q channels as shown:

$$
\begin{aligned}
\text{I channel} &= [AB/2\cos(\omega_s + \omega_c)t + AB/2\cos(\omega_s - \omega_c)t][\sin(\omega_c)t] \\
&= AB/4[\sin(\omega_s + 2\omega_c)t + \sin(\omega_s)t - \sin(\omega_s)t + \sin(\omega_s - 2\omega_c)t] \\
&= AB/4[\sin(\omega_s + 2\omega_c)t + \sin(\omega_s - 2\omega_c)t]
\end{aligned}
$$

$$
\begin{aligned}
\text{Q channel} &= [AB/2\cos(\omega_s + \omega_c)t + AB/2\cos(\omega_s - \omega_c)t][\cos(\omega_c)t] \\
&= AB/4[\cos(\omega_s)t + \cos(\omega_s + 2\omega_c)t + \cos(\omega_s - 2\omega_c)t\cos(\omega_s)t] \\
&= AB/4[\cos(\omega_s + 2\omega_c)t + \cos(\omega_s)t + \cos(\omega_s)t + \cos(\omega_s - 2\omega_c)t] \\
&= AB/2\cos(\omega_s)t + AB/4\cos(\omega_s + 2\omega_c)t + AB/4\cos(\omega_s - 2\omega_c)t
\end{aligned}
$$

This shows all of the signal being received in the Q channel, $\cos(\omega_s)$, and none in the I channel. Most of the time this will be a split, and the magnitude and phase will be determined when combining the I and Q channels. This quadrature technique uses carrier recovery loops when receiving a suppressed carrier waveform. For example, Costas loops, which are commonly used in direct sequence systems, use this technique to recover the carrier and downconvert the waveform into I and Q data streams.

3.11 Splitting Signals into Multiple Bands for Processing

Some receivers split the incoming signal into multiple bands to aid in processing. This is common with intercept-type receivers that are trying to determine what an unknown signal is. If the signal is divided up in multiple bands for processing, then the signal is split up but the noise remains the same if the bandwidth has not changed. Therefore, the SNR will degrade each time the signal is split by at least 3 dB. However, if the signal is coherent and is summed coherently after the split, then there is no change in the SNR except some loss in the process gain. Thus, careful examination of the required SNR and the type of processing used with this type of receiver will aid in the design.

3.12 Phase Noise

Phase noise is generally associated with oscillators or phase-locked loops (PLLs). Phase noise is the random change in the phase or frequency of the desired frequency or phase of operation. The noise sources include thermal noise, shot noise, and flicker noise or $1/f$ noise.

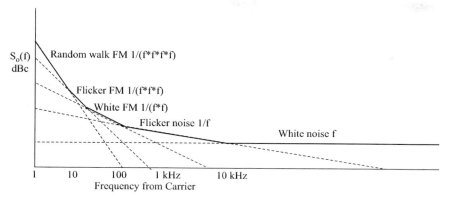

Note: $S_o(f)$ = power spectral density (radians^2/Hz) in dB down from carrier dBc/Hz (Log-Log Scale)

Figure 3-6   Phase noise analysis.

The different types of phase and frequency noise and the causes that generate each of the processes and definitions are shown in Figure 3-6. The noise beyond about 10 kHz is dominated by white noise, which is generally referred to as *kTB* noise since it is derived using Boltzmann's constant, temperature, and bandwidth. The 1/*f* noise or flicker noise is the dominant noise source from about 100 Hz to 10 kHz. This noise is produced by noisy electrical parts, such as transistors and amplifiers, and choosing a good LNA can help. The next type of noise, which is the dominant noise from about 20 Hz to 100 Hz, is white FM noise, caused by passive resonator devices such as cesium and rubidium. White FM noise appears to change the frequency with respect to white noise and is designated as 1/*ff*. The next type of noise is the physical resonance mechanism or the actual parts in oscillators and is close to the carrier, about 5 Hz to 20 Hz. This type of noise is known as flicker FM and is designated as 1/*fff*. The closest noise to the actual carrier frequency is 1/*ffff* noise, which is caused by vibration, shock, temperature, and other environmental parameters. This is called random-walk FM and is very difficult to measure since it is so close to the carrier and requires very fine resolution.

3.13 Bandwidth Considerations

The bandwidth of the transmitted and received signal can influence the choice of IF bands used. The IF bandwidth is selected to prevent signal foldover. In addition, the IF bandwidth needs to reduce the higher frequencies to prevent aliasing. Also, the bandwidths need to be selected to relax design constraints for the filters.

For example, a direct sequence spread bandwidth is approximately 200 MHz wide. Since the total bandwidth is 200 MHz, the center RF needs to be greater than twice the 200 MHz bandwidth so that the unwanted sidebands can be filtered. An RF of 600 MHz is chosen because of the availability of parts. This also provides sufficient margin for the filter design constraints. If the bandwidth is chosen so that the design constraints on the filters require the shape factor, or the frequency response of the filter, to be too large, then the filters may not be feasible to build.

Bandwidth plays an important function in the SNR of a system. The narrower the bandwidth, the lower the noise floor, since the noise floor of a receiver is related to *kTBF*, where *B* is the bandwidth.

3.14 Filter Constraints

The shape factor is used to specify a filter's roll-off characteristics (Figure 3-7). The shape factor (SF) is defined as

$$SF = f(-60\text{dB})/f(-3 \text{ dB})$$

where

$f(-60$ dB$)$ = frequency that is 60 dB down from the center of the passband
$f(-3$ dB$)$ = frequency that is 3 dB down from the center of the passband

For example,

$$SF = 20\text{MHz}/10 \text{ MHz} = 2$$

where

$f(-60$ dB$)$ = 20 MHz
$f(-3$ dB$)$ = 10 MHz

Note that these values may be evaluated at different levels for different vendors.

Typical achievable shape factors range from 2 to 12, although larger and smaller shape factors can be achieved. Shape factors as low as 1.05 can be achieved with crystal filters. Note that insertion loss generally increases as the shape factor decreases. Insertion loss ranges from 1.2 dB to 15 dB. Surface acoustic wave (SAW) filters achieve even a better shape factor at the expense of greater insertion loss and cost. They are also more prone to variations from vibration.

The percent bandwidth is used to specify the limitations of the ability to build a filter. The percent bandwidth is a percentage of the carrier frequency. For example, a 1% bandwidth for 100 MHz would be a 1 MHz bandwidth. This is usually specified as the 3 dB bandwidth.

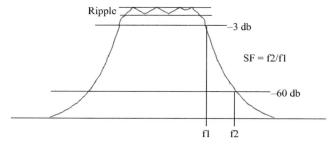

Figure 3-7   Typical filter response showing insertion loss and shape factor.

If the bandwidth is not specified, then the 3 dB bandwidth is used. Typical achievable bandwidths using crystal filters for 10 kHz to 200 MHz range from 0.0001 to 3%.

Filters are used primarily for out-of-band rejection. This helps reduce interference and adjacent channel interference for multiple users. Many different types of filters can be used depending on the type of roll-off, phase distortion, passband characteristics, and out-of-band sidelobes. Some filters may have good roll-off characteristics but may experience sidelobes in the out-of-band response.

Caution needs to be used for filters that may act differently at higher frequencies, where the capacitors may look inductive and the inductors may look capacitive. This is usually specified at the resonance frequency of the passive component. The self-resonance frequency (SFR) of a component is where the impedance of the capacitance and the impedance of the inductance cancel out to provide a low resistive impedance. The SRF should be much higher than the frequency at which the part is going to be used.

Another consideration in selecting a filter is the amount of amplitude variation in the passband, known as passband ripple. For example, Butterworth filters have low passband ripple, whereas Chebyshev filters specify the passband ripple that is desired.

Insertion loss is another key design consideration. Usually the higher order filters and steeper roll-offs have higher insertion losses that have to be accounted for in the design of the receiver. A typical filter with a specified insertion loss and shape factor is shown in Figure 3-7.

3.15 Group Delay

Group delay is the measurement of the delay of the frequency components through a device or system. This is an important consideration for digital communications. Digital signals and waveforms are equal to a sum of multiple frequencies with different amplitudes that are specified using the Fourier series. As these frequencies propagate through the device or system, if all of the frequency components have the same delay, then they are said to have constant group delay and the output digital signal will be the same as the input digital signal. However, if these frequencies have different delays through the device or system, which is nonconstant group delay, then they will not add up correctly on the output and the output digital signal will be distorted (Figure 3-8). This distortion is known as dispersion and will cause intersymbol interference (ISI). This is where one symbol or digital signal interferes with another symbol or digital signal because they are spread out in time due to dispersion.

Devices in a receiver such as filters do not have constant group delay, especially at the band edges (Figure 3-9). Some filters are better than others as far as group delay; for example, Bessel filters are known for their constant group delay. Constant group delay and linear phase are related since the group delay is the derivative of the phase. Thus, for linear phase, which means a constant slope, the group delay or the derivative of the constant slope is a constant number.

Some systems utilize group delay compensation to help mitigate the nonconstant group delay of a device or system. The basic technique measures the group delay/phase linearity of the device or system. Several components affect group delay, including filters, amplifiers, and narrowband devices. Once the measured group delay has been obtained, an all-pass phase linearizer/group delay compensator is designed (Figure 3-10).

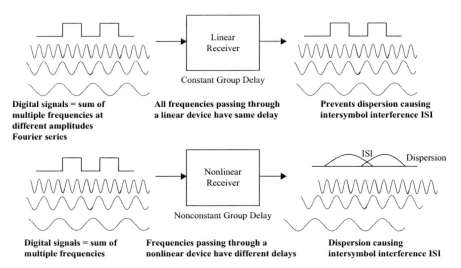

Figure 3-8   Group delay for digital signals.

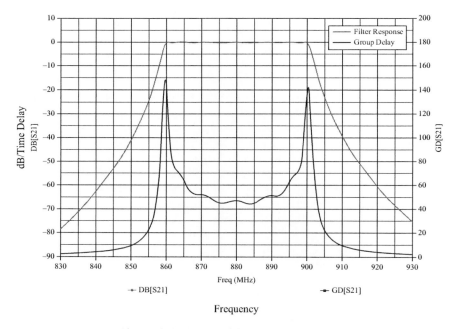

Figure 3-9   Group delay through a filter.

These compensators have no effect on magnitude, only phase, with their poles and zeros spaced equally and opposite of the $j\omega$ axis. The next step is to convolve the curves of the measured group delay and group delay compensator to produce a constant group delay for the system (Figure 3-11).

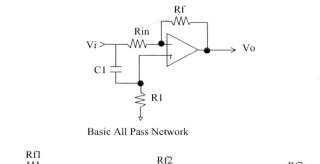

Basic All Pass Network

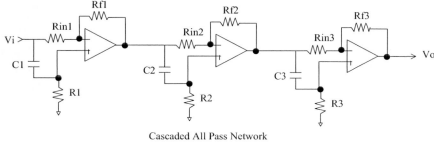

Cascaded All Pass Network

Figure 3-10   Op amp all pass group delay equalizer.

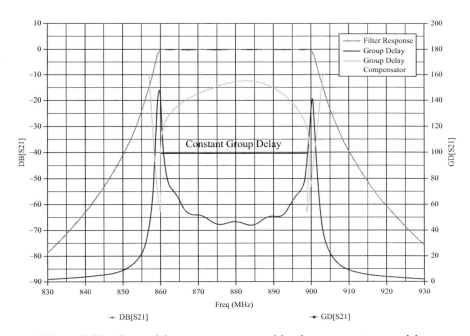

Figure 3-11   Group delay compensator resulting in a constant group delay.

3.16 Analog-to-Digital Converter

An ADC is a device that samples the analog signal and converts it to a digital signal for digital processing. It must sample at least as fast as the Nyquist rate, or twice as fast as the highest frequency component of the signal. One of the problems that occurs when an analog signal is digitized is the low resolution of the step size. The error associated with the step is called the quantization error.

As the signal becomes smaller in a linear ADC, the error becomes larger. Note that if the signal is large, the quantization errors in both the I and Q channels produce a small change in amplitude and phase. The phase error is just a few degrees, and the amplitude is a very small percentage of the actual amplitude, as shown in Figure 3-12. As for the smaller signal, quantization error in the I and Q channels produces a large error in phase.

Most ADCs are linear, which makes it harder to detect smaller signals. To improve the sensitivity of the ADC, a log amplifier is sometimes used before the ADC or an ADC with a log scale (μ-law) is used. This gives a higher response (gain) to lower level signals and a lower response (gain) to higher level signals and approximates the ideal ADC (Figure 3-12).

Another way to improve the sensitivity of lower level signals is to use a piecewise linear solution where there are multiple ADC stages with different scaling factors. Each finer resolution stage is set to cover the LSBs of the previous ADC. This method provides a large DR while still being a linear process. Also, a piecewise μ-law solution can be used to increase the DR of the ADC, but this would not be a linear solution.

3.17 Sampling Theorem and Aliasing

Nyquist states that an analog function needs to be sampled at a rate of at least two times the highest frequency component in the analog function to recover the signal. If the analog function is sampled less than this, aliasing will occur.

When a signal is sampled, harmonics of the desired signal are produced. These harmonics produce replicas of the desired signal (positive and negative frequencies) in the frequency domain at $1/T_S$ intervals. When the analog signals are undersampled, the harmonics of the repeated signal start to move into the passband of the desired signal and produce aliasing. Anti-aliasing or pre-aliasing filters can eliminate this problem for digital sampling and transmissions.

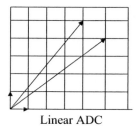
Linear ADC

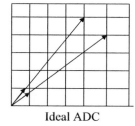
Ideal ADC

Ideal ADC percent error is reduced for lower level signals.
μ-law and linear approximation designed to approximate ideal ADC.

Figure 3-12   Quantization error for large and small signals using a linear versus an ideal ADC.

3.18 Anti-Aliasing Filter

A filter is required before any sampling function to prevent aliasing of the higher frequencies back into the desired signal bandwidth (Figure 3-13). This is required to filter the high frequencies above the Nyquist criteria, which states that the sample rate needs to be twice the highest frequency component in the desired signal and that all frequencies higher than that need to be filtered out. This filter is called an anti-aliasing filter. To illustrate the concept, if the highest frequency of the desired signal is 1 MHz, then the sample rate required is 2 million samples per second (Msps). If this frequency is sampled at this rate, the output of the sampler would be alternating plus and minus values at a 2 Msps rate (Figure 3-13). The reconstruction of the samples would produce a 1 MHz frequency. If there is an incoming signal at a higher frequency and it is sampled at the same rate (undersampled according to the Nyquist rate), then the output would produce a frequency that would appear as a lower frequency than the actual input frequency (Figure 3-13). This reconstructed frequency could fall into the desired signal bandwidth and interfere with the correct signal (Figure 3-14).

In some systems, this Nyquist undersampled technique is controlled and used to process desired signals at higher frequencies by undersampling. This allows for the sample rate of the detector to be slower and provides a means of processing much higher frequency signals with slower clocks. This can reduce cost and increase bandwidths for the detection process. For example, if the desired signal is the high-frequency signal in Figure 3-14, then the resultant aliased low-frequency signal would be the desired signal and is detected by this lower sample rate detector. However, careful design is needed so that unwanted signals do not alias and cause interference with the desired signal.

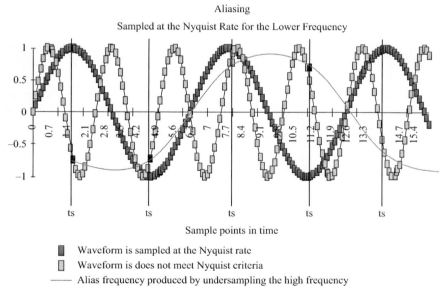

Figure 3-13   Graph showing the Nyquist criteria for sampling and aliasing.

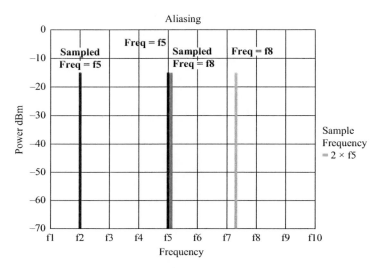

Figure 3-14   Graph showing the Nyquist criteria for sampling and aliasing.

3.19 Dynamic Range/Minimum Detectable Signal

There is a great deal of confusion in the definitions and measurements of MDS and DR. Care should be taken to ensure accurate analysis and to compare different systems with the same criteria. One of the main problems is that many systems today process signals in digital signal processing (DSP) integrated circuits, which makes comparisons with the analog-type systems difficult. MDS is a measurement of how well a receiver can detect a signal with respect to noise and is generally calculated at the ADC. DSP systems may not be able to evaluate the signal at the ADC, especially if spread spectrum systems are used and despreading is performed in the digital domain. Before determining the MDS for a particular system, careful analysis needs to be performed to provide the optimal place in the system to do the calculation. Also, the criteria for calculating the MDS need to be considered to evaluate each system's architecture. BER, tangential sensitivity (TSS), where the pulse is tangential to the noise, SNR, and others can be used as the criteria. To evaluate each system fairly, the same process should be used for comparison.

DR can be calculated using the minimum signal that the ADC can detect and the maximum voltage that the ADC can process. This assumes there is no saturation and the ADC can handle the full DR. For example, if the maximum voltage is 2 V, then the maximum 2 V will fill up the ADC with all "1"s. For each bit of the ADC, the DR is increased by approximately 6 dB/bit. The reason for this is that the DR of the voltage for each bit in the ADC is divided in half and then converted to dB:

$$10\log(\tfrac{1}{2})^2 = -6 \text{ dB}$$

For each ADC bit, the value of the bit could be high or low, which splits the decision by ½. For example, if there is a range from 0 V to 10 V on the input to the ADC and there is one bit in the ADC with a threshold set at 5 V, then the DR has been increased by ½. Now the

ADC can detect 5 V and 10 V instead of just 10 V. If the bit value is high, then the range is from 5 V to 10 V; if the bit value is low, then the range is from 0 V to 5 V. Since voltage levels are being evaluated, the ½ is squared. If there are two bits, then the range can be divided into four levels, and each level is ¼ the total DR, which calculates to -12 dB.

Consider, for example, a system with an ADC that can handle a maximum of 2 V and uses an 8-bit ADC. The 8-bit ADC provides 48 dB of DR which produces a minimum signal that is approximately -48 dB below 2 V. Therefore, the minimum signal in volts would be approximately 7.8 mV, with a DR equal to 48 dB.

The system noise, which establishes the noise floor in the receiver, is composed of thermal noise ($kTBF$) and source noise. The values of k, T, B, and F are usually converted to dB and summed together. Additional noise is added to the system due to the source noise, including LO phase noise, LO bleed-through, and reflections due to impedance mismatch. Consequently, the noise floor is determined by

$$\text{Noise floor} = \text{thermal noise} + \text{source noise}$$

The noise floor and the minimum signal at the ADC are used to calculate the SNR, and the MDS is calculated using given criteria for evaluation, such as the tangential sensitivity for pulsed systems.

The input noise floor for the system is used to determine the amount of gain that a receiver must provide to optimize the detection process. The gain required from the receiver is the amount of gain to amplify this input noise to the threshold level of the ADC's least significant bit (LSB). Therefore, the gain required (G_r) is

$$G_r \text{ (dB)} = [\text{noise floor at the ADC's LSB (dB)}] - [\text{input noise floor at the receiver (dB)}]$$

The DR of the receiver is the DR of the ADC unless additional DR devices are included, such as log amplifiers or AGC, to handle the higher level inputs. Many digital systems today rely on saturation to provide the necessary DR. The DR of the receiver depends on the portion of the receiver that has the smallest difference between the noise floor and saturation.

Dynamic range is another often misunderstood definition. The definition used here is the total range of signal level that can be processed through the receiver without saturation of any stage and within a set BER or defined MDS level. (However, some systems can tolerate saturation, which increases this dynamic). This assumes that the signal does not change faster than the AGC loops in the system. If the signal changes instantaneously or very fast compared to the AGC response, then another definition is required, instantaneous dynamic range (IDR). The IDR is the difference between saturation and detectability, given that the signal level can change instantaneously. The IDR is usually the DR of the ADC, since the AGC does not respond instantaneously.

To determine the receiver's DR, a look at every stage in the receiver is required. The NF does not need to be recalculated unless there is a very large amount of attenuation between stages or the bandwidth becomes wider with a good deal of gain. Generally, once the NF has been established, a look at the saturation point in reference to the noise level at each component can be accomplished. An even distribution of gains and losses is usually the best. A DR enhancer such as an AGC or a log amplifier can increase the receiver's DR but may not increase the IDR. However, devices such as log amplifiers can actually reduce the overall DR by adding noise to the desired signal.

Note that the DR can be reduced by an out-of-band large signal, which causes compression and can also mix in noise. The jamming signals should be tested and analyzed to determine the degradation of the receiver's DR.

3.20 Types of DR

Dynamic range in a receiver design can designate either amplitude or frequency. The two DRs are generally related and frequently only one DR is used, but the type of DR specified is given for a particular reason, for example, a two-tone test or receiver test.

3.20.1 Amplitude DR

If a design requires the output be within a given amplitude signal level, then the DR is generally given as the difference between the MDS and the maximum signal the receiver can handle (saturation). Often the 1 dB compression point provides a means of determining the maximum signal for the receiver. This point is where the output signal power is 1 dB less than what the linear output power is expected to be with a given input, which means there is some saturation in the receiver (Figure 3-15). Operating at this point results in 1 dB less power than what was expected.

If the DR needs to be improved, then an AGC or a log amplifier can be added. The amplitude DR is probably the most commonly used DR term. Compression and saturation are important to prevent distortion of the incoming signal, which generally increases the BER.

3.20.2 Frequency DR

Many systems are concerned with the frequency DR, which describes the ability of a receiver to distinguish different frequencies, whether it is between multiple desired frequencies or

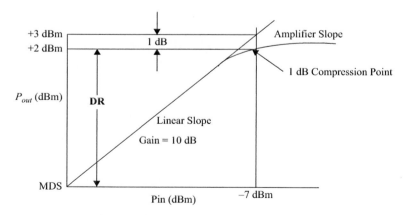

Figure 3-15 Amplitude DR using 1 dB compression point.

between the desired frequency and spurious signals or jammers. This is related to the 1 dB compression point DR, since when nonlinearities are present there are unwanted spurious signals generated which reduce the frequency DR (Figure 3-16). Two methods of analysis typically prove useful information when considering frequency DR.

3.20.3 Single-Tone Frequency DR

One method is to use a one-tone DR and calculate the spurs that will reduce the frequency DR because of this tone. This method is not used very often because generally there is more than one input signal.

3.20.4 Two-Tone Frequency DR

A better and more common way is to define the DR using a two-tone DR analysis. This approach takes two input signals and their generated spurious products and calculates their power levels using intercept points on the linear or desired slope (Figure 3-16).

The first-order spurs—f1 and f2—do not test nonlinearity, so this method is not used. The second-order spurs—2f1, 2f2, f1 * f2 = (f1 + f2, f1 − f2)—are caused by non-linearities and are at high power levels but can generally be easily filtered out. If they cannot be, since they are larger than third order, this could cause some problems due to the high power level of the spurs. The third-order spurs—3f1, 3f2, 2f1 * f2 = (2f1 − f2, 2f1 + f2), f1 * 2f2 = (f1 − 2f2, f1 + 2f2)—are also caused by nonlinearities. These spurs are all filtered out except for 2f1 − f2, f1 − 2f2, which are inband and are used to test the frequency DR.

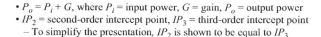

• $P_o = P_i + G$, where P_i = input power, G = gain, P_o = output power
• IP_2 = second-order intercept point, IP_3 = third-order intercept point
  – To simplify the presentation, IP_2 is shown to be equal to IP_3

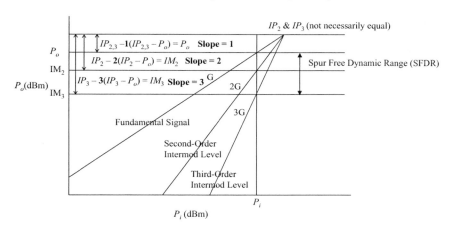

Figure 3-16 Frequency dynamic range—two-tone spur free dynamic range (SFDR).

The second-order intercept point helps determine the second harmonics of the two inputs (2×0, 0×2) and the product of the two signals (1×1). The second-order spurs are the highest power intermod products that are created due to a nonlinear portion of the receiver. However, if the bandwidth is limited to less than an octave, these products are eliminated and the third-order spurs (2×1, 1×2) become the strongest products in-band. These spurs generally fall into the operating band, especially when they are close together in frequency.

The third-order intercept point becomes the criteria for calculating the DR for the receiver (Figure 3-16). Caution must be used when performing the analysis because if the bandwidth is greater than an octave then the second-order intermods cannot be neglected.

3.21 Second- and Third-Order Intermodulation Products

The second- and third-order intercept points are used to determine the intermod levels in the system. The output power (in dB) plotted against the input power is a linear function up to the compression point. The slope of this line is constant with gain; that is, $P_o(\text{dBm}) = P_i(\text{dBm}) + G$ (dB). The slope of the second-order line is equal to twice that of the fundamental; that is, for every 1 dB of increase in output fundamental power, the second-order output signal power increases by 2 dB. The slope of the third-order line is equal to three times that of the fundamental; that is, for every 1 dB of increase in output fundamental power, the third-order output signal power increases by 3 dB. If the fundamental plot is extended linearly beyond the 1 dB compression point, then this line will intersect the second- and third-order lines. The point where the second-order curve crosses the fundamental is called the second-order intercept point. The point where the third-order curve crosses the fundamental is called the third-order intercept point and is generally not at the same location as the second-order intercept point. The intercept points are given in dBm. The intermod levels can then be calculated for a given signal level and can be graphically shown and calculated as shown in Figure 3-16.

The second-order intermod signal will be two times farther down the power scale from the intercept point as the fundamental. Therefore, the actual power level (in dB) of the second-order intermod is

$$IM_2 = IP_2 - 2(IP_2 - P_o)$$

where

$IM_2 =$ second-order intermod power level (in dBm)
$IP_2 =$ second-order intercept point
$P_o =$ output power of the fundamental signal

The third-order intermod signal will be three times farther down the power scale from the intercept point as the fundamental. Therefore, the actual power level (in dB) of the third-order intermod is

$$IM_3 = IP_3 - 3(IP_3 - P_o)$$

where

IM_3 = third-order intermod power level (in dBm)
IP_3 = third-order intercept point
P_o = output power of the fundamental signal

For example, suppose the second-order and third-order intercept points are at +30 dBm, the input power level is at −80 dBm, and the gain is at +50 dB. The signal output level would be at −30 dBm. The difference between the intercept points and the signal level is 60 dB. The following example shows what the intermod signal levels are.

Given

$IP_2 = IP_3 = +30$ dBm
$P_i = -80$ dBm
$G = 50$ dB

calculate

$$P_o = G * P_i = 50 \text{ dB} + -80 \text{ dBm} = -30 \text{ dBm}$$
$$IP_2 - P_o = IP_3 - P_o = +30 \text{ dBm} - (-30 \text{ dBm}) = 60 \text{ dB}$$
$$P_o = IP_{(2,3)} - 1(IP_{(2,3)} - P_o) = +30 \text{ dBm} - 1(60 \text{ dB}) = -30 \text{ dBm}$$
$$IM_2 = IP_2 - 2(IP_2 - P_o) = +30 \text{ dBm} - 2(60 \text{ dB}) = -90 \text{ dBm}$$
$$IM_3 = IP_3 - 3(IP_3 - P_o) = +30 \text{ dBm} - 3(60 \text{ dB}) = -150 \text{ dBm}$$

which results in

$IM_2 = -90$ dBm $- (-30$ dBm$) = -60$ dBc (60 dB down from P_o)
IM_2 is generally easy to filter out because second order products are generally not in the passband
$IM_3 = -150$ dBm $- (-30$ dBm$) = -120$ dBc (120 dB down from P_o)

Since the second-order signals can be generally filtered out, the third-order products limit the system's DR.

3.22 Calculating Two-Tone Frequency DR

A method of calculating DR is by comparing input power levels shown in Figure 3-17. This example is for third-order frequency DR. The input power that causes the output power (P_o) to be right at the noise floor is equal to P_1. This is the minimum input signal that can theoretically be detected. The input power (P_3) is the power that produces third-order intermod products at the same P_o as what is at the noise floor. An increase in input power will cause the third-order spurs to increase above the noise. (Note that the spurs will increase in power faster than the desired signal level as input power is increased.)

The following equations are simple slope equations:

$$(P_o - IP_3)/[P_1 - (IP_3 - G)] = 1$$
$$P_1 = P_o - G$$
$$(P_o - IP_3)/[P_3 - (IP_3 - G)] = 3$$
$$P_3 = (P_o + 2IP_3 - 3G)/3$$

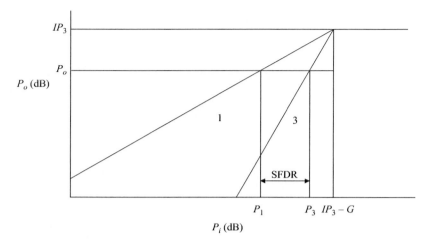

Figure 3-17   Frequency dynamic range analysis.

The difference in the input powers gives the DR of the system:

$$DR = P_3 - P_1 = (P_o/3 + 2IP_3/3 - G) - (P_o - G) = \tfrac{2}{3}(IP_3 - P_o)$$
$$P_o = \text{noise floor} = -174 + 10\log B + NF + G - G_p$$
$$DR = \tfrac{2}{3}(IP_3 + 174 - 10\log B - NF - G + G_p)$$

where

    DR = dynamic range
    IP_3 = third-order intercept point
    B = bandwidth
    NF = noise figure
    G = gain
    G_p = process gain

Before the detector, the receiver NF bandwidth uses the smallest IF bandwidth for the calculation.

The previously given analysis gives the theoretical DR, but for the usable DR, a SNR needs to be specified. This requires the input power for the desired signal to increase sufficiently to produce the desired SNR (Figure 3-17). Since the desired spur level is still at the noise floor, this is a direct subtraction (in dB) of the desired SNR:

$$DR_a = P_3 - P_{1a} = \tfrac{2}{3}(IP_3 + 174 - 10\log B - NF - G + G_p) - SNR$$

This is the third-order spurious-free dynamic range (SFDR).

From the same analysis, the second-order usable DR is

$$DR_a = \tfrac{1}{2}(IP_2 + 174 - 10\log B - NF - G + G_p) - SNR$$

To calculate the IP3 for an amplifier, the two-tone test is used. The power output per tone is designated as P_{tone}. The level of the highest intermodulation (intermod) product is also specified or measured and is designated as intermodulation distortion (IMDs) in dBc (referenced to the carrier). Therefore, the equation to calculate IP3 for both the input (IIP3) and the output (OIP3) is

$$\text{OIP3} = \text{P}_{out} + \text{dBc}/2\text{: dBc is the highest level of the IMDs}$$

$$\text{IIP3} = \text{OIP3} - \text{gain (dB)}$$

Here are two examples in calculating the OIP3 and the IMD level:

Example 1:

Given: Ptone = +5 dBm, dBc = 30 dBc
Solve for OIP3:
OIP3 = 5 dBm + 30 dBc/2 = 5dBm + 15 dB = 20 dBm

Example 2:

Given: OIP3 = 20 dBm, Ptone = +5 dBm
Solve for IMD level:
20dBm = 5 dBm + XdBc/2 = (20 dBm − 5 dBm)*2 = 30 dBc

3.23 System DR

One concern in receiver system analysis is the determination of where the minimum DR occurs. To easily determine where this is, a block diagram of the receiver with the noise level and the saturation level is developed. This can be included as a parameter in the link budget. However, instead of cluttering the link budget, a separate analysis is usually performed. The outputs of the devices are listed with the noise level, the saturation level, and the calculated amplitude DR. This is shown in a spreadsheet in Table 3-1.

The procedure is to list all of the power levels and noise levels at every point in the system. The estimated NF is established at the LNA, which in turn estimates the relative noise floor. Note that the bandwidth of the devices is being used initially. If the bandwidth is

Table 3-1 Receiver dynamic range calculations

RECEIVER ANALYSIS FOR OPTIMUM DYNAMIC RANGE					
Receiver	Gain/Losses	Sat. (dBm)	Noise (dBm)	IDR (dB)	DR (dB)
RF BW (KHz)(KTB)	100.00		−124		
RF Components	2.00	50.00	−124	174.00	174.00
LNA Noise Figure (int.)	3.00		−121.00		
LNA (Out)	30	7	−91.00	98.00	98.00
Filter (Out)	−2	50	−93.00	143.00	143.00
Mixer (Out)	−10	0	−103.00	103.00	103.00
IF Amp (Out)	50	7	−53.00	60.00	60.00
IF Bandwidth (kHz)	10		−63.00		
IF Filter Loss	−2	50	−65.00	115.00	115.00
Baseband Mixer (Out)	−10	0	−75.00	75.00	75.00
ADC (bits)	8	−27	−75.00	48.00	48.00

narrowed down the line, then the noise is lowered by $10\log B$. The top level is calculated by using either the 1 dB compression point or the third-order intercept point.

For the 1 dB compression point, the DR is calculated by taking the difference of the relative noise floor and this point. For the third-order intercept point method, the DR is calculated by taking ⅔ of the distance between the relative noise floor and the intercept point (see the previous equations). The minimum distance is found and from this point the gains are added (or subtracted) to determine the upper level input to the system. The difference between the input relative noise and the upper level input is the DR (Figure 3-16). The overall input third-order intercept point is found by adding ½ the input DR (refer back to the ⅔ rule, where ½ of ⅔ is ⅓, so the upper level input is ⅔ of the intercept point).

Adding an AGC extends the DR of the receiver generally by extending the video and detection circuitry, as shown in Table 3-2. To determine the required AGC, two main factors need to be considered: the dynamic range of the ADC and the input variations of the received signals.

The DR of the ADC is dependent on the number of bits in the ADC. For an 8-bit ADC, the usable DR with a SNR of 15 dB is approximately 30 dB (theoretical is approximately 6 dB/bit = 48 dB). This establishes the range of signal levels that can be processed effectively. If the input variations of the received signals are greater than this range, then an AGC is required. If the IDR needs to be greater than the ADC range, then a feed-forward AGC needs to be implemented to increase IDR. A feed-forward AGC uses a part of the input signal to adjust the gain before the signal is received at the AGC amplifier, so there is no time associated with the response of the gain. The amount of feed-forward AGC required is calculated by subtracting the ADC range from the input signal range.

The feedback AGC extends the DR but does nothing to the IDR, as shown in Table 3-2. The IDR tests the range in which the receiver can handle an instantaneous change (much quicker than the AGC response time) in amplitude on the input without saturation or loss of detection.

Table 3-2 Receiver dynamic range calculations using feedback AGC

- Does not improve instantaneous dynamic range IDR
- Improves overall dynamic range DR

RECEIVER ANALYSIS FOR OPTIMUM DYNAMIC RANGE WITH AGC					
Receiver	**Gain/Losses**	**Sat. (dBm)**	**Noise (dBm)**	**IDR (dB)**	**DR (dB)**
RF BW (KHz)(KTB)	100.00		−124		
RF Components	2.00	50.00	−124	174.00	174.00
LNA Noise Figure (int.)	3.00		−121.00		
LNA (Out)	30	7	−91.00	98.00	98.00
Filter (Out)	−2	50	−93.00	143.00	143.00
Mixer (Out)	−10	0	−103.00	103.00	103.00
AGC	30				
IF Amp (Out)	50	7	−53.00	60.00	90.00
IF Bandwidth (kHz)	10		−63.00		
IF Filter Loss	−2	50	−65.00	115.00	145.00
Baseband Mixer (Out)	−10	0	−75.00	75.00	105.00
ADC (bits)	8	−27	−75.00	48.00	78.00

The IDR can be extended by using a feed-forward AGC or a log amplifier. The problem with a log amplifier is that the IDR is extended as far as amplitude, but since it is nonlinear the frequency IDR may not improve and might even worsen. The feed-forward AGC is faster than the signal traveling through the receiver. Therefore, the IDR is improved for both amplitude and frequency and can be done linearly.

3.24 Tangential Sensitivity

For pulse systems, since the MDS is hard to measure, another parameter, tangential sensitivity, is shown in Figure 3-18. The signal power is increased until the bottom of the pulse noise is equal to the top of the noise level in the absence of a pulse. A tangential line is drawn to show that the pulse noise is tangential to the rest of the noise. With this type of measurement, a comparison between receivers can be made. The TSS measurement is good for approximately ±1 dB.

Dynamic range is critical in analyzing receivers of all types. Depending on the particular design criteria, both amplitude and frequency DRs are important in the analysis of a receiver. Methods for determining DR need to be incorporated in the design and analysis of all types of receivers. This is essential in optimizing the design and calculating the minimizing factor or device that is limiting the overall receiver DR. Graphical methods for calculating DR provide the engineer with a useful tool for analyzing receiver design. Both the DR and the IDR need to be considered for optimal performance. AGC can improve the receiver's DR tremendously but generally does not affect IDR. Feed-forward AGC and a log amplifier improve the IDR. However, the log amplifier is nonlinear and only improves the amplitude IDR and actually degrades the frequency IDR. The MDS is hard to define and measure.

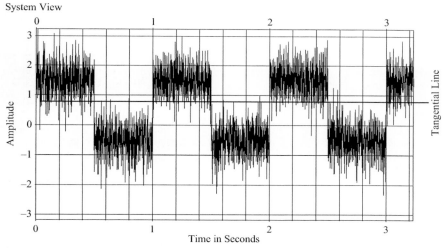

Used to measure sensitivity, need more SNR for reliable detection

Figure 3-18   Tangential sensitivity (TSS) for a pulse system.

An SNR is usually specified to help define the MDS. For pulse systems, the tangential sensitivity is utilized to improve the measurement of MDS and to compare receivers.

3.25 Digital Signal Processor

Once the analog signal is converted by the ADC to a digital signal, the digital signal processor finishes the reception by interpreting the data. Many receivers use DSP technology to accomplish most of the detection function. The RF/analog portion of the receiver is used only to downconvert the signal to a lower IF or baseband, spectrally shape the signal, and then convert it to the digital domain, where the remaining processing is digital. This enables the receiver to be configured for many receive applications and configurations for different waveforms by simply changing the software that drives the digital receiver.

Digital signal processing can do everything that analog processing can do, with the only limitation being the processor throughput required to sample fast enough for the frequency content of the signal. In conclusion, the sooner the signal can be processed digitally the better.

3.26 Summary

The receiver is an important element in the transceiver design. The receiver accepts the signal in space from the transmitter and amplifies the signal to a level necessary for detection. The LNA is the main contributor to the NF of the system. The superheterodyne is the most used receiver type and provides the most versatility by being able to apply a common IF. Saturation, compression, sensitivity, DR, reduction in unwanted spurious signals, and maximization of the SNR are the main concerns in designing the receiver. Mixers perform the downconversion process using spur analysis and selecting the correct mixer for the application. Two types of DR include amplitude, which is the most common way to express DR, and frequency DR, related to the two-tone third-order intercept point. Group delay plays an important role in digital communications, and careful consideration in the design will help reduce dispersion and ISI. Digital receivers perform most of the detection in data links today, and the ADC is used to translate an analog signal into the digital domain. The sooner the signal is in the digital domain, the better the receiver can optimize the detection process.

Reference

Tsui, J. B. *Microwave Receivers with Electronic Warfare Applications*. SciTech Publishing, NJ, 2005.

Problems

1. What is placed in the transceiver that prevents the transmitted signal from entering the receiver on transmit?
2. What is the gain required by the receiver for a system that has an 8-bit ADC, maximum of 1 V, NF of 3 dB, bandwidth of 10 MHz, and source noise of 4 dB?

3. What is the third-order SFDR of the system in problem 2 with a +20 dBm IP_3?
4. What is the usable DR of the system in problems 2 and 3 with a required SNR of 10 dB?
5. What is the NF without using an LNA with an IF NF of 3 dB and the total loss before the IF of 10 dB?
6. Calculate all intermodulation products up to third order for 10 MHz and 12 MHz signals.
7. What is the shape factor for a filter with 0 dBm at 90 MHz, −3 dBm at 100 MHz, and −60 dBm at 120 MHz?
8. Find the required sample rate to satisfy the Nyquist criteria if the highest frequency component is 1 MHz?
9. What would be the maximum amplitude and phase error on a linear ADC using 2 bits of resolution?
10. What are the advantages and disadvantages for oversampling a received signal?
11. Why is constant group delay important in digital communications? What is the result if the group delay is not constant?
12. What does LNA stand for? Why is its important?
13. What happens if a system samples the input signal at less than the Nyquist rate for the highest frequency content?
14. Why is each bit of an ADC equal to 6 dB?

CHAPTER 4

AGC Design and PLL Comparison

Automatic gain control (AGC) is used in a receiver to vary the gain to increase the dynamic range (DR) of the system. AGC also helps deliver a constant amplitude signal to the detectors with different radio frequency (RF) signal amplitude inputs to the receiver. AGC can be implemented in the RF section, the intermediate frequency (IF) section, in both the RF and IF portions of the receiver, or in the digital signal processing circuits. Digital AGCs can be used in conjunction with RF and IF AGCs. Most often the gain control device is placed in the IF section of the receiver, but placement depends on the portion of the receiver that limits the DR. The detection of the signal level is usually carried out in the IF section before the analog-to-digital converter (ADC) or analog detection circuits. Often the detection occurs in the digital signal processing (DSP) circuitry and is fed back to the analog gain control block. The phase-locked loop (PLL) is analyzed and compared with the AGC analysis, since both processes incorporate feedback techniques that can be evaluated using control system theory. The similarities and differences are discussed in the analysis. The PLL is used for tracking conditions and not for capturing the frequency or when the PLL is unlocked.

4.1 AGC Design

An AGC adjusts the gain of the receiver automatically. It is used to increase the DR of the receiver by adjusting the gain depending on the signal level; large signals need less gain, small signals need more gain.

There are different types of AGC designs, the most common of which is a feedback AGC, where the signal level is detected and processed, and a signal is sent back to the AGC amplifier or attenuator to control the gain. It uses the feedback of the received signal to adjust the gain. Another type of AGC is known as a feed-forward AGC. This type of AGC detects the signal level and adjusts the gain farther down the receiver chain. This type of AGC uses feed-forward, not signal feedback, and is used for high-speed control. Another type of control is dependent on time. This type of control is called a sensitivity time constant (STC) and is used in radar systems. For example, the change of gain is proportional to $1/t^2$ or other criteria that are dependent on the desired output and response. These types of gain control are shown in Figure 4-1.

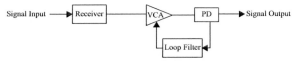

Feedback AGC: Uses feedback of the received signal to adjust the gain

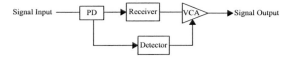

Feed forward AGC: Detects the signal level and adjust the gain ahead; for high speed control

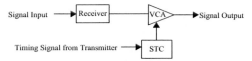

STC AGC: Used in radar systems; change the gain versus time; example 1/t2

Figure 4-1   Different types of AGCs used in receivers.

For the feedback AGC, the RF signal is downconverted and amplified, and the output is split and detected for use in the AGC to adjust the gain in the IF amplifiers. A voltage-controlled attenuator or variable gain amplifier, plus a linearizer, may be used for the actual control of the signal amplitude. The AGC voltage can be used to display the received power with an accuracy of about 0.5 dB by translating AGC volts to a received power equivalent which can be shown on a display or monitored by a personal computer (PC).

The requirements for AGC in a system are established by using several parameters to determine the amount of expected power fluctuations that might occur during normal operation. Some of the parameters that are generally considered include the variation in distance of operation, propagation loss variations such as weather and fading, multipath fluctuations, variations in the antenna pattern producing a variation in the gain/loss of the antenna in comparison to a true omnidirectional antenna, and expected power fluctuations because of hardware variations in both the transmitter and the receiver.

A typical AGC used in receivers consists of an AGC amplifier and a feedback voltage to control the gain. A voltage-controlled attenuator can be used in place of the AGC amplifier; however, the noise figure (NF) of the receiver can be changed if the attenuation is large. This feedback system can be designed using basic control system theory.

The basic model is shown in Figure 4-2a. This model is set up with the maximum gain from the AGC amplifier taken out of the loop as a constant and the feedback loop subtracts gain from this constant depending on the detected signal level.

The feedback system is redrawn to show the point at which the stability analysis is done, as shown in Figure 4-2b. This determines the stability of the AGC loop, and the feedback gain is then equal to unity, which makes it more convenient to perform the analysis using feedback control theory.

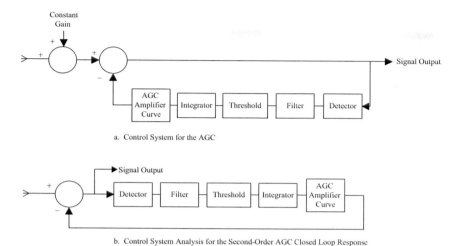

a. Control System for the AGC

b. Control System Analysis for the Second-Order AGC Closed Loop Response

Figure 4-2   AGC block diagrams used in the control system analysis.

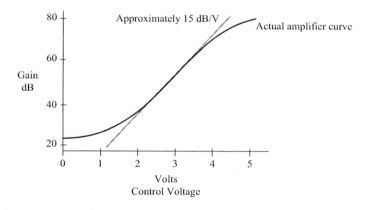

Figure 4-3   Typical gain versus control voltage for an AGC amplifier.

4.2 AGC Amplifier Curve

The first step in designing an AGC is to determine the amount of gain control needed. This may be given directly, or it can be calculated given the receiver's range of operation. Sometimes a compression ratio (range of the signal in versus range of the signal out) is provided. This will determine which AGC amplifier is needed.

Once the AGC amplifier is selected, the gain (dB) versus the control voltage (V) slope of the AGC amplifier is determined. The best way to do this is actual measurements in the lab, although some companies will furnish this curve in their data sheets. This slope is generally nonlinear. The AGC amplifier should be chosen so that the gain versus control voltage is as linear as possible, allowing the loop gain and noise bandwidth to remain stable as the control voltage is changed. An estimated linear slope can be applied depending on the design criteria. An example of an amplifier curve and an estimated linear slope is shown in Figure 4-3.

The slope is in dB/V and is used in the loop gain analysis. The estimated linear slope should be chosen where the AGC is operating most of the time. However, the AGC system should be designed so that it does not become unstable across the entire range of operation. The gain constant may be chosen at the steepest portion of the slope so that the loop gain will always be less than this gain value. This will cause the response time outside this approximation slope of the AGC to be longer (smaller bandwidth), which means the noise in the loop will always be less. Therefore, since the bandwidth can only get smaller, the reduction of desired modulation will always be smaller. Unless a slower response time presents a problem (as in the case of fading when a very slow AGC is used), these design criteria can be used in many situations.

4.3 Linearizers

Linearizers can be used to compensate for the amplifier's nonlinearity and create a more constant loop gain response over the range of the amplifier. A linearizer is a circuit that produces a curve response that compensates the nonlinear curve such that the sum of the two curves results in a linear curve (Figure 4-4). Note that the linear response is on a linear (V) versus log (P) scale.

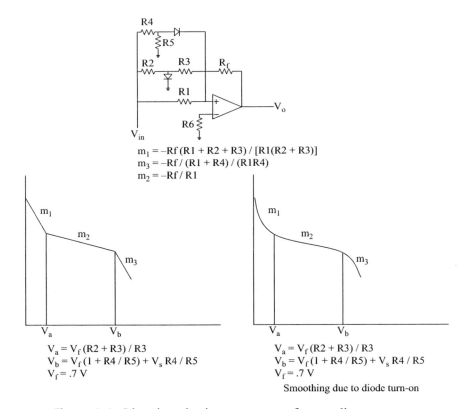

$$m_1 = -Rf\,(R1 + R2 + R3)\,/\,[R1(R2 + R3)]$$
$$m_3 = -Rf\,/\,(R1 + R4)\,/\,(R1R4)$$
$$m_2 = -Rf\,/\,R1$$

$V_a = V_f\,(R2 + R3)\,/\,R3$
$V_b = V_f\,(1 + R4\,/\,R5) + V_s\,R4\,/\,R5$
$V_f = .7\ V$

$V_a = V_f\,(R2 + R3)\,/\,R3$
$V_b = V_f\,(1 + R4\,/\,R5) + V_s\,R4\,/\,R5$
$V_f = .7\ V$

Smoothing due to diode turn-on

Figure 4-4 Linearizer circuit to compensate for a nonlinear response.

A linearizer can be designed by switching in diodes in a piecewise fashion, which alters the gain slope to compensate for the AGC amplifier gain curve. As the diodes are switched on, they provide a natural smoothing of the piecewise slopes due to the slope of the diode curve itself. This allows for fairly accurate curve estimations. Note that in Figure 4-4 the signs of the slopes have been neglected for simplicity. Care needs to be taken to ensure the correct sense of the loop to provide negative feedback. Linearizers can also be created and used in the digital processor, which is more accurate and is more immune to temperature variations and changes to the slopes of the desired curves.

4.4 Detector

The next step is the detector portion of the AGC. The sensitivity (gain) of the detector and the linearity of the curve (approaching a log of the magnitude-squared device) are important parameters in selecting the detector. The linear section of the detector can be chosen on the *square-law* portion of the curve (low-level signal) or the *linear* portion of the curve (high-level signal). The output power or voltage of the AGC is given or chosen for a particular receiver. The operating point can influence the choice of the detector used. The detector provides a voltage-out (V)/power-in (dBm) curve, or V/dB slope, which can be measured or given. The power in is an actual power level and is given in dBm. The change of power-in levels is a gain or loss given in dB. Therefore, the slope is a change of power-in levels and is given as V/dB not V/dBm. The ideal detector needs to be linear with respect to

$$V_o = \log[V_i^2] \text{ (log square law slope)}$$

The slope can be calculated if the detector is ideal:

$$(V + dV) - (V - dV)/[2\log(V + dV)/V - 2\log(V - dV)/V]$$

where

V = operating voltage level of the detector
dV = small variation of the operating voltage level

Note that if the detector is linear with respect to

$$V_o = \log[V_i] \text{ (log slope)}$$

the difference in the slope is a scale factor (2).

An amplifier stage with a gain of two following this detector would make it equal to the previous detector. A linearizer could be designed to compensate for either slope.

A linearizer design is shown in Figure 4-4. This method uses diode activation to set the break points of the curves, and the diode turn-on characteristics smooth the curves to a very close approximation of the desired curve.

The slope of a real detector is not linear. However, the slope is chosen so that it is linear in the region in which it is operating most of the time (Figure 4-5). This provides another part of the loop gain, given in V/dB.

The nonlinearity of the detector will change the instantaneous loop gain of the feedback system. Unless the system goes unstable, though, this is generally not a problem because the

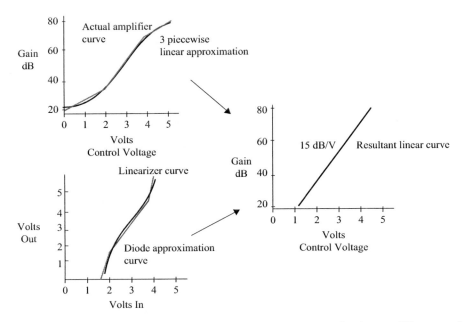

Figure 4-5 Using a linearizer curve to obtain a linear response for the amplifier control.

operating level to the diode is held constant by the AGC. Since the diode will always be operating in steady state at one point on the curve, the importance of linearization is not significant compared with the AGC amplifier linearization.

A linearizer can be used to help create a more linear slope from the detector so that the loop gain is more constant. If the modulating signal is large compared with the nonlinearity of the diode curve, there could be some distortion of the modulating signal. Also, there might be some distortion when the signal is on the edge of the AGC range of operation.

The resistor-capacitor (RC) time constant following the diode should be much larger than the period of the carrier frequency and much smaller than the period of any desired modulating signal. If $1/RC$ of the diode detector approaches the loop time response, then it should be included in the loop analysis. If $1/RC$ is greater than an order of magnitude of the loop frequency response, then it can be ignored, which is generally the case.

Since an AGC is generally not attempting to recover a modulating signal, as in amplitude modulation (AM) detection, failure-to-follow distortion is not considered. However, if the rate of change in amplitude expected is known, then this can be used as the modulating signal in designing for the response time. If a modulating signal (like a conical scan system produces to achieve angle information) is present, the total AGC bandwidth should be at least 10 times smaller to prevent the AGC from substantially reducing the conical scan modulating frequency. Here again, this depends on the design constraints. The modulation is actually reduced by $1/(1 + \text{loop gain})$. This is the sensitivity of the loop to the frequency change of the input.

A standard diode can be used to detect the signal. This is a common silicon diode, and the response of the diode is shown in Figure 4-6. Notice that this diode needs to be biased on, since the turn-on voltage is approximately 0.7 V for a silicon diode.

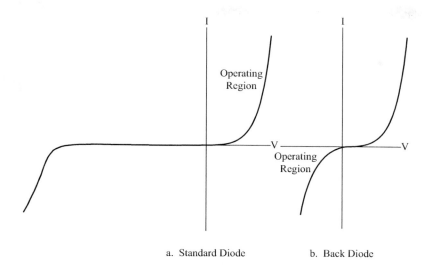

a. Standard Diode   b. Back Diode

Figure 4-6 Slope of some typical detectors.

Another type of diode that can be used to detect the signal is called a back diode. It provides very good sensitivity as a detector and uses the negative (back-biased) portion of the curve for detection (Figure 4-6). This detector operates well in the presence of low-level signals and requires no bias voltage for the diode since the operating region begins at zero.

The amplifier before the detector needs to be capable of driving the detector. A buffer amplifier can be used to supply the necessary current as well as to isolate the received signal from the loading effects of the detector. With a power divider and high-frequency detectors, this may not be a problem. The detector is matched to the system and the AGC amplifier output can drive the detector directly through the power divider.

4.5 Loop Filter

A loop filter needs to be designed to establish the frequency response of the loop and to stabilize the loop. The loop filter should follow the detector to prevent any high-frequency components from affecting the rest of the loop. A passive phase-lag network is used. The schematic consists of two resistors and a capacitor and is shown in Figure 4-7 along with the transfer function. This filter provides a pole and a zero to improve the stability of the loop. This filter is used to affect the attenuation and the roll-off point, since the phase shift for a phase lag filter generally has a destabilizing effect.

4.6 Threshold Level

A threshold needs to be set to determine the power level output of the AGC amplifier. A voltage level threshold in the feedback loop is set by comparing a stable voltage level with the output of the loop filter. This sets the power level out of the AGC amplifier. A threshold level circuit consists of an operational amplifier with a direct current (DC) offset summed in (Figure 4-7).

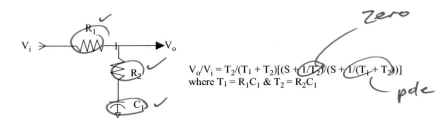

$V_o/V_i = T_2/(T_1 + T_2)[(S + 1/T_2)/(S + 1/(T_1 + T_2))]$
where $T_1 = R_1C_1$ & $T_2 = R_2C_1$

a. This is a lag filter providing a zero and a pole.

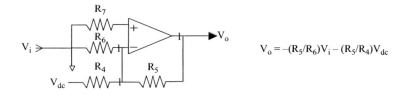

$V_o = -(R_5/R_6)V_i - (R_5/R_4)V_{dc}$

b. This is a threshold amplifier summing in the offset voltage which determines the AGC output level.

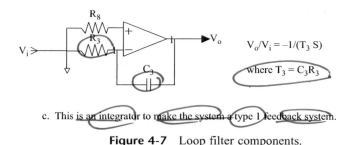

$V_o/V_i = -1/(T_3 S)$

where $T_3 = C_3R_3$

c. This is an integrator to make the system a type 1 feedback system.

Figure 4-7   Loop filter components.

In this case, if the output of the loop filter is less than the offset, then the output of the threshold device will be negative. If the loop filter output is greater than the offset, the output will be positive. This determines which way the gain is adjusted. The offset is equal to $-(R5/R4)V_{DC}$, as shown in Figure 4-7.

An additional gain is added to the loop and is selected for the particular AGC signal level and diode gain. This gain should be large enough to increase threshold sensitivity but small enough to prevent saturation of the loop amplifiers. The gain constant is defined as K_c and is equal to $-(R5/R6)V_1$, as shown in Figure 4-7.

4.7 Integrator

With just the phase-lag filter in the loop, the system is classified as a type 0 system, and there will be a steady-state error for a step response. If an integrator is included in the feedback loop, then the system is a type 1 system and the steady-state error for a step response is zero.

Therefore, an integrator is included in the loop. An integrator, using an operational amplifier, is shown in Figure 4-7. The output of the threshold device is integrated before controlling the gain of the AGC amplifier. With no signal present, the integrator will integrate to the rail of the operational amplifier (op amp) and stay there until a signal greater than the threshold is present. The AGC amplifier is held at maximum gain during this period of time, which is the desired setting for no signal or very small signals. The response time for the op amps to come out of saturation is negligible compared with the response time of the loop. If the voltage level out of the loop amplifiers is too large for the AGC amplifier, then a voltage divider or a voltage clamp should be used to protect the AGC amplifier. The DC gain of the integrator should not be limited by putting a resistor in parallel with the feedback capacitor or there will be a steady-state error. This makes the loop a second-order system and is easily characterized by using control system theory.

Most systems today use digital signal processing (DSP) techniques to detect the signal level and then provide feedback to control the gain of the amplifier or adjust the attenuator value. This is a feedback loop with the potential to oscillate, so careful design using control theory analysis is needed. This analysis includes the time delays, frequency responses, and gains through the digital circuitry path.

4.8 Control Theory Analysis

The open-loop transfer function including the integrator is

$$T_{sol} = \frac{K_a K_d K_c T_2 \left(S + \dfrac{1}{T_2} \right)}{T_3 (T_1 + T_2) S \left[S + \dfrac{1}{T_1 + T_2} \right]}$$

where

K_a = AGC amplifier loop gain
K_d = detector loop gain
K_c = gain constant
$T_1 = R_1 \times C_1$
$T_2 = R_2 \times C_1$
$T_3 = R_3 \times C_3$
T_{sol} = open-loop transfer function
$S = j\omega$ (assumes no loss)

The closed-loop response is

$$G(s)/(1 + G(s)) \text{ with } H(s) = 1$$

$$T_{scl} = \frac{K \left(S + \dfrac{1}{T_2} \right)}{S^2 \left(\dfrac{1}{T_1 + T_2} + K \right) S + \dfrac{K}{T_2}}$$

where

$$K = K_a K_d K_c (1/T_3)(T_2/[T_1 + T_2])$$

T_{scl} = closed-loop transfer function

From control system theory for the previous second-order system,

$$2\zeta\omega_n = K + 1/(T_1 + T_2)$$

and

$$\omega_n{}^2 = K/T_2$$

where

ω_n = natural frequency of the system

ζ = damping ratio

[handwritten: never seen this before]

The transfer functions and the block diagram of the analysis are shown in Figure 4-8.

[handwritten: NOTE] The natural frequency (ω_n) should be set at least 10 times smaller than any desired modulating frequency. The damping ratio is chosen so the system responds to a step function as fast as possible within a given percent of overshoot. With a 5% overshoot (PO), the minimum damping ratio is 0.707. This means the system is slightly underdamped. The damping ratio is chosen depending on the design criteria. If the overshoot is too high, the damping ratio should be changed. Once the damping ratio *[handwritten: made smaller]* and the natural frequency are chosen for a system, the time constants can be solved mathematically.

[handwritten: PO percentage overshoot]

Suppose the conical scan information modulation frequency of a radar system is equal to 30 Hz. The loop frequency response needs to be an order of magnitude less so that it does not interfere with the positioning information. Therefore, the loop frequency response is less than

[handwritten: 10x less or $\frac{30}{10} = 3$ Hz ✓]

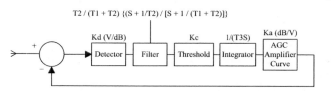

$$T2 / (T1 + T2) \{(S + 1/T2) / [S + 1 / (T1 + T2)]\}$$

Kd (V/dB) — Detector

Kc — Filter — Threshold

1/(T3S) — Integrator

Ka (dB/V) — AGC Amplifier Curve

Control system analysis for the second order AGC closed loop response.

$$\text{Open Loop Transfer Function} = Tsol = G(s) = Kd * \frac{T2}{(T1+T2)2} \frac{\left(S + \dfrac{1}{T2}\right)}{\left(S + \dfrac{1}{T1+T2}\right)} * Kc * \frac{1}{T3S} * Ka = \frac{KaKdKcT2\left(S + \dfrac{1}{T2}\right)}{T3(T1 + T2)S\left(S + \dfrac{1}{T1 + T2}\right)}$$

$$= \frac{K\left(S + \dfrac{1}{T2}\right)}{S\left(S + \dfrac{1}{T1 + T2}\right)}$$

$$\text{Closed Loop Transfer Function} = Tscl = G(s) / [1 + G(s)H(s)] = \frac{K\left(S + \dfrac{1}{T2}\right)}{S^2 + \left(\dfrac{1}{T1 + T2} + K\right)S + \dfrac{K}{T2}}$$

Figure 4-8   Transfer functions and block diagram used in the analysis of the AGC.

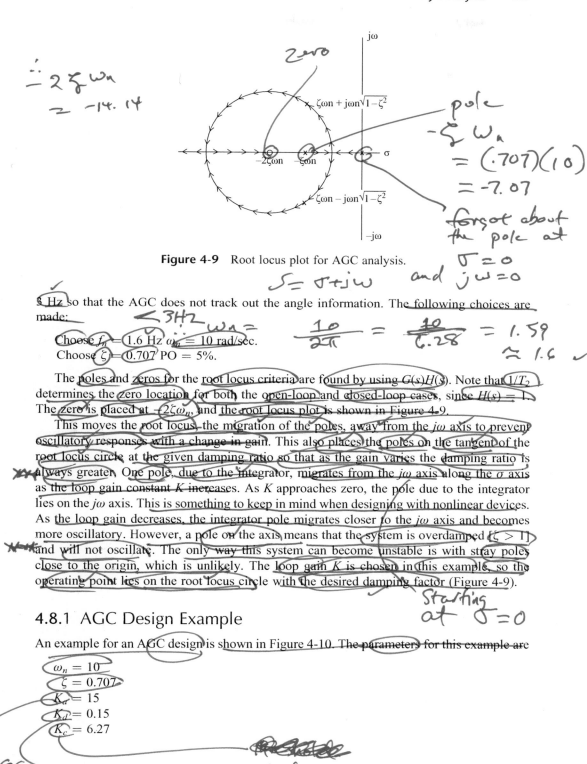

Figure 4-9   Root locus plot for AGC analysis.

[handwritten annotations:]
$-2\zeta\omega_n = -14.14$

zero

pole $-\zeta\omega_n = (.707)(10) = -7.07$

forgot about the pole at $\sigma = 0$ and $j\omega = 0$

$s = \sigma + j\omega$

3 Hz so that the AGC does not track out the angle information. The following choices are made:

[handwritten: < 3Hz $\omega_n = \frac{10}{2\pi} = \frac{10}{6.28} = 1.59 \approx 1.6$]

   Choose $f_n = 1.6$ Hz $\omega_n = 10$ rad/sec.
   Choose $\zeta = 0.707$ PO = 5%.

The poles and zeros for the root locus criteria are found by using $G(s)H(s)$. Note that $1/T_2$ determines the zero location for both the open-loop and closed-loop cases, since $H(s) = 1$. The zero is placed at $-2\zeta\omega_n$, and the root locus plot is shown in Figure 4-9.

This moves the root locus, the migration of the poles, away from the $j\omega$ axis to prevent oscillatory responses with a change in gain. This also places the poles on the tangent of the root locus circle at the given damping ratio so that as the gain varies the damping ratio is always greater. One pole, due to the integrator, migrates from the $j\omega$ axis along the σ axis as the loop gain constant K increases. As K approaches zero, the pole due to the integrator lies on the $j\omega$ axis. This is something to keep in mind when designing with nonlinear devices. As the loop gain decreases, the integrator pole migrates closer to the $j\omega$ axis and becomes more oscillatory. However, a pole on the axis means that the system is overdamped ($\zeta > 1$) and will not oscillate. The only way this system can become unstable is with stray poles close to the origin, which is unlikely. The loop gain K is chosen in this example, so the operating point lies on the root locus circle with the desired damping factor (Figure 4-9).

[handwritten: starting at $\sigma = 0$]

4.8.1 AGC Design Example

An example for an AGC design is shown in Figure 4-10. The parameters for this example are

   $\omega_n = 10$
   $\zeta = 0.707$
   $K_a = 15$
   $K_d = 0.15$
   $K_c = 6.27$

[handwritten: AGC amplifier detector   gain constant]

2ND ORDER AGC DESIGN:

Tclosed loop = [K(S + 1/T2)]/[S*S + (1/(T1 + T2) + K)S + K/T2]

Enter Constants:

Choose damping ratio z =	0.707	
Choose natural frequency fn =	10	Hz
AGC amplifier loop gain = Ka =	15	dB/Volt
Detector Loop Gain = Kd =	0.15	Volt/dB
Other component Gain = Kc =	10	Volt/Volt

Percent Overshoot = 4.325493 %
wn = 62.83185 rad/sec

Calculations:

The zero is set at 2zwn	T2 =	0.011256	sec
K = T2*wn*wn =	44.43554		
2zwn = K+1/(T1+T2)			
Solve for T1 = 1/(2zwn-K)-T2	T1 =	0.011262	sec
K = KaKdKc(1/T3)(T2/(T1+T2))			
T3 = KaKdKc(T2/(T1 + T2))/K	T3 =	0.253099	sec

Detector Design:

Enter RF source impedance = 5000 ohms
Detector gain = 0.15 Volts/dB

Detector time constant should be much faster than the AGC response time of the feedback loop or it needs to be included and much slower than the RF frequency.

RF frequency = 0.05 MHz
Time constant = 200 us
Choose C2 = 0.1 uf
R9 = 5000 ohms
R10 = 2 kohms

Enter Values:

Choose C1 =	1	uf
Choose C2 =	1	uf
Choose C3 =	1	uf

Offset Design:

Choose dBm out = 5 dBm
Enter coupler loss = 3 dB
Detector input = 2 dBm
Detector input = 1.407521 Vpeak
Detector output = 0.995268 Vrms
Amplifier output = 9.952679

Calculations:

R1 =	11.26246	kohms
R2 =	11.25565	kohms
R3 =	253.0992	kohms

Choose R5 = 5 kohms
Choose Vdc = 15 volts
R6 = 0.5 kohms
R4 = 7.535659 kohms

Etc.

R7 =	3.005689	kohms
R8 =	126.5496	kohms

Distortion on Signal:

Choose freq. of operation = 25 kHz
Frequency in radians = 157079.6
Tclosed loop = [K(S + 1/T2)]/[S*S + (1/(T1 + T2) + K)S + K/T2]
Numerator Real = 3947.842
Numerator Imag = 6979918
Denominator Real = -2.5E+10
Denominator Imag = 20928878
Closed Loop Mag = 0.000283
% affecting signal = 0.028289 %

Figure 4-10 An example of an AGC design.

[handwritten: the zero occurs at $\frac{1}{T_2}$ zero]

Choose the zero location:

$$T_2 = 1/(2\zeta\omega_n) = 1/((14.14)) = 0.0707$$

Using the previous root locus equations:

$$K = T_2\omega_n^2 = 0.0707(10)^2 = 7.07$$

$$2\zeta\omega_n = K + 1/(T_1 + T_2)$$

Solving for T_1:

$$T_1 = 1/(2\zeta\omega_n - K) - T_2 = 1/(14.14 - 7.07) - 0.0707 = 0.0707$$

Using the total loop gain expression to solve for T_3:

$$K = K_a K_d K_c (1/T_3)(T_2/[T_1 + T_2])$$

$$T_3 = K_a K_d K_c (T_2/[T_1 + T_2])/K = 15(0.15)(6.27)[(0.0707)/(0.0707 + 0.0707)]/7.07 = 1.0$$

Once the time constants are solved, the selection of the resistors and capacitors is the next step. Low-leakage capacitors are preferred, and the resistors should be kept small. This is to prevent current limiting of the source by keeping the current large compared with the current leakage of the capacitor. There are trade-offs here, depending on parts available and values needed.

Choose $C_1 = C_3 = 1\ \mu F$

$T_1 = R_1 C_1$, therefore $R_1 = T_1/C_1 = 0.0707/1\ \mu F = 70.7\ k\Omega$
$T_2 = R_2 C_1$, therefore $R_2 = T_2/C_1 = 0.0707/1\ \mu F = 70.7\ k\Omega$
$T_3 = R_3 C_3$, therefore $R_3 = T_3/C_3 = 1.0/1\ \mu F = 1\ M\Omega$

A block diagram showing the design of the AGC is shown in Figure 4-11.   *[handwritten: See p 116]*

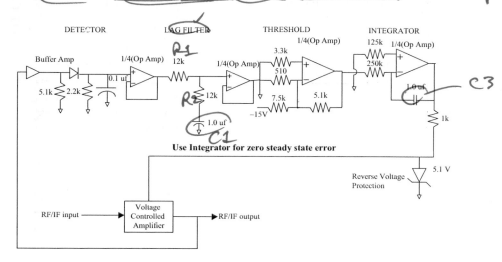

Figure 4-11   A block diagram of the AGC design example.

4.9 Modulation Frequency Distortion

The reduction of the modulating frequency (e.g., the conscan frequency information in the preceding example) is calculated by solving the closed-loop transfer function for the modulating frequency ω_m.

$$\omega_m = 188.5 \text{ rad/sec}$$
$$T_{closed\ loop} = 7.05(j188.5 + 14.14)/[(j188.5)^2 + 14.14(j188.5) + 100]$$
$$= 100 + j1329/(-35,432 + j2665) = 0.04\underline{/-90°}.$$

The 0.04 number means that for an input gain change of a signal level at the modulating frequency ω_m the AGC amplifier will change its gain by 0.04 of the input gain change. The modulating signal is reduced by 0.04 times the input gain change in dB.

For example, if the input signal level changes from 0 dBm to 10 dBm at the modulating frequency, the input gain change is 10 dB. The feedback loop response reduces the 10 dB change by a magnitude of, say, 0.04 or 0.4 dB. This means the AGC amplifier changes gain by 0.4 dB for an input change of 10 dB. However, since there is a phase shift of the feedback signal, the amplitude reduction will be less since it is associated with the cosine of the angle. In equation form:

$$\text{Signal input} = P_{in\ \text{dBm}} + A\cos(\omega_m t)_{\text{dB}}$$

$$\text{Signal output} = P_{in\ \text{dBm}} + A\cos(\omega_m t)_{\text{dB}} - HA\cos(\omega_m t + C)_{\text{dB}}$$

where

A = amplitude of the modulating signal (conscan)
H = amplitude through the feedback loop
C = phase shift through the loop
t = time

The superposition of the two cosine waves in the output equation results in the final gain (in dB) that is added to the power. We can also take the change in dB caused by the input modulation amplitude and phase, subtract the loop input and phase by converting to rectangular coordinates, and then convert back to polar coordinates. For example,

$$\text{Input} = 10 \text{ dB angle} = 0°$$
$$\text{Loop response} = \text{mag}(0.04 \times 10) = 0.4 \text{ dB angle} (-90° + 0°) = -90$$
$$\text{Total response} = (10 + j0) - (0 - j0.4) = 10 + j0.4 = 10.008 \text{ at an angle of } 2.3°$$

The magnitude changed by 0.008 dB and the phase of the modulating signal shifted by 2.3°.

Note that this assumes that the change in gain is within the DR of the AGC. If the gain change is outside of the AGC DR, then the magnitude of the loop response is 0.04 times only the portion of the AGC range that is being used.

To determine the resultant amplitude and phase for the general case:

$$H(\omega) = [H]e^{jC} = [H]\cos C + j[H]\sin C$$

Thus,

$$\{1 - H(\omega)\} = 1 - \{[H]\cos C - j[H]\sin C\}$$

The magnitude equals

$$\text{mag}[1 - H(\omega)] = \sqrt{\{1 + [H]^2 - 2[H]\cos C\}}$$

The angle equals

$$\tan^{-1}\{-[H]\sin C/(1 - [H]\cos C)\}$$

4.10 Comparison of the PLL and AGC Using Feedback Analysis Techniques

Phase-locked loops are important elements in the design of synthesizers and carrier recovery circuits. They are used in almost all receiver designs and have become a basic building block design tool for the engineer. This section simplifies the analysis of PLL stability and makes a comparison between AGC and the PLL.

4.11 Basic PLL

The basic PLL standard block diagram is shown in Figure 4-12. The operation of the PLL takes the input frequency and multiplies this frequency by the voltage-controlled oscillator (VCO) output frequency. If the frequencies are the same, then a phase error is produced. The nature of the PLL forces the phase of the VCO to be in quadrature with the phase of the incoming signal. This produces the sine of the phase error.

$$\cos(\omega t + a)\sin(\omega t + b) = {}^1/_2[\sin(a - b) + \sin(2\omega t + a + b)]$$

Using a low-pass filter to eliminate the sum term produces

$$ {}^1/_2[\sin(a - b)]$$

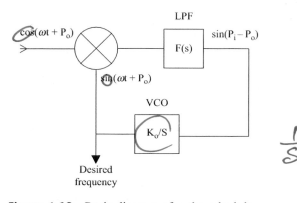

Figure 4-12   Basic diagram of a phase lock loop.

if $a = b$, then

$$\frac{1}{2}[\sin(0)] = 0$$

The zero voltage output sets the VCO to the correct steady-state value. If the voltage changes due to a change in input phase, the VCO changes until the phase error is approximately equal to zero. The low-pass filter removes higher frequency components and also helps establish the loop gain. The PLL tracks a phase change over time, and since the change of phase per unit time is frequency the PLL tracks frequency. Since it performs this task by converting the changing phase into a changing voltage that controls the VCO, the analysis is performed using phase as the parameter.

4.12 Control System Analysis

Phase-locked loops can be designed using basic control system theory since it is a feedback system. The analysis is very closely related to the AGC study. To analyze the stability of the PLL, the basic diagram is redrawn (Figure 4-13).

The block diagram for the PLL is almost identical to the AGC block diagram. The VCO contains an integrator and a gain constant, whereas the AGC amplifier only has a gain constant and the integrator is added as another block. The integrator included in the VCO produces a type 1 system, so for a step change in phase the PLL will have a steady-state error of zero. For the AGC, the added integrator provides a steady-state error of zero for a step change (in dB) of input power. If a zero steady-state error for a step in frequency is desired, another integrator needs to be included in the PLL. However, this chapter will limit the analysis to a second-order type 1 system.

type II System

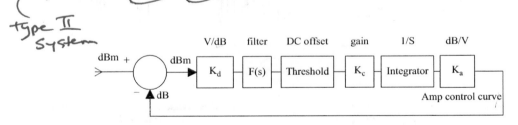

a. AGC Control System Analysis Block Diagram

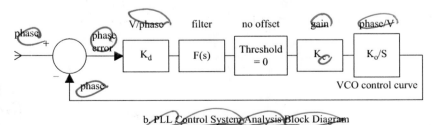

b. PLL Control System Analysis Block Diagram

Figure 4-13 Comparison between AGC and PLL feedback systems.

4.13 Detector

The phase detector or mixer in the PLL performs two operations (Figure 4-12). First, the detector takes the input phase and subtracts the feedback phase from the VCO to produce a phase error. Second, the detector converts the phase into voltage with a particular slope depending on the phase difference and the frequency of operation. This process for the AGC uses power (in dBm) for the input and subtracts the gain (in dB) to produce a power level error that is converted to voltage. The AGC detector's operation point is set at the desired dBm output, so the detector gain is the same around the operation point regardless of the input operating power. If the desired frequency is set for the PLL, then the same criteria apply. However, in many cases when a PLL is used, the desired operating frequency changes (as is the case for synthesizer design). Thus, the detector gain becomes a function of the input frequency or the input operating point. Care must be taken when selecting the gain constant for the VCO detector because it depends on the input frequency. There are approaches for linearizing the detector, but for this analysis the slope is chosen as a constant. If the frequency is constant, then the slope is easily chosen and the linearity is generally not a problem.

4.14 Loop Filter

The passive phase-lag network used for the loop stabilizing filter is identical for both the AGC and the PLL. The actual numerical values will be different depending on the loop requirements. The loop filter should follow the detector to prevent any high-frequency components from affecting the rest of the loop. The schematic consists of two resistors and a capacitor and is shown in Figure 4-7 along with the transfer function. This circuit provides a pole and a zero to improve the stability of the loop. The filter is used to affect the attenuation and the roll-off point since the phase shift for a phase-lag network generally has a destabilizing effect.

4.15 Loop Gain Constant

Additional gain is added to the loop and is selected for the particular operating levels of the PLL. The gain constant is defined as K, and is equal to $-R_s/R$, (Figure 4-7). The gain constant is used in both the PLL and the AGC cases and are different depending on each of their proper operating points. An added threshold is not required in the PLL since the desired phase difference is always zero, regardless of the desired frequency of operation.

4.16 Integrator

The integrator for the PLL is inherent in the loop, whereas the AGC integrator is added to the loop. The reason for this is because the PLL operates using phase as the parameter and the VCO actually delivers frequency. Therefore, the PLL forces the VCO to deliver phase that inherently adds an integrator because the integral of frequency is phase.

4.17 Conversion Gain Constant

The conversion in the AGC circuitry from volts to dBm produces a slope gain constant (dB/V) and is chosen considering the trade-off between stability and response time. The conversion in the PLL is a two-step process. First, the slope conversion constant converts from voltage to frequency, which is the constant for the VCO labeled K_o. Second, the inherent integration converts frequency into phase. The constant K_o for the PLL is not linear. This is also true for K_a in the AGC analysis because a linear approximation is used. A linearizer can be implemented for either the PLL or the AGC circuits.

4.18 Control Theory Analysis

The open-loop transfer function for the PLL is identical to the open-loop transfer function for the AGC except that the integrator is considered as $1/S$ instead of $1/(T_3S)$, as shown as follows:

Open-loop transfer function for AGC:

$$T_{sol} = \frac{K_a K_d K_c T_2 \left(S + \dfrac{1}{T_2}\right)}{T_3(T_1 + T_2)S \left[S + \dfrac{1}{T_1 + T_2}\right]}$$

Open-loop transfer function for PLL:

$$T_{sol} = \frac{K_o K_d K_c T_2 \left(S + \dfrac{1}{T_2}\right)}{(T_1 + T_2)S \left(S + \dfrac{1}{T_1 + T_2}\right)}$$

where

K_a = AGC amplifier loop gain
K_o = VCO gain
K_d = detector loop gain
K_c = gain constant
$T_1 = R_1 \times C_1$
$T_2 = R_2 \times C_1$
$T_3 = R_3 \times C_3$ *See p114*
T_{sol} = open-loop transfer function
$S = j\omega$ (for a lossless system)

The closed-loop response is

$$T_{scl} = \frac{K \left(S + \dfrac{1}{T_2}\right)}{S^2 + \left(\dfrac{1}{T_1 + T_2} + K\right)S + \dfrac{K}{T_2}}$$

where

$K = K_a K_d K_c (1/T_3)(T_2/[T_1 + T_2])$ for the AGC

$K = K_o K_d K_c (T_2/[T_1 + T_2])$ for the PLL

T_{scl} = closed-loop transfer function for both the AGC and PLL

From control system theory for the previously given second-order system

$$2\zeta\omega_n K + 1/(T_1 + T_2)$$

and

$$\omega_n^2 = K/T_2$$

where

ω_n = natural frequency of the system

ζ = damping ratio

This analysis applies to the condition of the loop being already phase locked and not capturing the signal.

The loop gain for the PLL is selected considering several factors:

- Stability of the loop
- Lock range and capture range
- Noise in the loop
- Tracking error

The wider the bandwidth, the larger the lock and capture ranges and also the more noise in the loop. The lock range is equal to the DC loop gain. The capture range increases with loop gain and is always less than the lock range. If the maximum frequency deviation or the maximum phase error desired is given, then the open-loop gain can be established. When designing a synthesizer, the crossover between the crystal reference and the VCO with respect to phase noise is chosen for best noise performance for the loop bandwidth. Once the frequency is chosen, the same stability analysis for the PLL can be performed as was done for the AGC circuit analysis.

The damping ratio is chosen so the system responds to a step function as fast as possible within a given percentage of overshoot. With a 5% overshoot, the minimum damping ratio is 0.707. This means the system is slightly underdamped. The damping ratio is chosen depending on the design criteria. If the overshoot is too high, the damping ratio should be changed. Once the damping ratio and the natural frequency are chosen for a system, the time constants can be determined mathematically. An example is shown as follows:

Determine loop bandwidth = 160 Hz
Choose ω_n = 1.0 krad/sec f_n = 160 Hz
Choose ζ = 0.707 PO = 5%

The poles and zeros for the root locus criteria are found by using $G(s)H(s)$. Note that $1/T_2$ determines the zero location for both the open-loop and closed-loop cases since $H(s) = 1$. The zero is placed at $-2\zeta\omega_n$, and the root locus plot is shown in Figure 4-9. This moves the root locus, the migration of the poles, away from the $j\omega$ axis to prevent oscillatory responses with a change in gain. It also places the poles on the tangent of the root locus circle at the given damping ratio so that the damping ratio is always greater as the gain varies. Due to the integrator, one pole migrates from the $j\omega$ axis along the σ axis as the loop gain constant K increases. As K approaches zero, the pole due to the integrator lies on the $j\omega$ axis. As the loop gain decreases, the integrator pole migrates closer to the $j\omega$ axis and becomes more oscillatory. However, a pole on the axis means that the system is overdamped ($\zeta > 1$) and will not oscillate. The only way this system can become unstable is with stray poles close to the origin. The loop gain (K) is chosen in this example so that the operating point lies on the root locus circle with the desired damping factor (Figure 4-9).

For the example, choose T_2:

$$T_2 = 1/(2\zeta\omega_n) = 1/1.414k = 707.2 \times 10^{-6}$$

Using the previous root locus equations:

$$K = T_2\omega_n{}^2$$
$$2\zeta\omega_n = T_2\omega_n{}^2 + 1/(T_1 + T_2)$$

Solving for T_1:

$$T_1 = [1/(2\zeta\omega_n - T_2\omega_n{}^2)] - T_2$$
$$T_1 = [1/\{1.414k - (707.2 \times 10^{-6})1000k\}] - 707.2 \times 10^{-6} = 707.6 \times 10^{-6}$$

For the AGC case, the constants are specified and T_3 is solved as shown:

$$K = K_aK_dK_c(1/T_3)(T_2/[T_1 + T_2])$$
$$T_3 = K_aK_dK_c(T_2/[T_1 + T_2])/K$$
$$K = T_2\omega_n{}^2$$
$$T_3 = K_aK_dK_c(T_2/[T_1 + T_2])/(T_2\omega_n{}^2)$$

For the PLL case, K_o and K_d are specified, and we solve for K_c:

$$K = K_oK_dK_c(T_2/[T_1 + T_2])$$
$$K_o = 20K_d = 0.90$$
$$K_c = K/[K_oK_d(T_2/[T_1 + T_1])]$$
$$K = T_2\omega_n{}^2$$
$$K_c = T_2\omega_n{}^2/[K_oK_d(T_2/[T_1 + T_1])]$$
$$= 707.2 \times 10^{-6}(1.0 \times 10^6)/[20(0.9)(707.2 \times 10^{-6})/(707.6 \times 10^{-6} + 707.2 \times 10^{-6})]$$
$$= 78.6$$

To complete the design, choose $C1 = 0.01\ \mu F$:

$$T_1 = R_1C_1, \text{ therefore } R_1 = T_1/C_1 = 707.6 \times 10^{-6}/0.01\ \mu F = 70.76\ k\Omega$$
$$T_2 = R_2C_1, \text{ therefore } R_2 = T_2/C_1 = 707.2 \times 10^{-6}/0.01\ \mu F = 70.72\ k\Omega.$$

The resultant transfer functions are

$$*T(\text{open loop}) = [707/S][S + 1.414k]/[S + 1.403k]$$
$$*T(\text{closed loop}) = 707[S + 1.414k]/[S^2 + 2.11kS + 1000k]$$

4.19 Similarities between the AGC and the PLL

The similarities between the PLL and the AGC are listed as follows. With very few exceptions the analysis is nearly identical:

- The PLL has a built-in integrator, part of the VCO when connected into the loop.
- AGC has an additional time constant due to the added integrator $1/(R_3C)$.
- They have the same open-loop transfer function.
- They have the same closed-loop transfer function.
- They are basically the same process since they are both feedback systems.

4.20 Feedback Systems, Oscillations, and Stability

Feedback systems are valuable in reducing errors and maintaining constant values over changing parameters. Feedback helps the system evaluate the error from the ideal setting and allows the system to adjust values to minimize the error.

However, using feedback systems can cause the system to oscillate. If an oscillator is desired, then the design criteria are set to provide oscillations using feedback. If an oscillator is not desired, then the design criteria need to be implemented to avoid oscillations that are inherent in feedback systems. These types of systems use negative feedback, since positive feedback would cause the system to oscillate or continue to increase in value. Even with negative feedback systems, there is the possibility for system oscillation. For a system to oscillate, it has to follow the Barkhausen criteria, which is 0 dB gain and 360° phase shift. A negative feedback system, since it is negative, which by definition has a 180° phase shift, requires only a 180° phase shift for the system to oscillate: 180° + 180° = 360°. If the negative feedback system has a 180° phase shift and 0 dB gain (gain of 1), then it satisfies the Barkhausen criteria and will oscillate.

Bode diagrams are used to provide an indication of the stability of a system and how much margin a system has from becoming unstable (Figure 4-14). This figure shows both the gain and phase as a function of frequency on a log scale. The gain margin is determined by

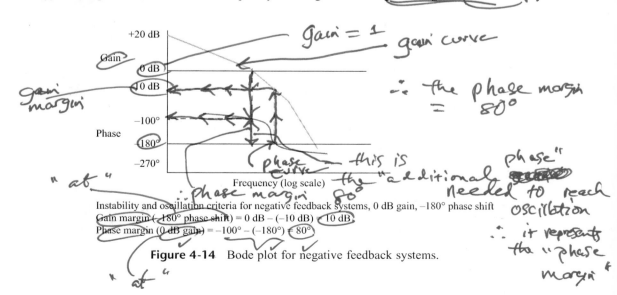

Instability and oscillation criteria for negative feedback systems, 0 dB gain, −180° phase shift
Gain margin (−180° phase shift) = 0 dB − (−10 dB) = 10 dB
Phase margin (0 dB gain) = −100° − (−180°) = 80°

Figure 4-14 Bode plot for negative feedback systems.

the amount of additional gain required at a $-180°$ (negative feedback) phase shift to equal 0 dB gain, which is the oscillation criteria. Since the gain is at -10 dB for a $-180°$ phase shift, the additional gain required to equal 0 dB is

$$\text{Gain margin} = 0 \text{ dB} - (-10 \text{ dB}) = 10 \text{ dB}$$

The phase margin is determined by the amount of additional phase shift required at 0 dB gain to equal a $-180°$ (negative feedback) phase shift, which is the oscillation criteria. Since the phase shift is at $-100°$ for a 0 dB gain, then the additional phase shift required to equal $-180°$ is

$$\text{Phase margin} = -100° - (-180°) = 80°$$

4.21 Summary

Automatic gain control is an important element in the design of all types of transceivers. AGC provides the necessary DR required for operation over varying distances and conditions. AGC can be analyzed using control system theory because it is a feedback system. The similarities between AGC and PLL are remarkable. They both can be analyzed using modern control system techniques. The steady-state error for both systems is zero for a step response. The step response for AGC is related to a power level change, and the step response for PLL is related to a phase change. If a zero steady-state error is required for a step in frequency, then another integrator needs to be added. Further study of PLL is required to answer various questions, such as capture and lock range, operation of the loop in a no-lock situation, and considerations concerning stray poles that may change the simple analysis. However, the idea of using these control system techniques for analyzing both AGC and PLL can be useful.

References

Bullock, S. R., and D. Ovard. "Simple Technique Yields Errorless AGC Systems." *Microwaves and RF*, August 1989.

Bullock, S. R. "Control Theory Analyzes Phase-Locked Loops." *Microwaves and RF*, May 1992.

Dorf, R. C. *Modern Control Systems*, 12th ed. Prentice Hall, 2010.

Gardner, F. M. *Phaselock Techniques*, 3d ed. John Wiley & Sons, 2005.

Problems

1. Name at least two RF devices that can be used for gain control.
2. What would be a reasonable value of the RC time constant, following the diode detector for the AGC, for a carrier frequency of 10 MHz and a desired modulating frequency of 1 MHz?
3. What does an integrator in the feedback loop do for the steady-state error?

4. Explain how the integrator achieves the steady-state error in problem 3.
5. If the input signal to a receiver changes suddenly from -70 dBm to -20 dBm, what is used to ensure a constant IF output of 0 dBm?
6. How does a nonlinear function affect the response of the AGC loop?
7. Why is a diode curve approximation much better that a piecewise linear approximation?
8. What is the main difference between an AGC and a PLL in the locked state?
9. What is the general reason that the AGC analysis is similar to the PLL analysis?
10. What state is the PLL assumed to be in the AGC/PLL analysis in this chapter?
11. What is the potential problem when using a feedback AGC?

CHAPTER 5

Demodulation

The demodulation process is part of the receiver process that takes the downconverted signal and retrieves or recovers the data information that was sent. This process removes the carrier frequency that was modulated in the transmitter to send the digital data and detects the digital data with the minimum bit error rate (BER). It also removes the high-speed code used to generate spread spectrum for jammer mitigation. This process was performed in the past using analog circuitry such as mixers and filters to remove the carrier frequency and spread spectrum codes, but today the process is incorporated in the digital circuitry using application-specific integrated circuits (ASICs), field-programmable gate arrays (FPGAs), and digital signal processing (DSP) integrated circuits. The basic system concepts are presented in this chapter, and the method of implementation is left to the designer.

5.1 Carrier Recovery for Suppressed Carrier Removal

The carrier frequency is modulated in the transmitter by a digital waveform consisting of digital data or digital data combined with a high-speed digital pseudo-noise (PN) code if spread spectrum is used. This process uses the digital waveform to shift the phase of the carrier according to the digital bit value. In either case, the digital waveform creates a modulated signal that suppresses the carrier frequency, which is difficult to detect. For analog signals, basic techniques such as a phased-lock loop (PLL) can be used to determine the carrier frequency. A PLL uses a phase comparison between the carrier frequency and the local oscillator (LO) to recover the frequency. However, the PLL has a problem with fast-changing phases (digitally modulated carrier), which causes it to lose lock. Therefore, to recover the carrier, the following methods, known as carrier recovery loops, are used:

- Squaring loop
- Costas loop
- Modified Costas loop

Once the carrier has been recovered, it is used to cancel the suppressed carrier in the modulated waveform, which produces the digital waveform. These loops are discussed in detail in the following sections. There are trade-offs for which type of carrier recovery loop should be used. A careful look at these parameters will provide the best type of carrier recovery loop for use in any given system design.

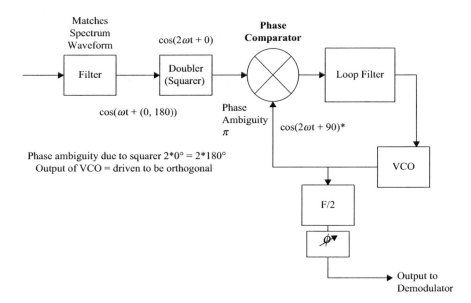

Figure 5-1   Squaring loop carrier recovery.

5.1.1 Squaring Loop

The squaring loop is employed for carrier recovery of a binary phase-shift keying (BPSK) direct sequence digital waveform. The advantage of the squaring loop is mainly its simplicity of implementation. The squaring loop requires minimal hardware, which makes it cost-efficient. A block diagram of a squaring loop is shown in Figure 5-1.

The squaring loop filters the incoming signal using a narrowband filter. The filter needs to be narrow enough for the spectrum of the input signal to be essentially constant. A filter matched to the waveform gives the best signal-to-noise ratio (SNR) and should be used in the intermediate frequency (IF) section before the squarer. This filter also reduces interference from other out-of-band signals and interference. The shape of the optimum filter is a sin$x$/x impulse response in the time domain (square wave function in the frequency domain). However, this filter is only theoretical and is not implementable. Therefore, for this application, a simple two-pole Butterworth filter can be used to approximate the ideal matched filter. The low-pass filter is designed to approximate the roll-off of the incoming waveform using a time period of T seconds, where T is one BPSK symbol interval. An alternative is to use an integrate-and-dump circuit, which also operates as a good matched filter.

The filtered signal is then fed into a doubler or squarer that basically squares the incoming signal. Squaring a direct sequence BPSK signal eliminates the modulation, since the modulated signal is phase shifted either 0° or 180°. Squaring the signal doubles both the frequency and phase, resulting in an output of twice the frequency with no phase shift:

$$S = A\cos(\omega t + \varphi), \text{ where } \varphi = 0°, 180°$$
$$S^2 = A^2(\cos^2(\omega t + \varphi) = A^2/2(1 + \cos[2(\omega t + \varphi)]) = A^2/2 + A^2/2\cos(2\omega t + 2\varphi)$$
$$= A^2/2 + A^2/2\cos(2\omega t + 0°, 360°))$$

Therefore,

$$S^2 = A^2/2 + A^2/2\cos(2\omega t + 0)$$

The output is now an unsuppressed carrier at twice the frequency and thus can use a PLL to determine the frequency. Since the PLL detected frequency is at twice the desired frequency, the output frequency is divided by two to achieve the fundamental frequency and the carrier is recovered.

However, the divider creates an ambiguity since it divides whatever the phase is of the actual carrier by two. For example, a change in one cycle (2π radians) results in a change of phase of π. This means that if the frequency shifts one cycle, the phase is shifted $180° = 360°/2$. Since the PLL does not recognize that there was a cycle slip, an ambiguity needs to be accounted for. This creates a potential problem in the data because of the possible phase reversal. Another example: suppose the frequency locks up at $\cos(2\omega t + (180°))$. When it is divided by two, the carrier frequency is $\cos(\omega t + (90°))$, which means that the carrier is 90° out of phase of the locked carrier. This phase needs to be adjusted out.

The output of the squaring loop is phase shifted to obtain a cosine wave, since the output of the PLL generates a sine wave and is multiplied with the incoming signal to eliminate the carrier and obtain the data. If the signal is 90° phase shifted with respect to the incoming signal, then the output would be zero:

$$(A\sin\omega t)(d(t)A\cos\omega t) = A/2d(t)(\sin 0 + \sin 2\omega t) = A/2d(t)\sin 0$$
$$= 0, \text{ since } \sin 2\omega t \text{ is filtered}$$

The phase shift is critical, may drift with temperature, and will need to be adjusted if different data rates are used. This is one of the main drawbacks to the squaring loop method of carrier recovery. Some other carrier recovery methods do not have this concern, though. The phase in question is the phase relationship between the input signal and the recovered carrier at the mixer where the carrier is stripped off.

Another disadvantage to the squaring loop approach to carrier recovery is that the filter for the squaring loop at the IF is good for only one data rate. If the data rate is changed, the filter needs to be changed. Some systems require that the data rate be variable, which makes the squaring loop less versatile than some other methods. A variable filter can be designed, but it is more complex than a fixed filter.

Squaring the signal gives rise to a direct current (DC) term and a tone at twice the carrier frequency. A possible disadvantage to the squaring loop is that it needs to operate at twice the frequency. For most cases this is not a problem, but higher frequencies can drive up the cost of hardware.

Another configuration of the squaring loop is shown in Figure 5-2. This configuration was developed for analog squaring loops, since it is generally easier to multiply frequencies than it is to divide frequencies in the analog domain. Since it is more difficult to divide frequencies in the analog domain, this configuration is another approach to achieve the same solution. Another advantage of this configuration is that the voltage-controlled oscillator (VCO) operates at a lower frequency, which is typically a cost savings. This configuration may not be an advantage in the digital domain since it is easier to divide using simple digital techniques than it is to multiply.

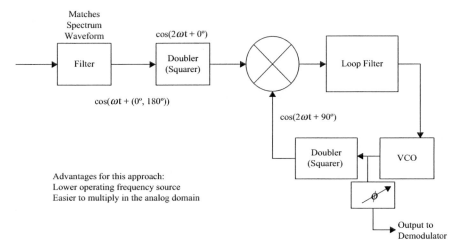

Figure 5-2 Another configuration of the squaring loop.

If higher order phase-shift keying (PSK) modulation schemes are used, then higher order detection schemes need to be implemented. For example, if a quadrature PSK (QPSK) modulation waveform is used, then the demodulation process should square the signal twice or raise the signal to the fourth power. Sometimes this circuit is referred to as a times-4 device, which is really a misnomer. The signal is not multiplied by 4, but the frequency is four times higher. Since with QPSK there are four phase states at $0°$, $90°$, $180°$, and $-90°$, it requires a four power, or squaring the waveform twice, to eliminate the phase shift. The following example shows squaring the input two times with the amplitudes and DC terms ignored for simplicity:

$$(\cos(\omega t + (0°, 90°, -90°, 180°)))^2 = \cos(2\omega t + (0°, 180°, -180°, 360°))$$
$$(\cos(2\omega t + (0°, 180°, -180°, 360°)))^2 = \cos(4\omega t + (0°, 360°, -360°, 360°))$$
$$= \cos(4\omega t + 0°)$$

The frequency is recovered by dividing the resultant frequency by four.

The description of the squaring loop can be applied for all the higher order loops, keeping in mind that the functions need to be modified for the higher order operations. The difference is how many times the signal needs to be squared. For an n-phase signal, $2^a = n$, where a is the number of times the signal needs to be squared. For example, an 8-PSK signal would be $2^3 = 8$. Therefore, the 8-PSK signal would need to be squared three times to eliminate the phase ambiguity. The higher order PSK demodulation processes can become complex and expensive due to the higher frequency components that need to be handled. Also, there are losses in the signal amplitude every time it is squared. These losses are called squaring losses, and they degrade the ability to detect the signal.

Once the carrier is recovered, it is used to strip off the phase modulated carrier to retrieve the digital modulated signal. This is accomplished in the demodulator, or mixer, and is then integrated over a bit or chip time to achieve the digital signal. A large positive number from

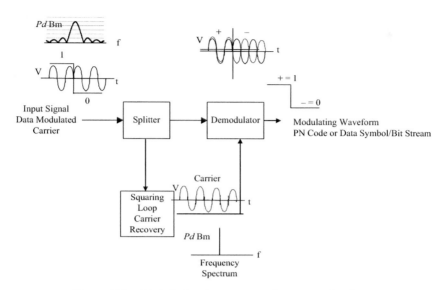

Figure 5-3 Digital signal recovery using a squaring loop.

the multiply and integrate is a digital "1," and a large negative number from the multiply and integrate is a digital "0"; thus, the digital signal is retrieved (Figure 5-3). As noted before, this process can all be accomplished in the digital domain, which improves the reliability and performance.

5.1.2 Costas Loop

The Costas loop is generally the preferred method of recovering the carrier frequency. It is used extensively in communication systems and does not require phase adjustment. It is basically a quadrature demodulation phase-locked loop (PLL) and contains the complete process that eliminates the carrier frequency. This does not have the squaring loop phase ambiguity; however, it does have an ambiguity on which quadrature path contains the data, and in some systems the quadrature outputs are combined to mitigate this ambiguity. Another way to eliminate this ambiguity is to use a modified version of the Costas loop, known as a hard-limited Costas loop, which will be discussed shortly.

The standard Costas loop does not use a squaring loop, so the double-frequency signal is not a part of the process. Therefore, the higher frequencies are not produced. Instead, the Costas loop uses basically two PLLs, one of which is in-phase (I) and the other is in quadrature (Q) as shown in Figure 5-4.

The outputs of each PLL are multiplied and low-pass filtered to eliminate the sum terms so that the loop tracks sin(2P), where P is the phase error. The low-pass filters are matched filters for the symbol or bit rate. To eliminate the need for an analog multiplier, a hard-limited Costas loop can be utilized. The matched filtering is done at baseband and the output is used to generate the error.

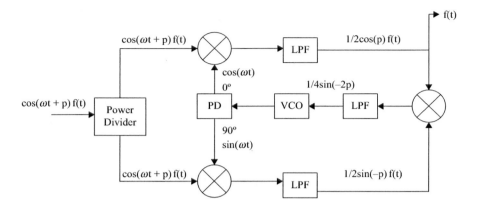

As p approaches zero the loop is locked. Note that the input could
have been a sin instead of a cos and the bottom leg would have
contained f(t).

Figure 5-4 Costas loop used for carrier recovery.

The phase shift, as in the squaring loop, is not critical, since the feedback forces the phase to be correct. Since the feedback is forcing the error to be zero, the error channel is driven to zero, but either channel can be the data channel. A modified Costas loop is a method to force the data to a known channel.

The matched filters in the Costas loop are good for only one data rate. If the data rate is changed, then the matched filters need to be changed. However, changing filters at baseband is much easier and more cost-effective than changing IF filters, as in the squaring loop. Note that the squaring loop baseband matched filters also need to be changed along with the IF filter.

5.1.3 Modified Costas Loop and Automatic Frequency Control Addition

Since the Costas loop is good only for a narrow bandwidth, an automatic frequency control (AFC) can be included to extend the pull-in range, which increases the bandwidth of the Costas loop. The standard Costas loop can be modified in other ways to improve the performance and versatility of this process. A hard-limited Costas loop or a data-aided loop can further improve the carrier recovery process. A hard-limited Costas loop has a hard limiter in one of the channels, which estimates the data pulse stream. By hard limiting one of the channels, that channel automatically becomes the data channel since it produces a sign change, and the other channel becomes the error channel, which is driven to zero. The multiplier—which is generally a drawback to Costas loops, since it is hard to build a multiplier in the analog domain—then can simply invert or noninvert the signal (Figure 5-5).

Using the data estimate, a noninverting amplifier or an inverting amplifier can be selected. This is much easier to implement in hardware. Commonly called a polarity loop, it strips the modulation off and leaves the phase error for the PLL. A data-aided loop uses the data estimation similar to the hard-limited Costas loop to improve the performance of the

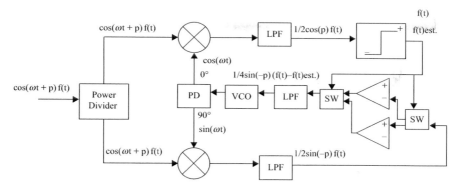

Figure 5-5   Hard-limited Costas loop for carrier recovery.

standard Costas loop. Many enhancements and variations can be made to the standard Costas loop, but a basic understanding of its operation provides the user with the ability to design a carrier recovery loop for a typical system, including modifications where improvement is desired.

5.2 Demodulation Process to Remove Spread Spectrum Code

Once the carrier is removed, the remaining digital waveform is a combination of the spread spectrum PN code and the digital data. To retrieve the digital data, the PN code needs to be removed or stripped off the digital waveform. There are basically two ways to demodulate this digital waveform for phase-shift keying (PSK) signals: a sliding correlator or a matched filter correlator. The method used depends on the type of modulation waveform and the corresponding demodulation process.

5.2.1 Sliding Correlator

A sliding correlator is used to remove the spread spectrum code on a spread spectrum system. This process incorporates a method of stripping off the PN code to retrieve the data information. The receiver has the same PN code generator and code as the transmitter so the technique is to align the incoming code with the received code. The correlator slides the code in time with the incoming code and looks for a correlation peak. When the peak is found, the system switches into lock condition and tracks the changes to keep the codes aligned. Once they are aligned, the codes are multiplied and integrated, which removes the PN code and retrieves the digital data. Since the code can be long before it repeats, some codes do not repeat for two years so an initial alignment needs to be performed. This requires additional bits to be sent that are not related to the data being sent. These bits are called the preamble, which consists of overhead bits that will lower the data rate by a certain percentage of the data messages sent. For continuous systems, these overhead bits are a very small percentage of the message. For pulsed-type systems or time division multiple access (TDMA) systems, these bits can be a large percentage of the message depending on the length of the pulse. The code alignment process can be accomplished by using the following techniques:

- Generating a short code for acquisition only.
- Using highly reliable clocks available to both the transmitter and receiver, such as rubidium, cesium, or global positioning system (GPS).
- Using an auxiliary subsystem that distributes system time to both parties.

In many system architectures, all of these techniques are used to simplify and accelerate the code alignment process. Once the code time is known fairly accurately, two steps are used to demodulate the PN code.

The first step is to determine if the code is lined up; the second step is to lock the code into position. The first step is achieved by switching a noncorrelated code into the loop and then into a sample-and-hold function to measure the correlation level (noise) output. Then the desired code is switched into the loop and into the sample-and-hold function to measure the correlation level (signal) output. The difference in these levels is determined, and the results are then applied to a threshold detector in case the search needs to be continued. If the threshold level is achieved, the second step is activated. This prevents false locking of the sidelobes. Note that a false lock generally occurs at half of the bit rate.

The second step is to maintain lock (also fine-tuning) by using a code lock tracking loop. This assumes that the codes are within ½ bit time. One type of code tracking loop is called an early-late gate. An early-late gate switches a VCO between an early code (½ bit time early) and a late code (½ bit time later). The correlation or multiplication of the early code and the input code and the late code and the input code produces points on the autocorrelation peak, as shown in Figure 5-6.

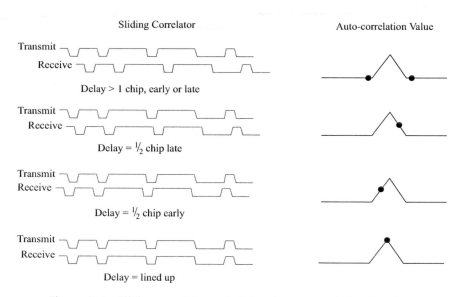

Figure 5-6   Sliding correlator early-late gate autocorrelation function.

The figure shows the code loop locked when the values of the autocorrelation function are equal. The nondelayed code is used for demodulation of the spread spectrum waveform. The peak of the autocorrelation function occurs when the nondelayed, aligned code is multiplied with the incoming signal and summed (Figure 5-6). These codes are mixed with the incoming signal, filtered, and detected (square-law envelope detected), and the output sign is changed at the dither rate. For example, the early code detector output is multiplied by -1, the late code detector output is multiplied by $+1$, and the final output is filtered by the loop filter, which takes the average of the two levels. This provides the error and controls the VCO to line up the codes so that the error is zero, which means the VCO is switching symmetrically around the correlation peak. Therefore, the early-late gate uses feedback to keep the code aligned. Sign is important in the design and hardware applications since it drives the direction of the oscillator. For example, if a negative number increases the frequency of the VCO or a positive number increases the VCO, this needs to be taken into account so the VCO is driven in the right direction given the sign of the error signal.

A delay-locked loop (DLL) is another form of a lock loop code. DLLs accomplish the same thing as early-late gates. They align the code to strip off the PN code using autocorrelation techniques. The main difference in the DLL compared with the early-late gate is that the DLL uses a nondelayed code and a code that is delayed by one chip. The DLL splits the incoming signal and mixes with the VCO output that is either shifted a bit or not. The main goal is to obtain the maximum correlated signal on the nondelayed code and the minimum correlated signal on the 1 chip delayed code. The DLL generally tracks more accurately (approximately 1 dB), but it is more complex to implement. Other forms of correlation that are used in communication systems include tau-dither loops (TDLs) and narrow correlators. Narrow correlators are used in GPS systems for higher accuracy and will be discussed in later chapters.

A basic code recovery loop is shown in Figure 5-7. The incoming code is a composite of both the data and the high-speed PN code from the transmitter to produce the spread spectrum signal. The same high-speed PN code resides in the receiver. The sliding correlator

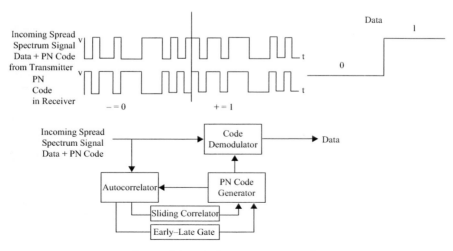

Figure 5-7   Code recovery example.

searches for the correlation peak by speeding up or slowing down the code clock, which aligns the codes. When these codes are aligned, multiplication is performed on each of the code bits and the integration of all the multiplication products results in a large negative number when the code is inverted, which represents a "0" data value in this example; the same process results in a large positive number when the code is not inverted, which represents a "1" data value. Thus, if the codes are aligned properly, the output is the retrieved digital data signal (Figure 5-7). The autocorrelator, or early-late gate in this example, compares the incoming digital waveform and correlates this digital signal with the internal PN code residing in the receiver. Using an early code and a late code or early-late gate and comparing these results provides the feedback to adjust the speed of the PN receiver clock to maintain code lock.

5.2.2 Pulsed Matched Filter

Another option to remove the spread spectrum PN code is to use an asynchronous process that combines the phase-shift times (bits or chips) using a matched filter or a correlator. This matched filter generally consists of a tapped delay line, with the tap spacing equal to the bit or chip duration, and provides a means to sum all of the phase shifts. This can be done either using analog tapped delay lines or digital filters. Most systems use digital means for implementing this type of demodulation scheme.

As the digital signal propagates through the matched filter, a correlation peak is generated at a particular time. The peak can be used as the retrieved data directly or in a pulse position modulation (PPM) scheme where the time position of the correlation peak is decoded as data.

The matched filter process is used for asynchronous detection of a spread spectrum signal. It can be used for pulsed systems where the arrival of the pulse provides the information for the system. The time of arrival (TOA) is used in conjunction with a PPM scheme to decode the sent data. This is a very useful for systems that do not transmit continuous data and that do not want to use extra bits (overhead bits) for synchronizing the tracking loops in a coherent system. This type of matched filter is a very simple process that sums the PN signals in such a way that the signal is enhanced, but it does not change the bandwidth.

An example of this type of matched filter correlator in the analog domain is an acoustic charge transport (ACT) device. It separates packets of sampled analog data with respect to delay, weights each packet, and then sums them together to receive a desired pulse. An analog version of a matched filter is shown in Figure 5-8.

This can be done digitally using finite impulse response (FIR) filters with the tap delay equal to the chip width and weights that are 1 or −1, depending on the code (Figure 5-9). The matched filter correlator provides a means of producing a TOA with a high SNR by combining all of the pulses that were sent into one time frame.

The matched filter correlator consists of a tapped delay line, with the delay for each of the taps equal to 1/chip rate. Each of the delay outputs are multiplied by a coefficient, usually ±1, depending on the code that was sent. The coefficients are the time reverse of the PN code values. For example,

Code values: 1, −1, 1, 1
Coefficient values: 1, 1, −1, 1

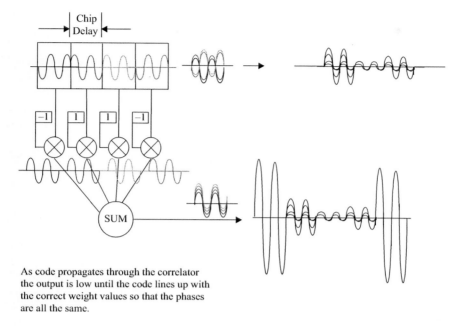

As code propagates through the correlator
the output is low until the code lines up with
the correct weight values so that the phases
are all the same.

Figure 5-8   Analog matched filter correlator.

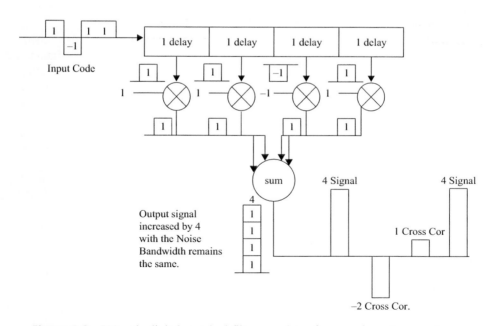

Figure 5-9   PN code digital matched filter correlator for spread spectrum systems.

As the signal is processed through the matched filter, the weights will correspond at one code delay in time, and the outputs of the resultants are all summed together to create a large signal (Figure 5-9).

This example shows a code that is 4 bits long. The output of the digital summation is a pulse with an amplitude of four times the input pulse amplitude. This provides the SNR necessary to detect the signal with the bandwidth and noise power unchanged.

Often digital systems will use a quadrature downconversion to strip off the carrier and provide both I and Q data to the matched filter. The matched filter can employ either a quadrature correlator or parallel matched filters to process the I and Q data.

To increase the process gain and jamming margin, cascaded PN code matched filters can be used. The more cascading of the match filters, the more process gain and jamming margin is achieved. An example of cascaded matched filters is shown in Figure 5-10. This simple example shows a four-stage matched filter, achieving a fourfold signal increase, then the output of this match filter is fed into another four-stage matched filter to provide a 16-fold signal increase (Figure 5-10).

If the code is longer and random, then passing a sine wave through it would result in superposition of many noninverted and inverted sine wave segments. When these segments are summed together in the matched filter, the average output would be close to zero amplitude, or at least reduced. The same thing applies with different codes that are passed through the matched filter, since they switch the phase of the carrier at the wrong places and times. The closer the code is to the pseudo-random code, the higher the output correlation peak will be. Note, however, that regardless of the code used, the bandwidth has not changed. The bandwidth is dependent on the chip rate or pulse width (PW), which is still two times the chip rate (or 2/PW) for double-sided, null-to-null bandwidth (Figure 5-11).

Frequently there is no carrier recovery since the process is asynchronous and the signal is demodulated. This method of detection is useful in pulsed systems or TDMA systems, where the overhead required to synchronize the system may overwhelm the amount of data to be sent. In other words, the overhead bits and time to synchronize the tracking loops reduces the amount of information that can be sent for a given transmission pulse width.

Once the pulse is recovered, this produces a pulse in time that establishes the TOA for that pulse. This can be applied to a PPM technique to demodulate the sent data.

5.3 Pulse Position Modulation

Pulse position modulation measures the TOA of the pulse received to determine where in time the pulse occurred. The data are encoded and decoded with reference to the time position of the pulse. The data bits represent a time slot on the time slot grid array. This is used with spread spectrum systems using PN codes, with matched filters to correlate PN codes. The integrated correlated signal produces a pulse, and the time position of the pulse provides data information.

The number of bits is dependent on the number of time slots; for example,

$$2^n \text{ bits} = \text{number of time slots}$$

If there are eight time slots, then there are 3 bits of information for each time slot:

$$2^3 \text{ bits} = 8 \text{ time slots}$$

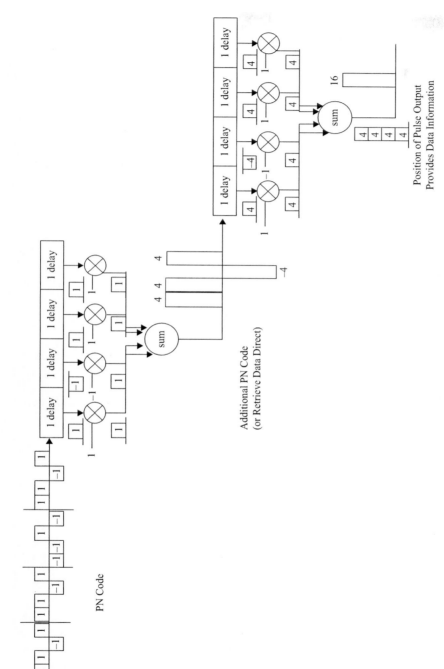

Figure 5-10 Cascaded PN code matched filters for increased process gain and margin.

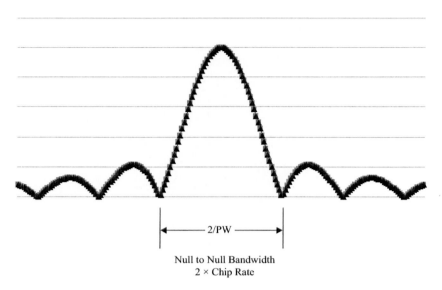

Figure 5-11   Sinc function squared for pulse modulation.

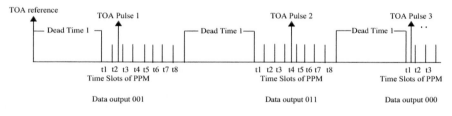

Dead time required for demodulation of the PN code to get a pulse
Dependent on the length of spread spectrum code

Figure 5-12   Absolute pulse position modulation for data encoding.

If absolute time is precisely known, then an absolute PPM demodulation process can be used. Therefore, the TOA referenced to absolute time provides the information. The amount of information this TOA pulse provides depends on the PPM grid or the number of possible time divisions in which the TOA could occur. For example, if there are eight possibilities for the TOA to occur, then 3 bits of information are produced on the TOA of one pulse (Figure 5-12).

The PPM grid is referenced to an absolute time mark, and the entire PPM grid changes only with absolute time variations. So the PPM grid is not dependent on the TOA pulses received except for the reference. Once the PPM structure is set up, it does not change. The dead time is allocated to ensure that the chips that make up the pulsed output have passed through the matched filter and do not interfere with other pulses. The dead times are the same lengths and do not vary in length between pulses.

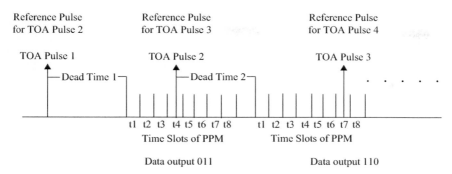

Figure 5-13 Differential pulse position modulation for data encoding.

Differential PPM also can be useful when absolute time is not known. This method relies on the difference in time between two received TOA pulses. Therefore, the first TOA pulse received contains no data but provides the reference for the following TOA pulse received. The time after a received TOA pulse is divided up into a PPM time slot grid, and when the following TOA pulse is received in that time slot grid then the time slot determines the sent data. For example, if the time on the grid is divided into eight time slots after receipt of a TOA pulse, then the next TOA pulse falls in one of the eight time slots and produces 3 bits of information (Figure 5-13). The third TOA pulse is mapped into the PPM grid, with the second TOA pulse taken as the reference, and so on, as shown in Figure 5-13.

Dead time is also required for this PPM scheme. However, the dead time spacing between the PPM grids is contingent on the location of the received reference pulse. For example, if a TOA pulse is received in time slot 1 of the PPM grid, it is used as a reference for the next TOA pulse. The dead time starts from the time of reception of the reference TOA pulse. If the same TOA pulse is received in time slot 8 of the PPM grid and is used as the reference for the next pulse, the absolute time of the PPM grid for the next pulse will be delayed more than if time slot 1 was used as the reference. The average would produce a higher throughput of data compared with absolute PPM, since the dead time starts immediately following a received TOA pulse. For the absolute time PPM, the grid for all pulses is at the maximum and does not change with the data used.

A comparison between absolute and differential PPM with respect to missed pulses is shown in Figure 5-14. With absolute PPM, if a pulse is missed or appears in another time slot, then the errors will be for that time slot only. However, if differential PPM is used, then a TOA pulse that is missed or falls into the wrong time slot would cause an error not only for that particular time slot but also for the next TOA pulse position because the time slot grid is establish by the previous TOA pulse that was in error (Figure 5-14). Therefore, differential systems have a possibility of having twice as many errors as the absolute system. This is true for all types of differential systems and will be discussed in other applications.

5.4 Code Division Encoding and Decoding

Another type of demodulation scheme using the matched filter concept is code division encoding, which contains a matched filter for every code sequence used. Each of the TOA

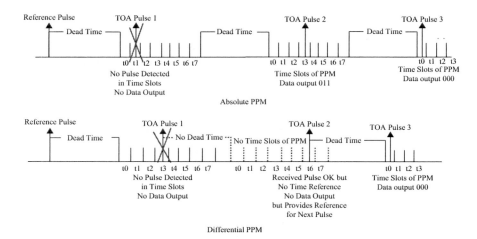

Figure 5-14   Comparison between absolute and differential PPM.

pulses has a different code that depends on the sent data. For example, if there are eight different code sets and eight different matched filters, an output on code 1 decodes to data of 000, indicating that code 1 was sent. If there is an output of the fourth matched filter, then the data decodes to 011, which means that code 4 was sent. The major drawback to this type of demodulation is the amount of hardware required to build multiple matched filters.

5.5 Coherent versus Differential Digital Modulation and Demodulation

Coherent modulation and demodulation techniques maintain a constant state throughout the signal. To achieve this, it requires a very stable source and phase locking techniques. It is generally more costly and more complex.

On the other hand, differential modulation and demodulation techniques are less complex and less expensive. They need to be stable only between state changes (phase/amplitude) of adjacent symbols. Thus, stability is critical just for a short period of time. However, these methods experience more bit errors and require more SNR compared with coherent systems and apply to all types of digital modulation.

A diagram showing the differences between coherent and differential systems with respect to the number of bit errors is shown in Figure 5-15. For coherent systems, if there is a bit error, it affects only the bit that was in error. For differential systems, there is the possibility of having twice as many errors. If the system receives one bit error, then the next bit could be in error, since it is dependent on the previous bit value and monitors the amount of change to the bit (Figure 5-15).

As mentioned earlier, a major advantage of the differential system is the simplicity and cost of implementing a demodulation scheme. For example, a simple differential binary phase-shift keying (DBPSK) system can be demodulated by using a delay and multiply, as shown in Figure 5-16. The delay is equal to one bit time, and the multiplication and

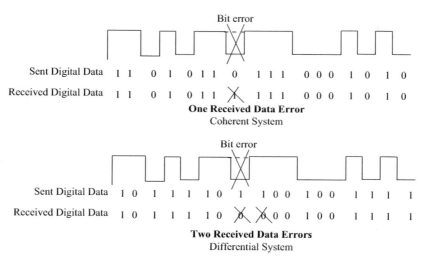

Figure 5-15   Coherent versus differential systems with respect to bit errors.

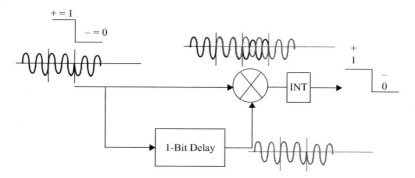

Figure 5-16   Differential demodulation using delay and multiply.

integration produces the bit that was sent. When the signal is in phase, then the delay and multiply detection integrates to a large positive number; when the signal is out of phase, the signal integrates to a large negative number, thus retrieving the differentially encoded bit stream (Figure 5-16).

5.5.1 Coherent Demodulation

Coherent demodulation is generally used for continuous systems or long-pulse systems that can afford the overhead required to enable the tracking loops in the system. This overhead includes the bits that need to be sent so that the tracking loops can start to track the incoming signal. These bits are wasted because the coherent data have not been demodulated if the tracking loops are not tracking the data yet. Therefore, if there is an output, it will contain a tremendous amount of errors. For short-pulse systems, this overhead can take up most of the signal. Every time a pulse is sent out, the tracking loops have to reacquire the signal. There are many ways to demodulate the incoming signal and to implement the tracking loops for each part of the system.

The demodulation process requires three basic functions to retrieve the sent data:

- Recover the carrier, since the digital modulation results in a suppressed carrier and the carrier is recovered to remove it from the incoming signal.
- Remove spreading spectrum coding, if using spread spectrum techniques; generally done using a despreading correlator or sliding correlator.
- Align and synchronize the sample point for sampling the data stream at the optimal SNR point, which requires lining up the bits with the sample time using a bit synchronizer.

5.6 Symbol Synchronizer

Once the carrier is eliminated and the spread spectrum PN code is removed, the raw data remain. However, due to noise and intersymbol interference (ISI) distorting the data stream, a symbol synchronizer is needed to determine what bit was sent. This device aligns the sample clock with the data stream so that it samples in the center of each bit in the data stream. An early-late gate can be used, which is similar to the early-late gate used in the code demodulation process. The latter requires integration over the code length or repetition to generate the autocorrelation function, but with the bit synchronizer, the integration process is over a symbol or bit if coding is not used. The early and late gate streams in the bit synchronizer are a clock of ones and zeros since there is no code reference and the data are unknown. Therefore, the incoming data stream is multiplied by the bit clock, set at the bit rate, with the transitions early or later than the data transitions, as shown in Figure 5-17.

This example shows the early gates and late gates are off from the aligned signal by ¼ symbol or bit, which is a narrower correlator compared with the standard ½. This provides more accuracy for the sample point. The integrated outputs are shown along with the aligned integration for reference. When the integrated outputs are equal in peak amplitude, as shown in the example, then the bit synchronizer is aligned with the bit. So the point between the early gate and late gate, which is the center of the aligned pulse, is the optimal point to sample and recover the data. This point provides the best SNR of the received data, which is the center of the eye pattern (Figure 5-18). Once these data are sampled at the optimal place, a decision is made to determine whether a "1" or a "0" was sent. This bit stream of measured data is decoded to produce the desired data.

5.7 The Eye Pattern

The eye pattern describes the received digital data stream when observed on an oscilloscope. Due to the bandwidth limitations of the receiver, the received bit stream is filtered, and the transitions are smoothed. Since the data stream is pseudo-random, the oscilloscope shows both positive and negative transitions, thus forming a waveform that resembles an eye (Figure 5-18).

The four possible transitions at the corner of the eye are low to high, high to low, high to high, and low to low. Observation of the eye pattern can provide a means of determining the

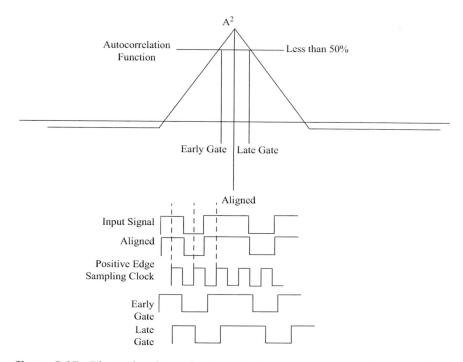

Figure 5-17   Bit synchronizer using the early-late gate narrow correlator technique.

performance of a receiver. The noise on the eye pattern and the closing of the eye can indicate that the receiver's performance needs to be improved or that the signal from the transmitter needs to be increased. The eye pattern starts to close with the amount of distortion and noise on the signal (Figure 5-18). The bit synchronizer samples the eye pattern at the center of the eye where the largest amplitude and the highest SNR occur. As the noise increases, it becomes much more difficult to determine the bit value without bit errors (Figure 5-19).

5.8 Digital Processor

Once the digital data stream has been sampled and it is determined that a "0" or a "1" was sent, the digital data stream is decoded into the data message. The digital processor is generally responsible for this task and also the task of controlling the processes in the receiver as well as demodulation of the received signal. In many systems today, the digital processor, in conjunction with specialized DSP integrated circuits, is playing more of a key role in the demodulation process. This includes implementation of the various tracking loops, bit synchronizing, carrier recovery, and decoding of the data, as mentioned already. The digital processor ensures that all of the functions occur at the necessary times by providing a control and scheduling capability for the receiver processes.

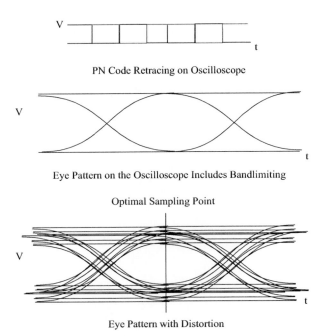

PN Code Retracing on Oscilloscope

Eye Pattern on the Oscilloscope Includes Bandlimiting

Optimal Sampling Point

Eye Pattern with Distortion

Figure 5-18   Eye pattern as seen on an oscilloscope.

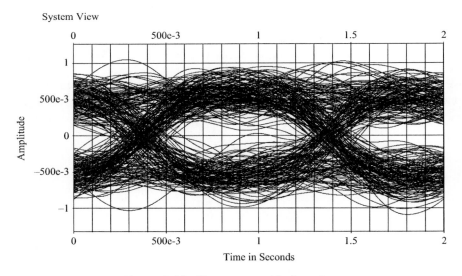

Figure 5-19   Eye pattern with distortion.

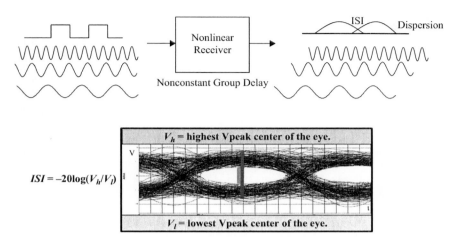

Figure 5-20   Intersymbol interference is the ratio of the highest excursions to the lowest excursions of the received waveform at the center of the eye.

5.9 Intersymbol Interference

Intersymbol interference is the amount of interference due to dispersion of the pulses that interfere with the other pulses in the stream. Dispersion occurs when there are nonlinear phase responses to different frequencies or a nonconstant group delay, which is the derivative of the phase (Figure 5-20).

Since pulses are made up of multiple frequencies according to the Fourier series expansion of the waveform, these frequencies need to have the same delay through the system to preserve the pulse waveform. Therefore, if the system has a constant group delay, or the same delay for all frequencies, then the pulse is preserved. If a nonconstant group delay is causing different delays for different frequencies, then these frequencies are added together to form a pulse waveform where the pulse is dispersed or spread out. This leads to distortion to adjacent pulses and thus creates ISI. To obtain a value of the amount of ISI, the eye pattern, as mentioned earlier, can be used. The ISI is determined by the following and shown in Figure 5-20:

$$ISI = 20\log(V_h/V_l)$$

where

V_h = highest peak voltage when measured at the center of the eye of the eye pattern
V_l = lowest peak voltage when measured at the center of the eye of the eye pattern

The measured amount of ISI in a receiver determines the amount of degradation in the ability to detect the desired signals and leads to increased BERs in the system. This can be included in the implementation losses of the system, described in Chapter 1. Sometimes this ISI is specified in the link budget as a separate entry.

5.10 Scrambler/Descrambler

Scramblers and descramblers are used to prevent a long stream of either "1"s or "0"s in the data or code. The scrambler uses an additional PN code selected with short runs of "1"s and "0"s which is combined with the code or data sequence. This is accomplished by performing a modulo-2 addition to the digital signal in the transmitter. This code is demodulated by the descrambler in the receiver with a matched filter or sliding correlator. The code used for the scrambler/descrambler contains no data and adds cost and complexity. Using this technique prevents drifts of VCOs due to large DC values caused by these long periods when the code does not change state.

5.11 Phase-Shift Detection

Spread spectrum systems are sometimes used for covert systems. This is known as electronic countermeasures (ECM) and provides decreased vulnerability to detection. Electronic counter-countermeasure (ECCM) receivers are designed to detect these types of signals. One of the methods to detect a BPSK waveform, as mentioned earlier, is to use a squaring, doubler, or ×2 (doubling the frequency) to eliminate the phase shift as follows:

$$[A\cos(\omega t + 0°, 180°)]^2 = A^2/2\cos(2\omega t + 0°, 360°) = A^2/2\cos(2\omega t)$$

Note that $2 \times 0°$ is $0°$ and $2 \times 180°$ is $360°$, which is equal to $0°$. This is the basis for squaring, to eliminate the phase shift modulation so that the resultant signal is a continuous wave (CW) spectral line instead of a sinc function with a suppressed carrier. The CW frequency is then easily detected, since this squaring process despreads the wideband signal.

The end result is a spectral line at twice the carrier frequency. This basic principle for detecting a BPSK data stream allows the ECCM receiver to obtain the frequency of the signal being sent by dividing the output frequency by two for BPSK. If BPSK is used, the null-to-null bandwidth, which is equal to twice the chip rate, allows the ECCM receiver to calculate the chip rate of the BPSK signal. The sidelobes are monitored to ensure that they are half of the width of the main lobe, which equals the chip rate. If they are not, then this provides additional information to the ECCM receiver that the signal is not BPSK and might be another form of modulation.

QPSK and offset QPSK (OQPSK) are detected by what is known as a times-4 (×4) detector, which quadruples the input signal to eliminate the phase ambiguities. OQPSK has all the same absolute phase states but is not allowed to switch more than 90°, which eliminates the 180° phase shifts. However, the criteria depend on only the absolute phase states and how they are eliminated. The following shows the results of squaring the signal twice. The first squaring function produces

$$[A\cos(\omega t + 0°, 90°, 180°, -90°)]^2 = A^2/2\cos(2\omega t + 0°, 180°, 360°, -180°)$$
$$= A^2/2\cos(2\omega t + 0°, 180°) \text{ plus DC offset filtered out}$$

Squaring again produces

$$\left[A^2/2\cos(2\omega t + 0°, 180°)\right]^2 = A^4/8\cos(4\omega t + 0°, 360°) \text{ plus a DC term filtered out}$$
$$= A^4/8\cos(4\omega t)$$

Note, since the possible phase states are 0°, 180°, 90°, and −90°, squaring them would only give $2 \times 0° = 0°$, $2 \times 180° = 360° = 0°$, $2 \times 90° = 180°$, and $2 \times -90° = -180° = 180°$. Therefore, the problem has been reduced to simple phase shifts of a BPSK level. One more squaring will result in the same as described in the previous BPSK example, which eliminates the phase shift. Thus, quadrupling or squaring the signal twice eliminates the phase shift for a quadrature phase-shifted signal.

The resultant signal gives a spectral line at four times the carrier frequency. This is the basic principle behind the ECCM receiver in detecting a QPSK or OQPSK data stream and allows the ECCM receiver to know the frequency of the signal being sent by dividing the output frequency by four for both QPSK and OQPSK waveforms.

For minimum shift keying (MSK), there is a sinusoidal modulating frequency proportional to the chip rate in addition to the carrier frequency along with the phase transitions. One way to generate classic MSK is to use two BPSK in an OQPSK-type system and sinusoidally modulate the quadrature channels at a frequency proportional to the bit rate before summation. In other words, the phase transitions are smoothed out by sinusoidal weighting. By positioning this into a times-4 detector as before, the resultant is

$$[A\cos(\omega t + 0°, 90°, 180°, -90°)(B\cos\pi t/2T)]^4$$
$$= [AB/2(\cos(\omega \pm \pi/2T)t + 0°, 90°, 180°, -90°)))]^4$$

The phase ambiguities will be eliminated because the signal is quadrupled so that the phase is multiplied by 4. This results in 0°, so these terms are not carried out.

Squaring the equation first produces the sums and differences of the frequencies. Assuming that the carrier frequency is much larger than the modulating frequency, only the sum terms are considered and the modulating terms are filtered out. For simplicity, the amplitude coefficients are left out. Therefore, squaring the above with the conditions stated results in

$$\cos(2\omega \pm \pi/T)t + \cos(2\omega)t$$

Squaring this equation with the same assumptions result in quadrupling MSK:

$$\cos 4\omega t + \cos(4\omega \pm 2\pi/T)t + \cos[(4\omega \pm \pi/T)t$$

There will be spectral lines at 4 times the carrier, 4 times the carrier ±4 times the modulating frequency, and 4 times the carrier ±2 times the modulating frequency. Therefore, by quadrupling the MSK signal, the carrier can be detected. Careful detection can also produce the chip rate features since the modulating frequency is proportional to the chip rate.

5.12 Shannon's Limit

In 1948, C. E. Shannon developed a theory of how much information could be theoretically sent over a channel. The channel could be any medium used to send information, such as wire, optic cable, or freespace. The Shannon limit for information capacity is

$$C = B\log_2(1 + SNR)$$

where

$\quad C = $ information capacity of a channel (bps)
$\quad B = $ bandwidth (Hz)
$SNR = $ signal-to-noise ratio

Note: SNR is the actual value of the SNR, which is unitless and is not converted to dB. So if the SNR is given in dB, the value needs to be reverted to the actual SNR.

Since it is more common to use logarithms in base 10 instead of base 2, a conversion is required as follows:

$$\log_2 X = \log_{10}X/\log_{10}2 = 3.32\log_{10}X$$

Therefore, Shannon's limit becomes

$$C = 3.32B\log_{10}(1 + SNR)$$

Many different coding schemes can send several bits of information for every symbol sent. For example, QPSK sends two bits of information per symbol of transmission. Shannon's limit provides the absolute maximum information that can be sent no matter how many bits can be put into a symbol.

5.13 Summary

The demodulation process is an important aspect in the design of the transceiver. Proper design of the demodulation section can enhance the sensitivity and performance of data detection. Two types of demodulation can be used to despread and recover the data. The matched filter approach simply delays and correlates each delay segment of the signal to produce the demodulated output. This process includes the use of PPM to encode and decode the actual data. Another demodulation process uses a coherent sliding correlator to despread the data. This process requires alignment of the codes in the receiver, which is generally accomplished by a short acquisition code. Tracking loops, such as the early-late gate, align the code for the dispreading process. Carrier recovery loops, such as the squaring loop, Costas loop, and modified Costas loop, provide a means for the demodulator to strip off the carrier. A symbol synchronizer is required to sample the data at the proper time in the eye pattern to minimize the effects of ISI. Finally, receivers designed for intercepting transmissions of other transmitters use various means of detection depending on the type of phase modulation utilized.

References

Haykin, S. *Communication Systems*, 5th ed. Wiley, 2009.

Holmes, J. K. *Coherent Spread Spectrum Systems*. Wiley & Sons, 1982, pp. 251–267.

Keate, C. R. "The Development and Simulation of a Single and Variable Data-Rate Data-Aided Tracking Receiver." Ph. D. dissertation, 1986.

Torrieri, D. *Principles of Spread-Spectrum Communication Systems*, 2d ed. Springer, 2011.

Problems

1. Given a code of 1010011, what would the weights of the pulse matched filter need to be assuming ×1 is the weight for the first stage of the matched filter and ×7 is the weight for the last stage of the matched filter?
2. Given a BPSK signal, show how a squaring loop eliminates the phase ambiguity.
3. What is the null-to-null bandwidth of a spread spectrum signal with a chipping rate of 50 Mcps?
4. How does the bandwidth change in (a) the pulsed matched filter and (b) the sliding correlator matched filter?
5. Explain what is meant by intersymbol interference.
6. What is an eye pattern?
7. Where on the eye pattern is the best place for sampling the signal for best performance? Where is the worst place?
8. Using the squaring loop approach for BPSK detection, what would be needed in (a) an 8-PSK detector, (b) a 16-PSK detector?
9. What would be the minimum possible phase shifts of each of the waveforms in problem 8?
10. Since MSK can be considered as a frequency-shifted signal, what would be another way of detection of MSK?
11. What is the advantage of MSK in reference to sidelobes?
12. What are the two basic ways to demodulate a PSK digital waveform?
13. What is a method used to maintain lock using a sliding correlator?
14. What two ways are used to eliminate the carrier from the PSK digital waveform?
15. What are the two parameters that establish Shannon's limit?

CHAPTER 6

Basic Probability and Pulse Theory

To achieve a better understanding of digital communications, some basic principles of probability theory need to be discussed. This chapter provides an overview of theory necessary for the design of digital communications and spread spectrum systems. Probability is used to calculate the link budget in regards to the error and required signal-to-noise (SNR) ratio and to determine whether a transceiver is going to work and at what distances. This is specified for digital communications systems as the required E_b/N_o.

6.1 Basic Probability Concepts

The central question concerning probability is whether something is going to occur, whether it is an error in the system, or the probability that multipath will prevent the signal from arriving at the receiver. The probability that an event occurs is described by the probability density function (PDF) and is defined as

$$f_x(x)$$

Probability can be expressed as a percentage, for example, a 10% chance that a value is present or that an event has occurred, or the probability of occurrence (Figure 6-1). The integral of the PDF is equal to $F_x(x)$ and is referred to as the cumulative distribution function (CDF). This integral of the PDF over $\pm\infty$ equals 1, which represents 100%.

$$F_x = \int_{-\infty}^{+\infty} x f_x(x) dx = 1 = 100\%$$

For example, if there is a 10% chance of getting it right, by default there is a 90% chance of getting it wrong. The CDF is shown in Figure 6-2.

Often the PDF is mislabeled as the CDF. For example, if something has a Gaussian distribution, the curve that comes to mind is the PDF, as shown in Figure 6-1. This is not the cumulative distribution function.

Another term used in probability theory is called the expected value $E[X]$. The expected value is the best guess as to what the value will be. To obtain the expected value,

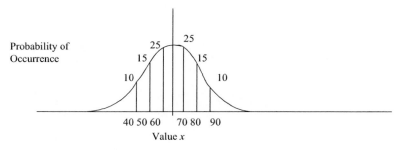

Figure 6-1   Probability density function for a Gaussian process.

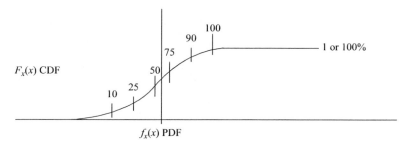

Figure 6-2   Cumulative distribution function for a Gaussian process.

the signal values are multiplied by the percentage or the PDF and then summed or integrated as shown:

$$E[x] = \int x f_x(x) dx$$

Using a discrete example, suppose the density function is symmetrical. To plug in some numbers:

x	$f_x(x)$	$x f_x(x)$
1	0.1(10%)	0.1
2	0.2(20%)	0.4
3	0.4(40%)	1.2
4	0.2(20%)	0.8
5	0.1(10%)	0.5

$$\sum x f_x(x) = E[X] = 3.0$$

The expected value is equal to the mean value or mean (m_x). Therefore,

$$m_x = E[x] = \int x f_x(x) dx = 3.0$$

6.2 The Gaussian Process

The Gaussian process or distribution is probably the most common distribution in analyzing digital communications. There are many other types of distributions, such as the uniform distribution, which is used for equal probability situations (e.g., the phase of a multipath signal), and the Rayleigh distribution, which is used to characterize the amplitude variations of the multipath. The Gaussian distribution, however, is used most frequently. It is called the normal distribution or bell-shaped distribution because of its common use and because it produces a PDF bell-shaped curve. Most often noise is characterized using a Gaussian distribution for transceiver performance and SNR evaluations. The PDF for a Gaussian process is defined as follows:

$$f_x(x) = \frac{1}{\sigma\sqrt{2\pi}} e^{-\frac{(x-m_x)^2}{2\sigma^2}}$$

where

σ = standard deviation
m_x = mean

This establishes a curve called the distribution function. This is really the density function, a slight misnomer. Note that it is not the cumulative distribution function as mentioned earlier. The cumulative distribution is defined as

$$F_x(x) = \frac{1}{2}\left(1 + erf\ \frac{x}{\sigma\sqrt{2}}\right)$$

where

erf = error function
σ = standard deviation

The mean is assumed to be zero.

The erf value can be calculated or a lookup table can be used with an approximation if x is large. The cumulative distribution function is used to calculate the percentage that the error is within a given range, such as 1σ or 2σ.

For example, what is the probability that x is between $\pm2\sigma$ or a 2σ variation or 2 times the standard deviation. The cumulative distribution is used, and substituting -2σ for x in the previous equation yields

$$F_x(x) = \frac{1}{2}\left(1 + erf\ \frac{-2}{\sqrt{2}}\right) = \frac{1}{2}\left(1 + erf(-\sqrt{2})\right)$$

Note that $erf(-x) = -erf(x)$. Therefore, $erf(-2^{\frac{1}{2}}) = -erf(1.414) = -0.954$ (Table 6-1).

Finishing the calculations:

$$F_x(x) = \tfrac{1}{2}(1 + (-0.954)) = 0.023$$

This represents the probability of the function at -2σ or less, as shown in Figure 6-3. Therefore, assuming Gaussian distribution with zero mean, the probability of the function being a $+2\sigma$ or greater is the same.

Table 6-1   Error function, $erf(x)$

	Hundredths digit of x									
x	0	1	2	3	4	5	6	7	8	9
0.0	0.00000	0.01128	0.02256	0.03384	0.04511	0.05637	0.06762	0.07886	0.09008	0.10128
0.1	0.11246	0.12362	0.13476	0.14587	0.15695	0.16800	0.17901	0.18999	0.20094	0.21184
0.2	0.22270	0.23352	0.24430	0.25502	0.26570	0.27633	0.28690	0.29742	0.30788	0.31828
0.3	0.32863	0.33891	0.34913	0.35928	0.36936	0.37938	0.38933	0.39921	0.40901	0.41874
0.4	0.42839	0.43797	0.44747	0.45689	0.46623	0.47548	0.48466	0.49375	0.50275	0.51167
0.5	0.52050	0.52924	0.53790	0.54646	0.55494	0.56332	0.57162	0.57982	0.58792	0.59594
0.6	0.60386	0.61168	0.61941	0.62705	0.63459	0.64203	0.64938	0.65663	0.66378	0.67084
0.7	0.67780	0.68467	0.69143	0.69810	0.70468	0.71116	0.71754	0.72382	0.73001	0.73610
0.8	0.74210	0.74800	0.75381	0.75952	0.76514	0.77067	0.77610	0.78144	0.78669	0.79184
0.9	0.79691	0.80188	0.80677	0.81156	0.81627	0.82089	0.82542	0.82987	0.83423	0.83851
1.0	0.84270	0.84681	0.85084	0.85478	0.85865	0.86244	0.86614	0.86977	0.87333	0.87680
1.1	0.88021	0.88353	0.88679	0.88997	0.89308	0.89612	0.89910	0.90200	0.90484	0.90761
1.2	0.91031	0.91296	0.91553	0.91805	0.92051	0.92290	0.92524	0.92751	0.92973	0.93190
1.3	0.93401	0.93606	0.93807	0.94002	0.94191	0.94376	0.94556	0.94731	0.94902	0.95067
1.4	0.95229	0.95385	0.95538	0.95636	0.95830	0.95970	0.96105	0.96237	0.96365	0.96490
1.5	0.96611	0.96728	0.96841	0.96952	0.97059	0.97162	0.97263	0.97360	0.97455	0.97546
1.6	0.97635	0.97721	0.97804	0.97884	0.97962	0.98038	0.98110	0.98181	0.98249	0.98315
1.7	0.98379	0.98441	0.98500	0.98558	0.98613	0.98667	0.98719	0.98769	0.98817	0.98864
1.3	0.98909	0.98952	0.98994	0.99035	0.99074	0.99111	0.99147	0.99132	0.99216	0.99248
1.9	0.99279	0.99309	0.99338	0.99366	0.99392	0.99418	0.99443	0.99466	0.99489	0.99511
2.0	0.99532	0.99552	0.99572	0.99591	0.99609	0.99626	0.99642	0.99658	0.99673	0.99688
2.1	0.99702	0.99715	0.99728	0.99741	0.99753	0.99764	0.99775	0.99785	0.99795	0.99805
22	0.99814	0.99822	0.99831	0.99839	0.99846	0.99854	0.99861	0.99867	0.99874	0.99880
2.3	0.99886	0.99891	0.99897	0.99902	0.99906	0.99911	0.99915	0.99920	0.99924	0.99928
2.4	0.99931	0.99935	0.99938	0.99941	0.99944	0.99947	0.99950	0.99952	0.99955	0.99957
2.5	0.99959	0.99961	0.99963	0.99965	0.99967	0.99969	0.99971	0.99972	0.99974	0.99975
2.6	0.99976	0.99978	0.99979	0.99980	0.99981	0.99982	0.99983	0.99984	0.99985	0.99986
2.7	0.99987	0.99987	0.99988	0.99989	0.99989	0.99990	0.99991	0.99991	0.99992	0.99992
2.8	0.99992	0.99993	0.99993	0.99994	0.99994	0.99994	0.99995	0.99995	0.99995	0.99996
2.9	0.99996	0.99996	0.99996	0.99997	0.99997	0.99997	0.99997	0.99997	0.99997	0.99998
3.0	0.99998	0.99998	0.99998	0.99998	0.99998	0.99998	0.99998	0.99999	0.99999	0.99999
3.1	0.99999	0.99999	0.99999	0.99999	0.99999	0.99999	0.99999	0.99999	0.99999	0.99999
3.2	0.99999	0.99999	0.99999	1.00000	1.00000	1.00000	1.00000	1.00000	1.00000	1.00000

The probability of the signal being outside $\pm 2\sigma$ is 0.0456. Thus, the probability of x being inside these limits is $1 - 0.0456 = 0.954$, or 95.4%. This is the same result as simply taking the previously given *erf* function, $erf(2)^{1/2} = 0.954$, or 95.4%. For a -1σ error, the error function of -0.707 is -0.682. The distribution function is 0.159, as shown in Figure 6-3. So for $\pm 1\sigma$ the probability of being outside these limits is 0.318. The probability of being within these limits is $1 - 0.318 = 0.682$, or 68.2%. Given the range of x, the probability that the solution falls within the range is easily calculated.

Note that if the number of samples is not infinite, then the Gaussian limit may not be accurate and will degrade according to the number of samples that are taken.

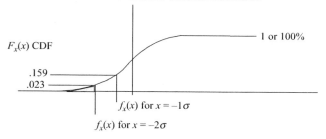

Cumulative Distribution Function for Gaussian Distribution

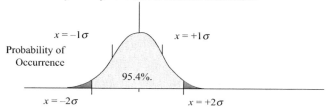

Probability Density Function for Gaussian Distribution

Figure 6-3   Probability within a given range using Gaussian distribution.

6.3 Quantization and Sampling Errors

Since all digital systems are discrete and not continuous, the digitizing process creates errors between the digital samples. This error is created because a digitizing circuit samples the signal at regular intervals, resulting in a sequence of numerical values that simulates a continuous signal. If a signal is changing continuously with time, then an error is caused by the estimation of the signal between the sample points. If the sample rate is increased, this estimation error is reduced.

There are basically two types of errors going from the analog signal to the digital signal: quantization error and sampling error. Quantization error occurs in the resolution of the analog signal due to discrete steps. This is dependent on the number of bits in the analog-to-digital converter (ADC): the more bits, the less quantization error. The value of the analog signal is now discrete and is either at one bit position or another. An example of quantization error is shown in Figure 6-4.

Sampling error comes about because the samples are noncontinuous samples. Basically the sampling function samples at a point in time and extrapolates an estimate of the continuous analog signal from the points. This is dependent on the sample rate.

For example, radar range is based on the time of arrival (TOA) of the pulse that is sent and returned. This is the basic principle that ranging and tracking radars work on when calculating the range of an aircraft from the ground station. One source of range error is called sampling error. If the arrival pulse is sampled, the sampling error is based on the distance between the sample pulses; halfway on either side is the worst-case sampling error, as shown in Figure 6-5.

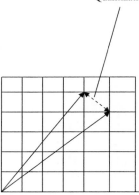

Figure 6-4 Quantization error in the ADC.

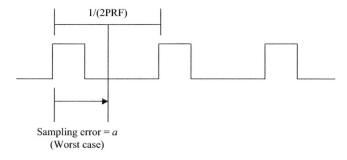

Figure 6-5 Sampling error showing worst-case conditions.

Since the probability that this occurrence is equal across the worst-case points, it is referred to as being uniform or having a uniform distribution. The variance for a uniform distribution is

$$\sigma^2 = a^2/3$$

where distribution varies between $+a$ and $-a$.

The standard deviation is equal to the square root of the variance:

$$\sigma = \sqrt{\frac{a^2}{3}}$$

For example, if the clock rate is 50 MHz, the time between pulses is 20 nsec, so a, or the peak deviation, is equal to 10 nsec. Therefore, the standard deviation is equal to

$$\sigma = \sqrt{\frac{10^2}{3}} = 5.77 \ ns$$

If a number of samples are taken and averaged out, or integrated, then the standard deviation is reduced by

$$\sigma_{ave} = \frac{\sigma}{\sqrt{n_{samples}}}$$

If the number of samples that are to be averaged is 9, then the standard deviation is

$$\sigma_{ave} = \frac{5.77 \; ns}{\sqrt{9}} = 1.92 \; ns$$

To determine the overall error, the distribution from each source of error needs to be combined using a root sum square (RSS) solution. This is done by squaring each of the standard deviations, summing, and then taking the square root of the result for the final error. For example, if one error has a $\sigma = 1.92$ and another independent error has a $\sigma = 1.45$, then the overall error is

$$Total \; Error \; \sigma = \sqrt{1.92^2 + 1.45^2} = 2.41 \; ns$$

This combination of the uniform distributions results in a normal or Gaussian distribution overall, in accordance with the central limit theorem. This is one of the reasons that the Gaussian or normal distribution is most commonly used. Therefore, even though each error is uniform and is analyzed using the uniform distribution, the resultant error is Gaussian and follows the Gaussian distribution for analysis. However, this assumes independent sources of error. In other words, one source of error cannot be related or dependent on another source of error.

6.4 Probability of Error

The performance of the demodulation either is measured as the bit error rate (BER) or is predicted using the probability of error (P_e) as shown for binary phase-shift keying (BPSK):

$$BER = Err/TNB$$

$$POE = \frac{1}{2} erfc \left(\frac{E_b}{N_o} \right)^{\frac{1}{2}} = P_{fa}$$

where

Err = number of bit errors
TNB = number of total bits
P_e = probability of error
E_b = energy in a bit
N_o = noise power spectral density (noise power in a 1 Hz bandwidth)
$erfc$ = $1 - erf$ = complimentary error function

The BER is calculated by adding up the number of bits that were in error and dividing by the total number of bits for that particular measurement. BER counters continuously calculate this ratio. The BER could be called the bit error ratio since it is more related to a ratio than it is to time. An instrument used to measure or calculate BER is a bit error rate tester (BERT).

Normally, the probability of error (P_e) is calculated from an estimated E_b/N_o for a given type of modulation. The probability of getting a zero when actually a one was sent for a BPSK modulated signal is

$$P_{e1} = \frac{1}{2} erfc((E_b/N_o)^{\frac{1}{2}})$$

The probability of getting a one when a zero was sent is

$$P_{e0} = \tfrac{1}{2}erfc((E_b/N_o)^{\frac{1}{2}})$$

The average of these two probabilities is

$$P_e = \tfrac{1}{2}erfc((E_b/N_o)^{\frac{1}{2}})$$

Note that the average probability is the same as each of the individual probabilities. This is because the probabilities are the same and are independent (i.e., either a one was sent and interpreted as a zero or a zero was sent and interpreted as a one). The probabilities happen at different times. The P_e for other systems are as follows:

$$\text{Coherent FSK: } P_e = \tfrac{1}{2}erfc((E_b/2N_o)^{\frac{1}{2}})$$

$$\text{Noncoherent FSK: } P_e = \tfrac{1}{2}\exp(-E_b/2N_o)$$

$$\text{DPSK (noncoherent): } P_e = \tfrac{1}{2}\exp(-E_b/N_o)$$

$$\text{Coherent QPSK and coherent MSK: } P_e = \tfrac{1}{2}erfc((E_b/N_o)^{\frac{1}{2}}) - \tfrac{1}{4}erfc^2((E_b/N_o)^{\frac{1}{2}}).$$

Note: The second term can be eliminated for $E_b/N_o \gg 1$

A spreadsheet with these plotted curves is shown in Figure 6-6. Note that the coherent systems provide a better P_e for a given E_b/N_o, or less E_b/N_o is required for the same P_e. Also note that coherent BPSK and coherent QPSK are fairly close at higher signal levels and diverge with small E_b/N_o ratios.

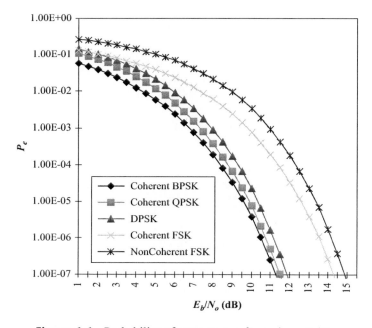

Figure 6-6 Probability of error curves for a given E_b/N_o.

6.5 Probability of Detection and False Alarms

Pulsed systems, such as pulse position modulation (PPM) systems and radars, use probability of detection and probability of false alarms since data are not sent on a continuous basis. They operate on whether or not a pulse was detected correctly and whether or not noise or other signals caused the system to detect a signal that was not the true signal (false alarm). It is not just a matter of whether a zero or a one was sent but whether anything was sent at all. Therefore, errors can come from missed detection or false detection.

The time between false alarms (T_{fa}) is dependent on the integration time, which determines the number of pulses integrated (N_{pi}). The number of pulses integrated is divided by the probability of false alarm (P_{fa}) as shown:

$$T_{fa} = N_{pi}/P_{fa}$$

Thus, the longer the integration time, the fewer false alarms will be received in a given period of time.

For a pulsed system, there will be times when nothing is sent, yet the detection process needs to detect whether or not something was sent. There is a threshold set in the system to detect the signal. If nothing is sent and the detection threshold is too low, then the probability of false alarms or detections will be high when there is no signal present. If the detection threshold is set too high to avoid false alarms, then the probability of detecting the signal when it is present will be too low. If the probability of detection is increased to ensure detection of the signal, then there is a greater chance that we will detect noise, which increases the probability of a false alarm. A trade-off specifies where to set the threshold. Notice that the probability of detection describes whether or not the signal was detected. The probability of false alarm refers to whether or not noise was detected. This should not be confused with whether a zero or a one was detected, which is evaluated using the probability of error. A curve showing the PDFs of one-sided noise and signal plus noise is shown in Figure 6-7.

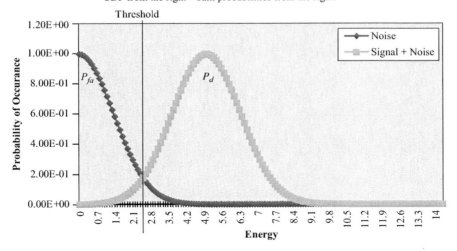

- Use cumulative distribution function to determine the probabilities of these one sided noise and S+N gaussian probability density functions.
- CDF from the right – sum probabilities from the right.

Figure 6-7   Probability of detection and false alarms curves.

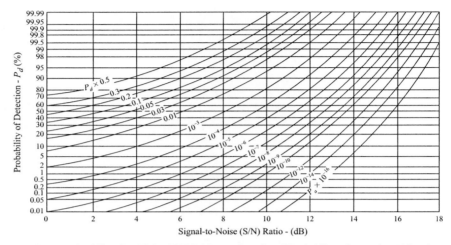

Nomograph of Signal-to-Noise (S/N) Ratio as a Function of Probability of Detection (P_d) and
Probability of False Alarm Rate (P_a)

Figure 6-8   Probability of detection and false alarms versus SNR.

Another useful graph in determining what threshold to select is given in Figure 6-8. The probabilities of detection and false alarms are shown with respect to the level of the SNR. The energy (E_p) is selected to provide the desired probability of detection and probability of false alarm. It is contained in the pulse (E_p). The cumulative distribution functions are used to calculate these parameters. For example, if the threshold is selected at a value of 2.4 (Figure 6-7), then the cumulative distribution function from 2.4 to $+\infty$ (area under the signal + noise curve from 2.4) is the probability of detection. The probability of false alarm is the cumulative distribution function from 2.4 to $+\infty$ (area under the noise curve from 2.4). This point, where the two Gaussian curves intersect, provides a good point to select, and the trade-offs between detection and false alarms are performed to optimize a design to meet the requirements for the system.

The cumulative distribution function for Gaussian distribution can be used since these processes are Gaussian. Since the P_{fa} and P_d are integrating the function from 2.4 to $+\infty$, and the total distribution function is equal to unity, then the distribution function is equal to

$$F_x(x) = 1 - \frac{1}{2}\left(1 + erf\,\frac{x}{\sigma\sqrt{2}}\right) = \frac{1}{2}\left(1 - erf\,\frac{x}{\sigma\sqrt{2}}\right)$$

The complementary error function is equal to

$$erfc = 1 - erf$$

Substituting the *erfc* in the previous equation produces

$$F_x(x) = \frac{1}{2}\left(erfc\,\frac{x}{\sigma\sqrt{2}}\right)$$

The variance for both P_{fa} and P_d is

$$Var = \frac{1}{2}(E_p/N_o)$$

The standard deviation or σ is the square root of the variance:

$$Std\,Dev = \sqrt{\frac{1}{2}\frac{E_p}{N_o}}$$

Substituting the standard deviation into the equation gives

$$F_x(x) = \frac{1}{2}erfc\left(\frac{x}{\sqrt{\frac{E_p}{N_o}}}\right) = P_{fa}$$

Since P_d contains an offset of the actual standard deviation from zero, this offset is included in the equation as follows:

$$F_x(x) = \frac{1}{2}erfc\left(\frac{x}{\sqrt{\frac{E_p}{N_o}}}\right) - \sqrt{\frac{E_p}{N_o}} = P_d$$

P_d is the probability of detecting one pulse at a time. If 10 pulses are required for a system to receive and each of these is independent and has the same probability for detection, then the probability of detecting all 10 pulses is $(P_d)^{10}$ for the 10 independent probabilities for each pulse sent. Error correction is frequently used to provide an increased probability of detection of each of the pulses. In many systems, the probability of missing a pulse needs to be determined and can be calculated by probability theory using the binomial distribution function (BDF).

6.6 Pulsed System Probabilities Using the BDF

The BDF is used to calculate the probability of missing a pulse, as already mentioned. For example, suppose that the probability of detection is 99% for one pulse according to the criteria for probability of detection as described in Section 6.5. If a message consists of 10 pulses, then the probability of getting all 10 pulses is $0.99^{10} = 0.904$, or 90.4%. The probability of missing one pulse is determined by using the binomial theorem as follows:

$$p(9) = \begin{bmatrix}10\\9\end{bmatrix}p^9(1-p)^{(10-9)} = \begin{bmatrix}10\\9\end{bmatrix}(.99)^9(.01)^1 = \frac{10!}{(10-9)!9!}(.99)^9(.01)^1 = 9.1\%$$

Therefore, the percentage of errors that result in only 1 pulse lost out of 10 is

$$Percent\ (one\ pulse\ lost) = 9.1\%/(100\% - 90.4\%) = 95\%.$$

If there are errors in the system, 95% of the time they will be caused by one missing pulse. By the same analysis, the percent of the time that the errors in the system are caused by two pulses missing is 4.3%, and so forth, for each of the possibilities of missing pulses. The total possibilities of errors are summed to equal 100%.

6.7 Error Detection and Correction

Error detection and correction are used to improve the integrity and continuity of service for a system. The integrity is concerned with how reliable the received information is. In other words, if information is received, how accurate are the data, and is it data or noise? Error detection can be used to inform the system that the data are false.

Continuity of service is concerned with how often the data are missed, which disrupts the continuity of the data stream. Error correction is used to fill in the missing or corrupted data to enhance the continuity of service.

Many techniques are used to perform both error detection and error correction. When designing a transceiver, the requirements need to be known as well as the amount of overhead the system can tolerate. Note that error detection and error correction require extra bits to be sent that are not data bits. So the amount of error detection and correction can significantly reduce the data capacity of a system, which is a trade-off in the system design. The extra bits are often referred to as overhead.

6.7.1 Error Detection

There are several different methods of detecting errors in a system and informing the receiver that the information received from the transmitter is in error. If an error is detected, then the receiver can either discard the information or send a request to the transmitter to resend the information.

Three main detection schemes are commonly used for error detection in digital communication systems today: parity, checksum, and cyclic redundancy check (CRC).

6.7.2 Error Detection Using Parity

Parity is used to determine if a message contains errors. The parity bit is attached to the end of a digital message and is used to make the number of "1"s that are in the message either odd or even. For example, if the sent message is 11011 and odd parity is being used, then the parity bit would be a "1" to make the number of "1"s in the message an odd number. The transmitter would then send 110111. If even parity is used, then the parity bit would be a "0" to ensure that the number of "1"s is even, so the sent digital signal would be 110110. If the digital message is 11010, then the sent digital signal would be 110100 for odd parity and 110101 for even parity.

To generate the parity bit, a simple exclusive-OR function of all the bits in the message including an extra bit produces the correct parity bit to be sent. The extra bit equals a "1" for odd parity and a "0" for even parity. For example, if the message is 11011 and the system is using odd parity (extra bit equals 1), then the parity bit would be

$$1 \text{ XOR } 1 \text{ XOR } 0 \text{ XOR } 1 \text{ XOR } 1 \text{ XOR } 1(\text{extra bit}) = 1$$

XOR is an exclusive-OR function in the digital domain. Consequently, the digital signal that is sent for this message would be 110111.

The exclusive-OR function is best described in the form of a truth table. When two bits are XORed, the truth table yields the resultant bit value. An XORed truth table is as follows:

Bit 1		Bit 2	Resultant Bit Value
0	XOR	0	0
0	XOR	1	1
1	XOR	0	1
1	XOR	1	0

The receiver detects an error if the wrong parity is detected. If odd parity is used, all of the received bits are exclusive-ORed, and if the result is a "1" then there were no errors detected. If the result is a "0," then this constitutes an error, known as a parity error. If even parity is used, then a "0" constitutes no errors detected and a "1" means that there is a parity error. For example, using odd parity, the receiver detects 110011 and by exclusive-ORing all the received bits, the output is a "0," which means that the system has a parity error.

One of the major drawbacks to this type of error detection is multiple errors. If more that one error occurs, then this method of error detection does not perform very well. If two errors occur, it would look as if no errors occurred. This is true for all even amounts of error.

6.7.3 Error Detection Using Checksum

Another method of detecting errors in a system is referred to as checksum. As the name connotes, this process sums up the digital data characters as they are being sent. When it has a total, the least significant byte of the sum total is appended to the data stream that is being sent.

The receiver performs the same operation by summing the input digital data characters and then comparing the least significant byte with the received least significant byte. If they are the same, then no error is detected. If they are different, then an error is detected. This is called a checksum error. For example, if the message contained the following characters

$$110, \ 011, \ 111, \ 011, \ 110, \ 001, \ 111, \ 110$$

then the digital sum would be 100, **111**. The least significant byte is **111**, which is attached to the message, so the final digital signal sent is

$$110, 011, 111, 011, 110, 001, 111, 110, \mathbf{111}$$

The receiver would perform the same summing operation with the incoming data without the checksum that was sent. If the least significant byte is 111, then there is no checksum error detected. If it is different, then there would be a checksum error. Note that the receiver knows the length of the transmitted message.

6.7.4 Error Detection Using CRC

One of the best methods for error detection is known as a CRC. This method uses division of the data polynomial $D(x)$ by the generator polynomial $G(x)$, with the remainder being generated by using an XOR function instead of subtraction. The final remainder is truncated to the size of the CRC and is attached to the message for error detection. The data polynomial is multiplied by the number of bits in the CRC code to provide enough place holders for the CRC. In other words, the remainder has to be large enough to contain the size of the CRC.

The polynomial is simply a mathematical expression to calculate the value of a binary bit stream. For example, a bit stream of 10101 would have a polynomial of $X^4 + X^2 + X^0$. The numeric value is calculated by substituting $X = 2$: $2^4 + 2^2 + 2^0 = 21$. Therefore, if there is a "1" in the binary number, the value it represents is included in the polynomial. For example, the first one on the left in the aforementioned digital word has a value of 16 in binary. Since $2^4 = 16$, this is included in the polynomial as X^4.

The receiver performs the same operation as before using the complete message, including the attached error detection bits, and divides by the generator polynomial. If the remainder is zero, then no errors were detected. For example, suppose we have a data polynomial, which is the data to be sent, and a generator polynomial, which is selected for a particular application:

$$D(x) = X^6 + X^4 + X^3 + X^0 \quad 1011001$$
$$G(x) = X^4 + X^3 + X^2 + X^0 \quad 11101$$

To provide the required placeholders for the CRC, $D(x)$ is multiplied by X^4, which extends the data word by four places:

$$X^4(X^6 + X^4 + X^3 + X^0) = X^{10} + X^8 + X^7 + X^4 \quad 10110010000$$

Note, this process is equivalent to adding the number of zeros as placeholders to the data polynomial since the generator polynomial is X^4 (4 zeros). The division using XOR instead of subtraction is shown in Table 6-2a.

Therefore, the CRC would be 0110 and the digital signal that would be transmitted is the data plus the CRC, which equals 10110010110.

The receiver would receive all the data including the CRC and perform the same function as shown in Table 6-2b. Since the remainder is zero, the CRC does not detect any errors. If there is a remainder, then there are errors in the received signal. Some of the standard CRCs used in the industry are listed in Table 6-3.

6.7.5 Error Correction

Error correction detects and also fixes errors or corrects the sent data. One form of error correction is to use redundancy. Even though it is not actually correcting the error that is sent, this method involves sending the data more than once. Another form of error correction, which is really a true form of error correction, is forward error correction (FEC).

Table 6-2a   CRC Generator using division with XOR instead of subtraction

```
              1 1 1 1 1 1 0
      1 1 1 0 1 ⟌1 0 1 1 0 0 1 0 0 0 0
              1 1 1 0 1
              0 1 0 1 1 0
                1 1 1 0 1
                0 1 0 1 1 1
                  1 1 1 0 1
                  0 1 0 1 0 0
                    1 1 1 0 1
                    0 1 0 0 1 0
                      1 1 1 0 1
                      0 1 1 1 1 0
                        1 1 1 0 1
                        0 0 0 1 1 0
                              CRC
```

Table 6-2b   CRC Error Detection using division with XOR instead of subtraction

```
                      1 1 1 1 1 1 0
          1 1 1 0 1 | 1 0 1 1 0 0 1 0 1 1 0
                      1 1 1 0 1
                      0 1 0 1 1 0
                        1 1 1 0 1
                        0 1 0 1 1 1
                          1 1 1 0 1
                          0 1 0 1 0 0
                            1 1 1 0 1
                            0 1 0 0 1 1
                              1 1 1 0 1
                              0 1 1 1 0 1
                                1 1 1 0 1
                                0 0 0 0 0 0
```

No Errors

Table 6-3   A list of common CRC generator polynomials

- CRC-5 USB token packets $G(x) = x^5 + x^2 + x^0$
- CRC-12 Telecom Systems $G(x) = x^{12} + x^{11} + x^3 + x^2 + x^1 + x^0$
- CRC-ANSI $G(x) = x^{16} + x^{15} + x^2 + x^0$
- CRC-CCITT $G(x) = x^{16} + x^{12} + x^5 + x^0$
- IBM-SDLC $= x^{16} + x^{15} + x^{13} + x^7 + x^4 + x^2 + x^1 + x^0$
- IEC TC57 $= x^{16} + x^{14} + x^{11} + x^8 + x^6 + x^5 + x^4 + x^0$
- CRC-32 IEEE Standard 802.3 $G(x) = x^{32} + x^{26} + x^{23} + x^{22} + x^{16} + x^{12} + x^{11} + x^{10} + x^8 + x^7 + x^5 + x^4 + x^2 + x^1 + x^0$
- CRC-32C (Castagnoli) $G(x) = x^{32} + x^{28} + x^{27} + x^{26} + x^{25} + x^{23} + x^{22} + x^{20} + x^{19} + x^{18} + x^{14} + x^{13} + x^{11} + x^{10} + x^9 + x^8 + x^6 + x^0$
- CRC-64-ISO ISO 3309 $G(x) = x^{64} + x^4 + x^3 + x^1 + x^0$
- CRC-64 ECMA-182 $G(x) = x^{64} + x^{62} + x^{57} + x^{55} + x^{54} + x^{53} + x^{52} + x^{47} + x^{46} + x^{45} + x^{40} + x^{39} + x^{38} + x^{37} + x^{35} + x^{33} + x^{32} + x^{31} + x^{29} + x^{27} + x^{24} + x^{23} + x^{22} + x^{21} + x^{19} + x^{17} + x^{13} + x^{12} + x^{10} + x^9 + x^7 + x^4 + x^1 + x^0$
- Others not listed

6.7.6 Error Correction Using Redundancy

Error correction using redundancy means correcting the errors by having the transmitter resend the data. There are basically two concepts for this type of correction.

The first method requires no feedback on the sent message. The transmitter merely sends the data more than once to increase the probability that the message will be received. For example, if a burst signal occurs on the first time the message is sent, then by sending the message again the probability is increased that the receiver will receive the message.

The second method requires feedback from the receiver. The transmitter sends the message to the receiver, and if the receiver detects an error it sends out a request to the transmitter to resend the message. The process is referred to as automatic repeat request (ARQ).

Both of these methods take time, which slows down the data rate for a system, making these rather inefficient means of correction or error recovery. Also, if an error occurs in every message that is transmitted, then these methods are not a viable means of error correction.

6.7.7 Forward Error Correction

FEC not only detects errors but also makes an attempt to correct the error so that the transmitter does not have to send the message again. There are several types of FEC, from very simple methods that can correct one error to very complex methods that can correct multiple errors. A process known as interleaving can improve some error correction methods.

6.7.8 Interleaving

Large interfering signals may occur for a short period of time, or burst, causing several errors for that short period of time. Since most error correcting schemes only correct a small percentage of errors in a data stream, they are unable to correct several errors at one time. To minimize this problem, the data are spread out and interleaved with other data signals so that when a burst occurs, instead of affecting one data message with several errors, only a small number of errors are present in each data message. Therefore, when the received data are deinterleaved, the error correction for each message contains a minimal number of errors to correct.

To illustrate interleaving, we start out with a digital signal using "a" as a placeholder for the bit value in a message. An example would be

a a a a <u>a a a</u> a

Suppose a burst jammer occurred that was wide enough to cause three of these bits to be in error (the underlined bits). If three digital messages needed to be sent, interleaving could be performed to allow only 1 bit error in message "a":

Message a Message b Message c

a a a <u>a a a</u> a a b b b b b b b b c c c c c c c c

Interleaved Messages a, b, c

a b c <u>a b c</u> a b c a b c a b c a b c a b c

The jamming signal still caused three errors at the same point in time; however, it caused only one error per message. The deinterleaving process is used to sort out the bits, and then error correction is used to correct the bits in error for each message:

Received messages after deinterleaving

a <u>a</u> a a a a a a b <u>b</u> b b b b b b c <u>c</u> c c c c c c

Therefore, instead of losing message "a," one error is corrected in each of the messages, and the correct data are obtained. This method of interleaving and deinterleaving the data is used extensively in systems that are susceptible to burst jammers or burst noise.

6.7.9 Types of FEC

Two types of FEC are commonly used in communications, block codes and convolutional codes:

Block codes. These receive k information bits into a block encoder, which is memoryless (does not depend on past events). The block encoder maps these information bits into an n-symbol output block with a rate $R = k/n$ (Figure 6-9a). Block codes are specified as (n,k), where n is the number of bits in the output code word, and k is the number of information bits or data bits in the data word. For example, a block code of (7,4) would have 7 bits in the output code word and 4 bits in the data code word. Thus, each 4-bit information/data code word is mapped into a 7-bit symbol code word output. This means there are 2^k number of distinct 4-bit input messages, which also means that there are 2^k number of distinct 7-bit output code words. The output code is called a block code. The number of bits in the output code word is always greater than the number of bits in the data word: $n > k$.

For long block codes, 2^k with length n is complex and can take up a lot of memory. To reduce the amount of memory needed for these code words, linear block codes are used. They are easier to encode because they store only the generator matrix and do a linear combination of the generator matrix and the data word; in other words, they perform a modulo-2 sum of two linear block code words, which produces another linear block code word (Table 6-4).

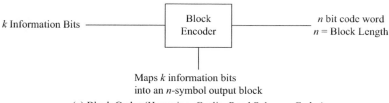

(a) Block Codes (Hamming, Cyclic, Reed Solomon Codes)

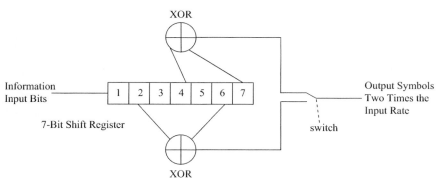

(b) Convolutional Code, Rate ½, Constraint Length 7

Figure 6-9   Block codes and convolutional codes for FEC.

Table 6-4   Linear systematic block code generator

	Generator Matrix	Message	Modulo-2	
Generator Matrix includes Identity Matrix →	(1000 011)	1	(1000011)	⎤ XOR
	(0100 110)	0		⎥
	(0010 111)	1	(0010111)	⎦
	(0001 101)	0		4 bits of data **1010**
		Systematic Code Word =	(1010 100) ←	3 parity bits **100**
Identity Matrix				
Messages		Codewords	(7,4)	
(0000)		(0000000)		
(0001)		(0001101)		
(0010)		(0010111)		
(0011)		(0011010)		
(0100)		(0100110)		
(0101)		(0101011)		
(0110)		(0110001)		
(0111)		(0111100)		
(1000)		(1000011)		
(1001)		(1001110)		
(1010)		**(1010 100)**		
(1011)		(1011001)		
(1100)		(1100101)		
(1101)		(1101000)		
(1110)		(1110010)		
(1111)		(1111111)		

Many codes are systematic and contain the actual information bits plus some parity bits. For a (7,4) linear block code, the code word contains the 4-bit data word plus 3 bits of parity (Table 6-4).

Error detection for linear systematic block codes calculates the syndrome. The syndrome is equal to the received matrix r times the transpose of the parity check matrix H^{-T}:

$$s = r * H^{-T}$$

If the syndrome is equal to zero, then there are no errors. Table 6-5 shows the process used to generate the syndrome.

An example showing the generation of the syndrome with both the no-error case and the case where there is an error occurring in the last bit is shown in Table 6-6. The syndrome is equal to (0,0,0) for the no-error case and is equal to (0,0,1) for the error case. This is using the same (7,4) linear block code as before. Therefore, the error is detected with the syndrome not equal to (0,0,0).

A measure of the ability to distinguish different code words is called the minimum distance. The minimum distance between the code words is the number of bits that are different. To calculate the minimum distance between code words, the code words are modulo-2 added together and the number of "1"s in the resultant is the minimum distance. For example,

$$\text{Code word 1} = 1100101$$
$$\text{Code word 2} = \underline{1000011}$$
$$\text{Modulo} - 2 = 0100110 = \text{three ''1''s}$$
$$\text{Minimum distance} = 3$$

Table 6-5   Syndrome generator for linear systematic block codes

		$\underline{\mathbf{H}}^{-T}$			
s = (s2, s1, s0) =		0 1 1			
(r6, r5, r4, r3, r2, r1, r0)	*	1 1 0		Modulo-2	
		1 1 1	=	s2 =	r5 + r4 + r3 + r2
		1 0 1		s1 =	r6 + r5 + r4 + r1
		1 0 0		s0 =	r6 + r4 + r3 + r0
		0 1 0		s2, s1, s0 = 0,0,0 no errors	
		0 0 1		Example: s2, s1, s0 = 0,0,1	

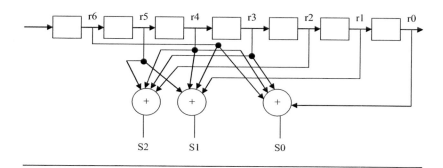

This is also equal to the minimum weight for linear block codes, three "1"s in code word 2. To ensure that the minimum weight is accurate, all code possibilities need to be examined to determine the fewest number of "1"s in a code word.

   The ability for an error detection method to detect errors is based on the minimum distance calculated. The error detection capability, or the number of errors that can be detected, is equal to

$$\text{Number of errors detected} = d_{\min} - 1$$

where $d_{\min}$ is the minimum distance. This example shows that the number of errors that can be detected is equal to

$$d_{\min} - 1 = 3 - 1 = 2$$

Even though the system is able to detect more errors equal to and greater than $d_{\min}$, it cannot detect all of the errors. Thus, undetected errors will pass through the system.

   The ability of an error correction method to correct errors is also based on the minimum distance calculated. The error correction capability, or the number of errors that can be corrected, is equal to

$$\text{Number of errors corrected} = (d_{\min} - 1)/2$$

This example shows that the number of errors that can be corrected is equal to

$$\text{Number of errors corrected} = (d_{\min} - 1)/2 = (3 - 1)/2 = 1$$

Table 6-6   Syndrome generator example, error detection for linear systematic block codes

Generator Matrix	Parity Matrix H		
(1000011)	**(0111100)**		
(0100110)	**(1110010)**		
(0010111)	**(1011001)**		
(0001101)			
Identity Matrix			
Code Word Received	$\mathbf{H}^{-T}$	**Modulo-2**	
(1010100) *	0 1 1	0 1 1	
	1 1 0	0 0 0	
	1 1 1 =	1 1 1	
	1 0 1	0 0 0	
	1 0 0	1 0 0	
	0 1 0	0 0 0	
	0 0 1	0 0 0	
		0 0 0	Equal to zero, no errors
Code Word Received with Error	$\mathbf{H}^{-T}$	**Modulo-2**	
(1010101) *	0 1 1	0 1 1	
	1 1 0	0 0 0	
	1 1 1 =	1 1 1	
	1 0 1	0 0 0	
	1 0 0	1 0 0	
	0 1 0	0 0 0	
	0 0 1	0 0 1	
		0 0 1	Not equal to zero, errors

An error correction method can perform a combination of both error correction and detection simultaneously. For example, if d_{min} is equal to 10, the combined ability of this method could correct three errors and detect six errors or correct four errors and detect five. Error correction and detection codes often specify the minimum distance in their description:

$$(7, 4, 3) = (n, k, d_{min})$$

An example of error correction is shown in Table 6-7. The error vector, which is the received vector that contains an error, is multiplied by the transpose of the parity check matrix H^{-T}. This generates three error equations equivalent to the syndrome calculated in Table 6-6, which is equal to (0,0,1). There are three equations and seven unknowns, so the possible solutions are equal to

$$2^{(7-3)} = 2^4 = 16$$

These equations are solved to find the solution with the most zeros. The solutions with the most zeros are known as coset error vectors. The example shows that if all the e values are set to zero except for $e_0 = 1$, then the coset error vector is equal to 0000001. To correct the

Table 6-7   Error vector and error correction example for linear systematic block codes

Error Vector	H-T				Coset, error vectors
$e_6, e_5, e_4, e_3, e_2, e_1, e_0$ *	0 1 1	0	e_6	e_6	1 0 0 0 0 0 0
	1 1 0	e_5	e_5	0	0 1 0 0 0 0 0
	1 1 1 =	e_4	e_4	e_4	0 0 1 0 0 0 0
	1 0 1	e_3	0	e_3	0 0 0 1 0 0 0
	1 0 0	e_2	0	0	0 0 0 0 1 0 0
	0 1 0	0	e_1	0	0 0 0 0 0 1 0
	0 0 1	0	0	e_0	**0 0 0 0 0 0 1**

$$e_5 + e_4 + e_3 + e_2 = 0$$
$$e_6 + e_5 + e_4 + e_1 = 0$$
$$e_6 + e_4 + e_3 + e_0 = 1$$

0   0   1

3 equations, 7 unknowns $(7 - 3 = 4)$, possible solutions $= 2^4$
Solve for a solution with the most zeros $e_6 = e_5 = e_4 = e_3 = e_2 = e_1 = 0$, $e_0 = 1$
Therefore, the Coset, error vector $= 0000001$ satisfies the three above equations with most zero
The correct code sent $= 1010100$
$$Vc = r + e = 1010101 + \mathbf{0000001} = 1010100 \text{ Corrected bit in code word}$$

code error, the received code is modulo-2 added to the coset error vector to correct the error:

$$\text{Received code} = 1010101$$
$$\text{Coset error vector} = \underline{0000001}$$
$$\text{Correct code} = 1010100$$

The probability of receiving an undetected error for block codes is calculated with the following equation:

$$P_u(E) = \sum_{i=1}^{n} A_i p^i (1 - p)^{n-i}$$

where

     $p = $ probability of error
     $i = $ number of "1"s in the code word
     $A_i = $ weight distribution of the codes (number of code words with the number of "1"s specified by i)

An example is shown in Table 6-8 for a (7,4) block code. Note that there is one code word with no "1"s, one code word with seven "1"s, seven code words with three "1"s, seven code words with four "1"s, and no code words with one, two, five, and six "1"s. Therefore, this example shows that the probability of an undetected error is equal to

$$P_u(E) = 7p^3(1 - p)^4 + 7p^4(1 - p)^3 + p^7$$

If the probability of error equals 10^{-2}, then the probability of an undetected error is

$$P_u(E) = 7(10^{-2})^3(1 - (10^{-2}))^4 + 7(10^{-2})^4(1 - (10^{-2}))^3 + (10^{-2})^7 = 6.8 \times 10^{-6}$$

Table 6-8 Probability of undetected error performance of block codes

$$P_u(E) = \sum_{i=1}^{n} A_i p^i (1-p)^{n-i}$$

Where A_i is the weight distribution of the code as shown below for a 7,4 code

Code words		
(0000000)	$i = $ # of '1's, $n = 7$ #bits of code	
(0001101)	A_i is the number of code words	
(0010111)	with the same number of '1's	
(0011010)	$A_0 = 1$	Not used, no ones
(0100110)	$A_1 = 0$	
(0101011)	$A_2 = 0$	
(0110001)	$A_3 = 7$	7 codes have 3 ones
(0111100)	$A_4 = 7$	7 codes have 4 ones
(1000011)	$A_5 = 0$	
(1001110)	$A_6 = 0$	
(1010100)	$A_7 = 1$	1 code has 7 ones
(1011001)		
(1100101)		
(1101000)		
(1110010)		
(1111111)		

If the probability of error equals 10^{-6}, then the probability of an undetected error is

$$P_u(E) = 7(10^{-6})^3(1-(10^{-6}))^4 + 7(10^{-6})^4(1-(10^{-6}))^3 + (10^{-6})^7 = 7 \times 10^{-18}$$

The Hamming code is another popular block code. The number of bits required for the Hamming code is

$$2^n \geq m + n + 1$$

where

$n = $ number of Hamming bits required
$m = $ number of data bits

Hamming bits are randomly place in the data message. For example, suppose a digital data stream is 1011000100010, which is 13 bits long. The number of Hamming bits required is

$$2^5 \geq 13 + n + 1$$

where $n = 5$. The Hamming bits slots are placed randomly in the data message as follows:

17   15   14   8   3

H 1 0 1 1 H 0 0 H 0 1 0 H 0 0 1 H 0

The bit values of the Hamming bits are calculated by exclusive-ORing the bit positions that contain a "1," which are 3, 8, 14, 15, 17:

$$00011 \ 3$$
$$01000 \ 8$$
$$01110 \ 14$$
$$01111 \ 15$$
$$\underline{10001} \ 17$$
$$\textbf{11011} \ \text{Hamming Code}$$

Consequently, the signal that would be sent out of the transmitter would be

H   H H   H   H

1101**1**1000**0**10**1**00**1**10

If an error occurs at bit position 15, which means that the "1" value at the transmitted signal at bit 15 is detected as a "0" in the receiver, then the following bit stream would be received:

1100<u>0</u>11000010100110

To detect and correct this error, at the receiver, all bits containing a "1" are exclusive-ORed:

$$00010 \ 2$$
$$00011 \ 3$$
$$00110 \ 6$$
$$01000 \ 8$$
$$01101 \ 13$$
$$01110 \ 14$$
$$10001 \ 17$$
$$\underline{10010} \ 18$$
$$01111 \ 15$$

The result, $01111 = 15$, is equal to the bit position where the error occurred. To correct the error, bit 15 is changed from a "0" to a "1."

The Hamming block code corrects only one error. Other higher level codes, such as the Reed-Solomon code, has the ability to correct multiple errors. Some other types of codes used for error correction include:

- Hsiao code. Uses a shortened Hamming code.
- Reed-Muller code. Used for multiple error corrections.
- Golay code. Used in space programs for data communications.

In addition, a very popular communications code is the cyclic code, which shifts one code word to produce another. This is a subset of linear block codes that uses linear sequential circuits such as shift registers with feedback to generate the codes. Since they are serially generated, they are able to be decoded serially. Reed-Solomon code is generated by cyclic codes and is used extensively for communications. Bose-Chaudhuri-Hocquenghem (BCH) code is another good example of a cyclic code. Even Hamming and Golay codes can be made to be cyclic codes.

Sometimes shortened codes are used. This is where the data are zero padded (make some data symbols zero at the encoder); they are not transmitted, but they are reinserted at the decoder.

Error correction and detection methods improve the system's ability to detect and correct errors to ensure more reliable communications. However, the cost of doing this is either more bandwidth to send more bits at a higher rate or a decrease in data throughput to send more bits at the same rate. These additional bits required for error correction and detection are often referred to as overhead. Along with the overhead concerns with error detection and correction, another caution is that for high error rates (10^{-2}) or more error correction may actually degrade performance.

Convolutional codes. These codes are also commonly used because of their simplicity and high performance. They are usually generated using a pseudo-noise (PN) code generator with different taps selected for the XOR functions and a clock rate that selects output symbols generally at a higher rate than the input symbols. For example, suppose the output rate is equal to twice the speed of the input information. This would produce a rate as follows:

$$R = \text{Input information bit rate/output symbol clock rate} = \frac{1}{2}$$

Also, for convolutional codes, the size of the shift register is called the constraint length. Therefore, if the shift register is seven delays long, then this example would be a convolutional code with rate ½ and constraint length 7 (Figure 6-9b). Many different rates are possible, but some of the more common ones are ½, ¾, and 7/8.

The more common methods of decoding these coding schemes use maximum likelihood sequence (MLS) estimation and the Viterbi algorithm, a modification of the standard MLS decoding scheme.

The MLS performs the following three basic operations:

1. It calculates the distances (the amount of difference, generally minimizes bit errors) between the symbol sequence and each possible code sequence.
2. It selects the code sequence that has the minimum distance calculated as being the correct code sequence.
3. It outputs this sequence as the MLS estimation of the received sequence.

Therefore, to retrieve the data information that was sent, the receiver performs the deinterleaving process and sorts out the data bits in the correct messages. Then a decoder, such as a Viterbi decoder, is used to perform the necessary error correction to receive reliable data.

6.7.10 Viterbi Decoder

A Viterbi decoder is a real-time decoder for very high-speed short codes, with a constraint length of less than 10. It is a major simplification of the MLS estimator. The basic concept is that it makes real-time decisions at each node as the bits progress through the algorithm. Only two paths reach a node in the trellis diagram, and the Viterbi algorithm makes a decision about which of the paths is more likely to be correct (Figure 6-10). Only the closest path to the received sequence needs to be retained.

Multi-h is a technique to improve detection of a continuous phase modulation system by changing the modulation index. For example, if the modulation index changes between ¼ and ½, often specified by {1,2/4}, then the modulation changes between $\pi/4$ and $\pi/2$.

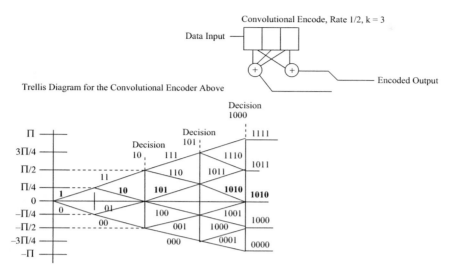

Figure 6-10   Viterbi decoder trellis diagram.

This means that the modulation changes from a $\pi/4$ shift to a $\pi/2$ shift, which improves the detection process by extending the ambiguity of the nodes of the trellis diagram. Therefore, the first merge point is twice as far on the multi-h trellis diagram (Figure 6-11). Extending the ambiguity points farther in the trellis diagram lessens the number of errors that can occur or the number of bit decisions.

The multi-h method improves spectral efficiency by providing more data throughput for a given bandwidth. The disadvantage is that the system becomes more complex. In addition, it has constraints that require

$$q \geq M^k$$

where

  $q =$ common denominator of modulation index (4 in the example)
  $M =$ number of phase states (binary in the example)
  $K =$ number of modulation states (2 in the example)

This example shows this meets the constraint for a good use of multi-h: $4 = 2^2$.

6.7.11 Turbo and Low-Density Parity Check Codes

Two other FEC codes used in data link communications are *turbo codes* and *low-density parity check* (LDPC) *codes*. Turbo codes were invented in 1993 by Claude Berrou and Alain Glavieux. They use two or more parallel convolutional encoders/decoders. These codes claim that they are approximately 0.5 dB away from Shannon's limit. They use parallel-concatenated convolutional codes (PCCCs). The disadvantage of turbo codes is the decoding delay, since iterations take time, and this may affect real-time voice, hard disk storage, and optical transmissions.

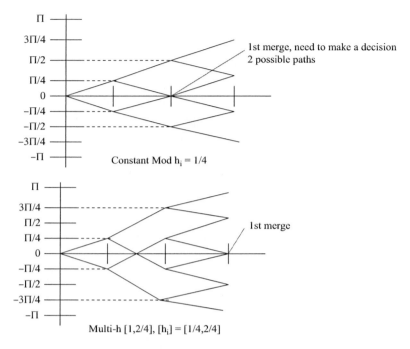

Figure 6-11   Trellis diagram for multi-h.

Turbo codes use a recursive systematic code (RSC). It is recursive because it uses feedback of the outputs back into the input. It is systematic because one of the outputs is exactly the same as the input. These two properties provide better performance over non-recursive, nonsystematic codes. The term turbo refers to turbo engines, which use exhaust to force air into the engine to boost combustion, similar to the function of the code, where it provides feedback from the output of the decoder to the input of the decoder.

The transmitted code word structure is composed of the input bits, parity bits from the first encoder using the input bits, and parity bits from the second encoder using interleaved input bits (Figure 6-12). The decoders use log-likelihood ratios as integers with reliability estimate numbers of -127 (0) to $+127$ (1). It uses an iterative process using both decoders in parallel.

A technique known as puncturing is also used. This process deletes some bits from the code word according to a puncturing matrix to increase the rate of the code. For example, puncturing can change the rate from ½ to ⅔ while still using the same decoder for these different rates. However, systematic bits (input bits) in general are not punctured. Puncturing provides a trade-off between code rate and performance.

Low-density parity check codes outperform turbo codes, with claims that the performance is approaching 0.1 dB away from Shannon's limit. The advantage of these codes is that the patents have expired so they are free for use. A possible disadvantage is high encoding complexity. These codes work to produce convolutional codes derived from quasi-cyclic block codes of LDPC.

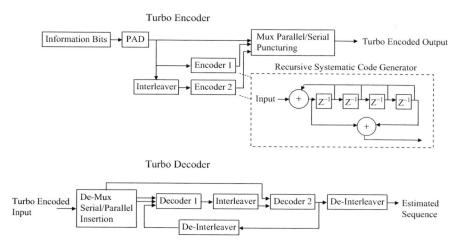

Figure 6-12   Basic turbo code diagram.

6.8 Theory of Pulse Systems

The easiest way to explain what is happening with a spread spectrum waveform is to look at a simple square wave. A square wave has a 50% duty cycle and is shown in Figure 6-13. This is a time domain representation of a square wave because it shows the amplitude of the signal with respect to time, which can be observed on an oscilloscope. The frequency domain representation of a square wave shows the amplitude (usually in dB) as a function of frequency, as shown in Figure 6-13. The frequency spectrum of a square wave contains a fundamental frequency, both positive and negative, and the harmonics associated with this fundamental frequency. The negative frequencies do not exist; they are there to help in analyzing the signal when it is upconverted. The fundamental frequency is the frequency of the square wave. If the corners of the square wave were smoothed to form a sine wave, it would be the fundamental frequency in the time domain. The harmonics create the sharp corners of the square wave. The sum of all these frequencies is usually represented by a Fourier series. The Fourier series contains all the frequency components with their associated amplitudes.

In a square wave, even harmonics are suppressed and the amplitudes of the odd harmonics form a sinc function, as shown in Figure 6-13. The inverse of the pulse width is where the nulls of the main lobe are located. For a square wave, the nulls are located right at the second harmonic, which is between the fundamental frequency and the third harmonic. Note that all the other nulls occur at the even harmonics, since the even harmonics are suppressed with a square wave (Figure 6-13).

If the pulse stream is not a square wave (50% duty cycle), then these frequency components move around in frequency. If the pulse width is unchanged and the duty cycle is changed, then only the frequency components are shifted around and the sinc function is unchanged. Also, the suppressed frequency components are different with different amplitude levels. For example, if the duty cycle is changed to 25%, then every other even harmonic (4, 8, 12) is suppressed (Figure 6-14).

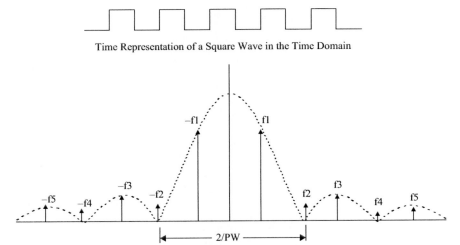

Time Representation of a Square Wave in the Time Domain

Frequency Spectrum of a Square Wave in the Frequency Domain—Mag Sinx/x

Figure 6-13 Time and frequency domain representations of a pulsed signal with 50% duty cycle.

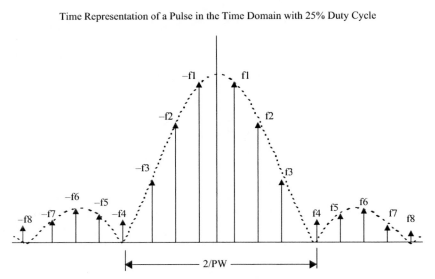

Time Representation of a Pulse in the Time Domain with 25% Duty Cycle

Frequency Spectrum of a Pulse in the Frequency Domain with 25% Duty Cycle—Mag Sinx/x

Figure 6-14 Time and frequency domain representations of a pulsed signal with 25% duty cycle.

Note that not all the even harmonics are not suppressed as they were for the 50% duty cycle pulse because they are not located in the nulls of the sinc pattern. Also, there are more and more frequencies in the main lobe as the duty cycle gets smaller, and if the pulse width does not change then the nulls will always be at the same place, at the inverse of the pulse width. If the pulse width changes, then the position of the nulls will change.

6.9 PN Code

To obtain an intuitive feel for what the spectrum of a PN coded signal is doing on a real-time basis, as the code is changing the pulse widths (variable number of "1"s in a row) and the duty cycle change (variable number of "−1"s in a row). Variations in the pulse widths cause the nulls of the sinc function to change (will not be any larger than 1/chip width). In addition, variations in the duty cycle cause the number of frequency components inside the sinc function to change. This results in a more continuous sinc function spectrum in the frequency domain, with the null of the sinc function being 1/chip width (Figure 6-15).

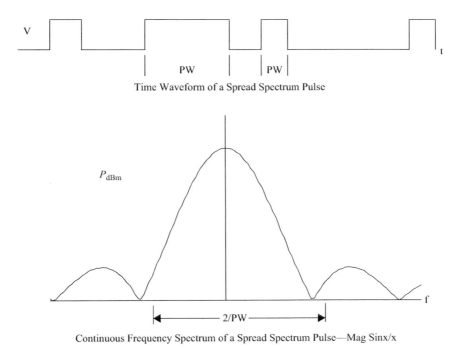

Time Waveform of a Spread Spectrum Pulse

Continuous Frequency Spectrum of a Spread Spectrum Pulse—Mag Sinx/x

Figure 6-15   Time and frequency domain representation of a spread spectrum continuous waveform.

6.10 Summary

A simple approach to understanding probability theory and the Gaussian process is provided to allow the designer to better understand the principles of designing and evaluating digital transmission. Quantization and sampling errors in the analysis and design of digital systems were discussed.

The probability of error, probability of detection, and probability of false alarms are the keys in determining the performance of the receiver. These errors are dependent on the received SNR and the type of modulation used. Error detection and correction were discussed, and several approaches were examined to mitigate errors. Theory on pulsed systems, showing time and frequency domain plots, provide knowledge and insight into the design and optimization of digital transceivers.

References

Haykin, S. *Communication Systems*, 5 ed. Wiley, 2009.

Lin, S., and Daniel J. Costello. *Error Control Coding*, 2nd ed. Prentice Hall, 2004.

Ross, S. *A First Course in Probability*, 8th ed. Pearson, 2009.

Problems

1. Using the fact that the integral of the probability density function is equal to 1, what is the percent chance of getting this problem wrong if the chance of getting it right is 37%?
2. What is the expected value of x if $f_x(x) = 0.4$ when $x = 1$ and $f_x(x) = 0.6$ when $x = 2$? What is the mean?
3. Is the answer in problem 2 closer to 1 or 2, and why?
4. What is the $E[x^2]$ in problem 2?
5. What is the variance in problem 2?
6. What is the standard deviation of problem 2?
7. What is the probability the signal will not fall into a 2σ value with a Gaussian distribution?
8. Given a system with too much quantization error, name a design change that can reduce the error.
9. What is the probability of receiving 20 pulses if the probability of detection is 0.98 for one pulse?
10. What is the probability that the error occurs because of one lost pulse in problem 9?
11. What is the difference between probability of error and BER?
12. What are the three basic types of error detection? Which is the best type?
13. What are the two basic types of error correction?
14. How does interleaving maximize the error correction process?

CHAPTER 7

Multipath

Multipath is a free-space signal transmission path that is different from the desired, or direct, free-space signal transmission path used in communications and radar applications. The amplitude, phase, and angle of arrival of the multipath signal interfere with the amplitude, phase, and angle of arrival of the desired or direct path signal. This interference can create errors in angle of arrival information and in received signal amplitude in data link communications. The amplitude can be larger or smaller depending on whether the multipath signals create constructive or destructive interference. Constructive interference is when the desired signal and the multipath signals are more in-phase and add amplitude. Destructive interference is when the desired signal and the multipath signals are more out of phase and subtract amplitude.

Angle of arrival errors are called glint errors. Amplitude fluctuations are called scintillation or fading errors. Therefore, the angle of arrival, the amplitude, and the phase of the multipath signal are all critical parameters to consider when analyzing the effects of multipath signals in digital communications receivers. Frequency diversity and spread spectrum systems contain a degree of immunity from multipath effects since these effects vary with frequency. For example, one frequency component for a given range and angle may have multipath that severely distorts the desired signal, whereas another frequency may have little effect. This is mainly due to the difference in the wavelength of the different frequencies.

7.1 Basic Types of Multipath

Multipath reflections can be separated into two types of reflections—specular and diffuse—and are generally a combination of these reflections. Specular multipath is a coherent reflection, which means that the phase of the reflected path is relatively constant with relation to the phase of the direct path signal. This type of reflection usually causes the greatest distortion of the direct path signal because most of the signal is reflected toward the receiver. Diffuse multipath reflections are noncoherent with respect to the direct path signal. The diffuse multipath causes less distortion than the specular type of multipath because it reflects less energy toward the receiver and usually has a noise-like response due to the random dispersion of the reflection. Both types of multipath can cause distortion to a system, which increases the error of the received signal and reduces coverage according to the link budget. Multipath effects are included as losses in the link budget (see Chapter 1). Specular reflection is analyzed for both a reflection off a smooth surface and a rough surface.

7.2 Specular Reflection on a Smooth Surface

Specular reflections actually occur over an area of the reflecting surface (which is defined as the first Fresnel zone, similar to the Fresnel zones found in optics). Most of the time the reflecting area is neglected and geometric rays are used. These rays obey the laws of geometrical optics, where the angle of incidence is equal to the angle of reflection, as shown in Figure 7-1.

For a strictly specular reflection (referred to as a smooth reflection), the reflection coefficient (ρ_o) depends on the grazing angle, the properties of the reflecting surface, and the polarization of the incident radiation. ρ_o assumes a perfectly smooth surface. This reflection coefficient is complex. A smooth reflection has only one path directed toward the receiver, as shown in Figure 7-1.

The magnitude of the "smooth" reflection coefficient (ρ_o) is plotted in Figure 7-2 as a function of the reflection coefficient and the grazing angle, with plots showing both horizontal and vertical polarization effects of the incident radiation. The graph shows that for vertical polarization near the pseudo-Brewster angle the reflection coefficient is very small, a phenomenon that is also observed in optics. This phenomenon can be used to minimize multipath reflections. For horizontal polarization, the reflection coefficient is fairly constant but starts dropping off as the grazing angle approaches 90°. Not only does the magnitude of the reflected radiation change, but the phase of the reflected radiation is also modified. A phase shift of the incident radiation can vary from near 0° to 180°, depending on the polarization of the incident radiation and the conductivity and dielectric constant of the reflecting surface.

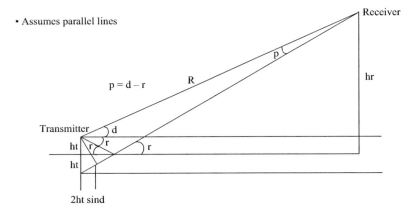

Specular analysis, angle of incident = angle of reflection.

d = arcsin[(hr − ht)/R]

r = arcsin[(hr + ht)/(R + 2ht sind)]

Figure 7-1   Single ray specular multipath analysis.

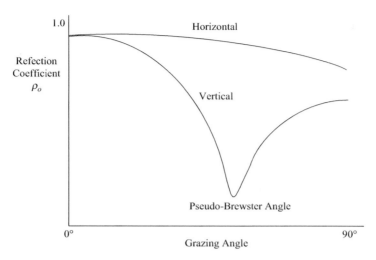

Figure 7-2   Reflection coefficient versus grazing angle for different antenna polarizations.

7.3 Specular Reflection on a Rough Surface

Generally the ideal reflecting case must be modified because reflecting surfaces (usually the earth's surface) are not perfectly smooth. The roughness of the reflecting surface decreases the amplitude of the reflection coefficient by scattering energy in directions other than the direction of the receiving antenna. To account for the loss in received energy, a scattering coefficient (ρ_s) is used to modify the smooth reflection coefficient. The smooth reflection coefficient is multiplied by the scattering coefficient, creating an overall modified reflection coefficient to describe a specular surface reflection on a rough surface. ρ_s is the root mean square (RMS) value. This coefficient can be defined as a power scattering coefficient:

$$\overline{\rho}_s^{\,2} = e^{[-(\frac{4\pi h \sin d}{\lambda})]^2}$$

where

   ρ_s = power scattering coefficient or reflection coefficient for a rough specular surface
   h = RMS height variation (normally distributed)
   d = grazing angle
   λ = wavelength

The roughness of the reflecting surface modifies the smooth reflection coefficient and produces a noncoherent, diffuse reflection. If the variations are not too great, then a specular analysis is still possible and ρ_s will be the scattering modifier to the smooth specular coefficient (ρ_o). This analysis can be used if the reflecting surface is smooth compared with a wavelength. The Rayleigh criterion is used to determine if the diffuse multipath can be neglected. It looks at the ratio of the wavelength of the radiation and the height variation of the surface roughness and compares it to the grazing angle or incident angle. The Rayleigh criterion is defined as follows:

$$h_d \sin d < \lambda/8$$

where

h_d = peak variation in the height of the surface
d = grazing angle
λ = wavelength

If the Rayleigh criterion is met, then the multipath is a specular reflection on a rough surface, and ρ_o and ρ_s are used to determine the multipath. If the Rayleigh criterion is not met, then the diffuse multipath coefficient ρ_d must be taken into account for complete multipath analysis.

7.4 Diffuse Reflection

Diffuse multipath has noncoherent reflections and is reflected from all or part of an area known as the *glistening surface* (Figure 7-3). It is also called this because in the optical (visible) equivalent the surface can sparkle when diffuse reflections are present. The boundaries of the glistening surface are given by

$$y = +/- \frac{X_1 X_2}{X_1 + X_2} \left(\frac{h_r}{X_1} + \frac{h_t}{X_2} \right) \sqrt{\beta_0{}^2 - \frac{1}{4} \left(\frac{h_r}{X_1} - \frac{h_t}{X_2} \right)^2}$$

where

β_0 = reflection angle
h_r = height of receiver antenna
h_t = height of transmitter antenna
X_1 = distance from the transmitter antenna base to the point of reflection
X_2 = distance from the receiver antenna base to the point of reflection

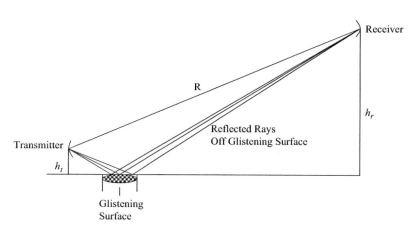

Figure 7-3   Multipath over a glistening surface.

Diffuse reflections have random amplitudes and phases with respect to the amplitude and phase of the direct path. The amplitude variations follow a Rayleigh distribution, while the phase variations are uniformly distributed from 0° to 360°. The phase distribution has zero mean (i.e., diffuse multipath appears as a type of noise to the receiver).

For diffuse reflection, a diffuse scattering coefficient is multiplied by ρ_o to obtain the diffuse reflection coefficient. Calculating the diffuse scattering coefficient (ρ_d) is more tedious than calculating the specular scattering coefficient. The area that scatters the diffuse multipath cannot be neglected, as was done in the specular multipath case. One procedure to determine ρ_d is to break up the glistening area into little squares, each square a reflecting surface, and then calculate a small (δ) diffuse scattering coefficient. The scattering coefficients are summed up and the mean is calculated, which gives the overall diffuse scattering coefficient (ρ_d). This can be converted to a power coefficient:

$$\rho_d^2 \cong \frac{1}{4\pi\beta_0^2} \int \frac{R^2 dS}{X_1^2 X_2^2} = \frac{1}{2\pi\beta_0^2} \int \frac{R^2 y dX}{(R-X)^2 X^2}$$

where

$R = $ range
$X = $ ground range coordinates

This is for a low-angle case. This diffuse scattering coefficient is derived based on the assumption that the reflecting surface is sufficiently rough that there is no specular multipath component present. The roughness factor has been defined as F_d^2 and modifies the diffuse scattering coefficient (ρ_d):

$$F_d^2 = 1 - \rho_s^2$$

A plot of the specular and diffuse scattering coefficients with respect to roughness criteria is shown in Figure 7-4. With a roughness factor greater than 0.12, the reflections become predominantly diffuse. The diffuse reflection coefficient amplitude reaches approximately 0.4, which is 40% of the amplitude of the smooth reflection coefficient, whereas the specular coefficient reaches 1.0, or 100% of the smooth reflection coefficient. Precautions need to be taken when using the plots in Figure 7-4 since ρ_d does not include the roughness factor F_d^2 and may cause significant errors in the analysis of actual values.

7.5 Curvature of the Earth

For most communications systems other than satellite links, the divergence factor D caused by the curvature of the earth can be taken as unity. However, if a particular scenario requires the calculation of the divergence factor, it is equal to

$$D = \lim_{f \to 0} \sqrt{\frac{A_r}{A_f}}$$

where

$A_r = $ area projected due to a round earth
$A_f = $ area projected due to a flat earth

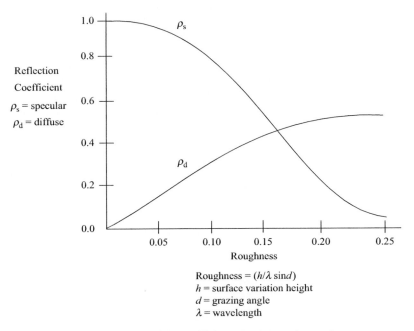

Figure 7-4 Reflection coefficient versus roughness factor.

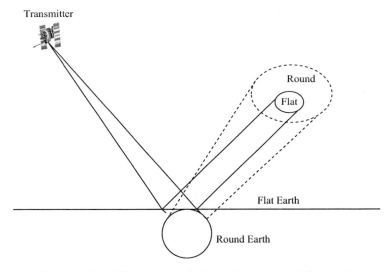

Figure 7-5 Different areas for both the round and flat earth.

A diagram showing the different areas that are projected for both the round and flat earth are shown in Figure 7-5. The curvature of the earth produces a wider area of reflection, as shown in the figure. The ratio of the areas is used to calculate the divergence factor. Note that for most systems the area difference is very small and generally can be neglected.

7.6 Pulse Systems (Radar)

Multipath can affect radar systems by interfering with the desired signal. Leading edge tracking can be used to eliminate a large portion of the distortion caused by multipath because the multipath arrives later in time than the direct path. In leading edge tracking, only the first portion of the pulse is processed, and the rest of the pulse, which is distorted due to multipath, is ignored. If the grazing angle is very small, then the time of arrival (TOA) for both direct path and multipath signals is about the same. For this case, the angle error between the direct path and multipath is very small. If the multipath is diffuse, then the multipath signal will appear as an increase in the noise floor of the direct path and a slight angle variation will occur due to the multipath mean value coming at a slightly different angle. If the multipath is specular, then the multipath signal will increase or decrease the actual amplitude of the direct path signal and also create a slight angle error. If one or more of the antennas are moving, then the specular reflection will also be a changing variable with respect to the phase of the multipath signal and the direct path signal, thus producing a changing amplitude at the receiver. As the grazing angle increases, the angle error also increases. However, the TOA increases, which means that leading edge tracking becomes more effective with larger grazing angles.

One other consideration is that the pulse repetition frequency (PRF) of the radar must be low enough so that the time delay of the multipath is shorter than the time between the pulses transmitted. This prevents the multipath return from interfering with the next transmitted pulse.

Another consideration is the scenario where a surface such as a building or other smooth surface gives rise to specular reflection with a high reflection coefficient where the multipath could become fairly large. For a stationary situation, this is a real problem. However, if one of the antennas is moving, then the multipath is hindered, depending on the velocity of the antenna and the area of the reflector, for only the time the angle of reflection is right. The processor for the receiver could do an averaging of the signal over time and eliminate some of these problems.

7.7 Vector Analysis Approach

Different approaches can be taken to calculate the resultant effects of the reflected energy. One method is vector addition. The reference vector is the direct path vector with a given amplitude and a zero reference angle. The coherent reflection vector, C, for specular reflection is now calculated with the phase referenced to the direct path vector:

$$\overline{C} = \overline{D}\rho_0\rho_s e^{j2\pi\frac{dR}{\lambda}}$$

where

    D = divergence factor
    ρ_0 = specular reflection coefficient
    ρ_s = scattering coefficient
    dR = path length difference
    λ = wavelength

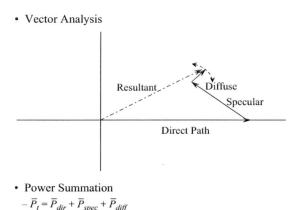

• Power Summation

$$-\bar{P}_t = \bar{P}_{dir} + \bar{P}_{spec} + \bar{P}_{diff}$$

Figure 7-6   Different approaches to analyzing multipath.

Determining an accurate vector for ρ_s is difficult. Note that ρ_s represents an average amplitude distributed value that is a function of the distribution of the surface.

The noncoherent vector (I) for diffuse reflection is calculated as

$$\bar{I} = \frac{\overline{D}\rho_0\overline{\rho_d}}{\sqrt{2}}(I_1 + jI_2)e^{j2\pi\frac{dR}{\lambda}}$$

where

$$dR = \text{path length difference}$$
$$D = \text{divergence factor}$$
$$I_1 + I_2 = \text{independent zero mean unit variance normalized Gaussian processes}$$

The vectors C and I are summed with the reference phasor to produce the resultant phasor, as shown in Figure 7-6.

Designing an accurate model for ρ_d is difficult. Observed test data from known systems can aid in the selection and verification of the scattering coefficients ρ_s and ρ_d.

7.8 Power Summation Approach

Another approach for determining the error caused by reflections is to sum the powers of each type of reflection with the direct power:

$$P_t = P_{dir} + P_{spec} + P_{diff}$$

where

$$P_{dir} = \text{direct power}$$
$$P_{spec} = \text{mean specular power}$$
$$P_{diff} = \text{mean diffuse power}$$

This approach uses power addition instead of vector addition. Mean power reflection coefficients are used. The resultant mean specular power is

$$\overline{P}_{diff} = \overline{P}_D \rho_0^2 \rho_s^2$$

where

P_D = power applied to the surface

The mean diffuse power is equal to

$$\overline{P}_{diff} = \overline{P}_D \rho_0^2 \rho_d^2$$

A divergence factor can be included to account for the curvature of the earth.

7.9 Multipath Mitigation Techniques

Some of the techniques used to reduce and mitigate multipath are the following:

- Leading edge tracking for radars. The multipath is delayed with respect to the direct path; therefore, the leading edge of the waveform contains the signal and not the multipath. This assumes a low PRF to prevent long multipath from affecting the next radar pulse. However, the longer the multipath, the smaller the signal amplitude, which will interfere less with the desired signal.
- Movement changes the specular multipath to provide an average error. Multipath is position sensitive, so movement will change the multipath nulls.
- Antenna design to prevent low-angle multipath. This technique is used for global positioning systems (GPS), where the desire is to have the mask angle or angle to the earth low but at the same time prevent multipath signals that are reflecting off the earth.
- Spread spectrum. Nulls caused by multipath are frequency dependent, so with a spread spectrum system spread across a wide band of frequencies, the multipath only affects a portion of the spectrum.
- Antenna diversity. Since multipath is dependent on position, antennas can be located at different positions so that as one antenna is in a multipath null the other antenna is not.

7.9.1 Antenna Diversity

Antenna diversity is a technique to help mitigate multipath. It requires the data link to have two or more antennas at the receiver. The signals in these different paths are selected to provide the best signal that is not interfered with by multipath. The theory is that when one antenna is in a multipath null, the other is not. This works very well with only two antennas with the antennas separated a short distance to prevent both antennas from being in the null (see Figure 7-7). This technique is used in many applications in wireless communications.

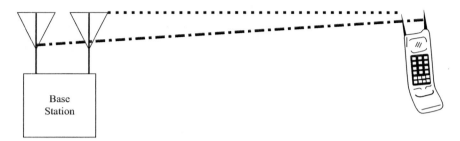

Figure 7-7   Antenna diversity to help mitigate multipath.

7.10 Summary

Multipath affects the desired signal by distorting both the phase and the amplitude. This can result in a lost signal or a distortion in the TOA of the desired signal. Multipath is divided into two categories: specular and diffuse. Specular multipath generally affects the system the most, resulting in more errors. Diffuse multipath is more noise-like and is generally much lower in power. The Rayleigh criterion is used to determine if the diffuse multipath needs to be included in the analysis. The curvature of the earth can affect the analysis for very long-distance multipath. One of the ways to reduce the effects of multipath is to use leading edge tracking so that most of the multipath is ignored. Some approaches for determining multipath effects include vector analysis and power summation. Several methods of multipath mitigation were discussed, including using multiple antennas for antenna diversity.

References

Barton, D. K. *Modern Radar System Analysis*. Artech, 1988.

Beckman, P., and A. Spizzichino. *The Scattering of Electromagnetic Waves from Rough Surfaces*. Pergamon Press, 1963.

Bullock, S. "Use Geometry to Analyze Multipath Signals." *Microwaves and RF*, July 1993.

Munjal, R. L. "Comparison of Various Multipath Scattering Models." Johns Hopkins University, December 8, 1986.

Skolnik, M. *Radar Handbook*, 3d ed. McGraw-Hill, 2008.

Problems

1. What is the difference between glint errors and scintillation errors?
2. Which type of multipath affects the solution the most? Why?
3. What is the effect of multipath on an incoming signal if the signal is vertically polarized at the pseudo-Brewster angle?

4. What is the effect of multipath on an incoming signal if the signal is horizontally polarized at the pseudo-Brewster angle?

5. What is the criterion for determining which type of multipath is present?

6. According to the Rayleigh criterion, what is the minimum frequency at which the multipath will still be considered specular for a peak height variation of 10 m and a grazing angle of 10°?

7. What is the divergence factor, and how does it affect the multipath analysis?

8. How do most radars minimize multipath effects on the radar pulses?

9. Graphically show the resultant vector for a reference signal vector with a magnitude of 3 at an angle of 10° and the smooth specular multipath signal vector with a reflection coefficient of 0.5 at an angle of 180° with respect to the signal vector.

10. What is the main difference between the vector summation approach to multipath analysis and the power summation approach? How does this approach affect the reflection coefficient?

11. How does antenna diversity help mitigate multipath?

CHAPTER 8

Improving the System against Jammers

The receiver is open to reception of not only the desired signal but also all interfering signals within the receiver's bandwidth, which can prevent the receiver from processing the desired signal properly (Figure 8-1). Therefore, it is crucial for the receiver to have the ability to eliminate or reduce the effects of the interfering signals or jammers on the desired signal.

This chapter discusses in detail three solutions to reduce the effects of jammers: a method to protect the system against pulse or burst jammers; an adaptive filter to reduce narrowband jammers such a continuous wave (CW); and a jammer reduction technique called a Gram-Schmidt orthogonalizer (GSO). Other techniques to reduce the effects of jammers are antenna siting, which mounts the antenna away from structures that cause reflections and potential jammers, and the actual antenna design to prevent potential jammers from interfering with the desired signal.

In addition, in some systems, the ability for another receiver to detect the transmitted signal is important. These types of receivers are known as intercept receivers. A discussion is presented on the various types of intercept receivers, and the advantages and disadvantages of each are evaluated.

8.1 Burst Jammer

One of the best jammers for direct sequence spread spectrum modulation is a burst jammer. A burst jammer is generally a high-amplitude narrowband signal relative to the desired signal and is present for a short period of time. Typical bursts range from a 0 dB to 40 dB signal-to-jammer ratio (SJR), with a duration of 0.5 μsec to 1000 μsec. An example of a burst jammer is shown in Figure 8-2.

The burst jammer affects the receiver as follows:

- The high amplitude of the burst saturates the automatic gain control (AGC) amplifiers, detectors, and the processor. The information is lost during the burst time and also during the recovery time of each of the devices.
- The burst can capture the AGC. With the burst present, the AGC voltage slowly increases, which reduces the gain of the amplifier. When the burst is gone, the amplifier gain is small, and the signal is lost until the AGC has time to respond, at which time the burst comes on again.

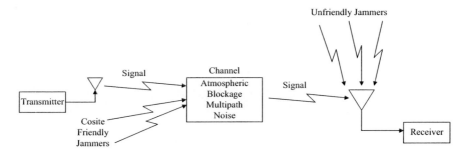

Figure 8-1   The receiver accepts both the desired signal and jammers.

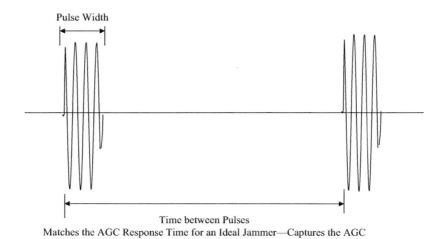

Figure 8-2   An example of a burst jammer.

One method of reducing the effects of a burst jammer is to use a burst clamp, which detects the increase in power at radio frequency (RF) and prevents the burst from entering the receiver. To determine the threshold for the power detector, the previous AGC voltage is used. The process gain and the bit error rate (BER) are also factors in determining the threshold desired. The duration thresholds are determined by the noise spikes for a minimum and the expected changes in signal amplitude for a maximum. The burst clamp performs four functions:

- It detects the power level to determine the presence of a burst.
- It switches out the input signal to the AGC amplifier.
- It holds the AGC voltage to the previous value before the burst occurred.
- It tells the decoder that a burst is present.

A detector log amplifier is used to determine the power level of the burst (Figure 8-3). The log amplifier compensates for the detector and gives a linear response of power in dB to volts out. The detected output is compared with the AGC voltage plus a threshold voltage to

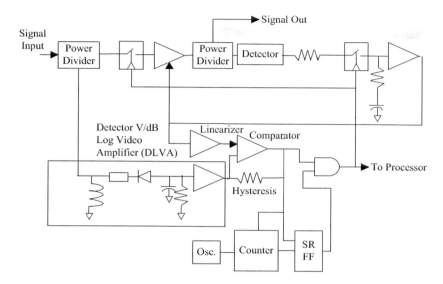

- Counter sets for longest expected burst.
- Counter resets FF after time out to close the switches again.
- SR FF goes high for one count cycle to allow capacitor to charge up.

Figure 8-3   Burst clamp used for burst jammers.

determine whether or not a burst is present. If a burst is present, the counter in the time-out circuit is enabled. This switches out both the intermediate frequency (IF) path and the AGC path, prevents the burst from continuing through the circuit, and holds the AGC voltage at the level set before the burst occurred. The counter time-out is set for the longest expected burst and then resets the flip-flop that closes the switches to allow the AGC voltage to build up again. This prevents the burst from locking up the AGC if there is a large change in signal level or when the transmitter is turned on. The speed of the circuitry is important so that the receiver will respond to quick bursts and not allow the AGC voltage to change due to a burst. The response of the IF amplifier is slow enough so that the detection circuitry has time to respond. Some considerations when designing the circuitry are as follows:

- Linearizing and matching over temperature.
- Response time of the burst clamp.
- False triggering on noise.
- Dynamic range of the detector.
- Detector amplitude dependent on frequency.
- Burst clamp saturation.
- Holding AGC voltage for long burst duration.

The response time needs to be fast enough so that the burst is not in the system longer than the error correction used in the system. The instantaneous dynamic range (IDR) of the system (amplitude) affects the soft decision (various thresholds are used in soft decisions), which can affect the BER regardless of the process gain.

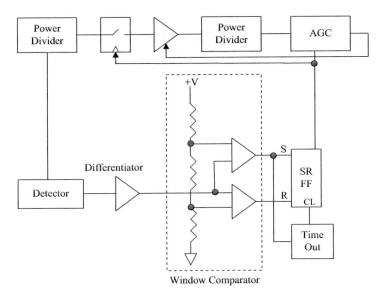

Figure 8-4   Slope detection burst clamp.

Another method for making a burst clamp is by using a slope detector, as shown in Figure 8-4. This design uses a differentiator to detect the presence of a burst. The window comparator looks at both the rising slope and falling slope of the burst. The rest of the circuitry is basically the same as before. The advantages of this type of clamp are as follows:

- Same threshold for all signal levels.
- No need for curve matching or linearizing.
- Insensitive to slow changes in noise level and signal level.

The disadvantages are as follows:

- Hard to detect the slope (detection error).
- Will not detect slow rise time bursts.
- Difficult to select the proper time constant.
- Slower detection process.

The placement of the burst clamp depends on the type of receiver used in the system. Most systems use a burst clamp at the IF and possibly one at RF. If the system has an AGC at the RF, then a burst clamp would be needed there also. The receiver either can have one detector at RF and use the sum of the RF and IF AGC voltages for the threshold or can use a detector at both places in the system. The latter would increase the sensitivity of the overall burst detection, depending on the receiver. Most detectors have a sensitivity of around −45 dBm.

8.2 Adaptive Filter

An adaptive notch filter that minimizes the effect of a narrowband jammer on data links can improve the transceiver. Most adaptive tapped delay line filters function only in baseband systems with bandwidths on the order of a few megahertz. To operate this filter over a wide bandwidth and at a high frequency, several problems must be overcome. Phase delays must be compensated for, and the quadrature channels must be balanced. Therefore, the actual performance of the filter is limited by the ability to make accurate phase delay measurements at these high frequencies.

In situations where the desired signal is broadband, there is a requirement to design a filtering system that can reduce narrowband jamming signals across a very wide band at a high center frequency without an extensive amount of hardware. The effect of the narrowband signals across a very large band can be reduced in several ways.

One such method is the use of spread spectrum techniques to obtain a process gain, which improves the SJR. Process gain is the ratio of the spread bandwidth to the rate of information sent. However, there are limitations on the usable bandwidth and on how slow the information can be sent in a given system. Coding techniques and interleaving can improve the system and reduce the effect of the narrowband jammer. However, the amount and complexity of the hardware required to achieve the necessary reduction limits the amount of coding that is practical in a system. If the bandwidth is already wide, spreading the signal may be impractical and the spread spectrum techniques will be of no value.

Passive notch filters placed in-line of the receiver can be used to reduce the unwanted signals, but with a large degradation in the wideband signal at the location of the notches. Also, prior knowledge of the frequency of each interferer is required and a notch for each undesired signal needs to be implemented. Since these notch filters are placed in series with the rest of the system, the overall group delay of a communication link will be altered.

Adaptive filters have been used for noise cancellation using a tapped delay line approach. The noise cancellation filter uses a separate reference input for the noise and uses the narrowband output for the desired signal. In this application, the desired signal is the wideband output, and the reference input and the signal input are the same. The reference signal goes through a decorrelation delay that decorrelates the wideband component, but the narrowband signal, because of its periodicity, remains correlated. When an adaptive filter is configured in this manner, it is called an adaptive line enhancer (ALE).

Adaptive filters, configured as ALEs, have been used to reduce unwanted signals, but they have been limited to relatively narrowband systems and at low frequencies. Adaptive filters can be used in broadband systems with a high carrier frequency, provided certain modifications are made to the system.

8.3 Digital Filter Intuitive Analysis

A finite impulse response (FIR) digital filter is shown in Figure 8-5. The signal enters the tapped delay line. Each output of the taps is multiplied by a weight value, and then they are all summed together. One way to look at this is that each tap is moving up and down according to the input signal and time. Therefore, the sum is equal to a point on the sine wave in time. At another point in time, the signal levels are different in the taps. However, the multiplication and resulting sum equal another point on the sine wave. This continues to happen until the output is a sine wave or the input is a sine wave with a delay. Note that other frequencies will not add up correctly with the given coefficients and will be attenuated.

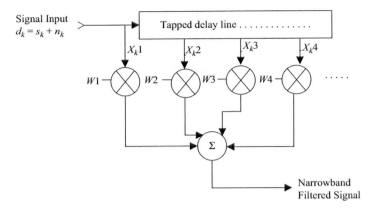

Figure 8-5 Digital finite impulse response filter.

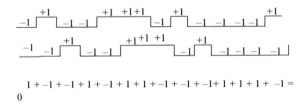

$1 + -1 + -1 + 1 + -1 + 1 + 1 + -1 + -1 + -1 + 1 + 1 + 1 + -1 =$
0

Figure 8-6 Correlation of a wideband PN code with a 1 chip delay.

An adaptive filter uses feedback to adjust the weights to the correct value for the input sine wave compared with the random signal, and in ALE the sine wave is subtracted from the input signal, resulting in just the wideband signal output. Note that there is a decorrelation delay in the ALE that slides the wideband signal in such a way that the autocorrelation of the wideband signal is small with a delayed version of itself. The narrowband signal has high autocorrelation with a delayed version of itself. This technique is used to reduce narrowband jamming in a wideband spread spectrum system. The wideband signal is not correlated with the delayed wideband signal, especially for long pseudo-noise (PN) codes. If a long PN code is used, then a delay greater than one chip of the code is not correlated. The longer the code, the less correlation for a delay greater than a chip (Figure 8-6). The correlator multiplies the code with the delayed version of the code, and the results are integrated. The integration value approaches zero as the code becomes longer.

8.4 Basic Adaptive Filter

A block diagram of a basic adaptive filter configured as an ALE is shown in Figure 8-7. The wideband input (d_k) consists of a narrowband signal (s_k) plus a wideband signal or noise (n_k). The composite signal is split, and one channel is fed to a decorrelation delay indicated by Z^{-D}, which decorrelates the wideband signal or noise. The other goes to a summing junction. The output of the decorrelation delay is delivered to a chain of delays, and the output of each

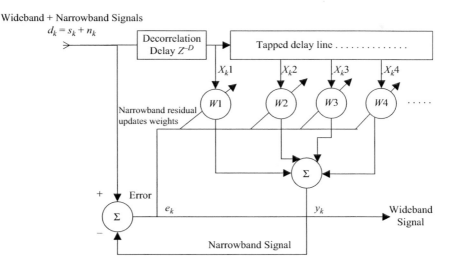

Figure 8-7   Basic adaptive filter configured as an ALE.

delay, $X_k(i)$, is multiplied by its respective weight values, $W_k(n)$. The weights are drawn with arrows to indicate that they are varied in accordance with the error feedback (e_k). The outputs of all the weights are summed together and produce the estimated narrowband spectral line (y_k). This narrowband signal is subtracted from the wideband plus narrowband input (d_k) to produce the desired wideband signal output, which is also the error (e_k). The name ALE indicates that the desired output is the narrowband signal (y_k). However, for use in narrowband signal suppression, the wideband or error signal (e_k) is the desired output.

Adaptive line enhancers are different from fixed digital filters because they can adjust their own impulse response. They have the ability to change their own weight coefficients automatically using error feedback, with no a priori knowledge of the signal or noise. Because of this ability, the ALE is a prime choice in jammer suppression applications where the exact frequency is unknown. The ALE generates and subtracts the narrowband interferer, leaving little distortion to the wideband signal. Also, one ALE can reduce more than one narrowband interference signal at a time in a given bandwidth. The adaptive filter converges on the superposition of the multiple narrowband jammers. The adaptive filter scheme has the ability to adapt in frequency and amplitude to the interferer in a specified bandwidth, so exact knowledge of the interferer is not necessary. If the interferer changes in frequency in the given band, the filter will automatically track the change and reduce the narrowband signal.

8.5 Least Mean Square Algorithm

The adaptive filter works on the principle of minimizing the mean square error (MSE), $E[e_k^2]$, using the least mean square (LMS) algorithm (Figure 8-8). The input to the filter is

$$d_k = s_k + n_k(\text{scalar})$$

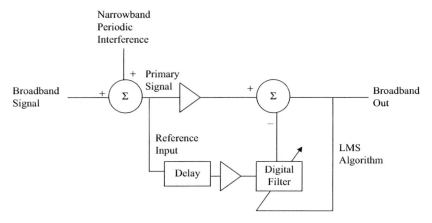

Figure 8-8   Adaptive filter using LMS algorithm.

where

s_k = narrowband signal (scalar)
n_k = noise or wideband signal (scalar)

The ALE is designed to minimize the MSE:

$$\text{MSE} = E[e_k^2] = E[(d_k - W_k^T X_k^2)] = E[d_k^2 - 2d_k W_k^T X_k + (W_k^T X_k^2)]$$
$$= E[d_k^2] - 2E[d_k X_k^T]W_k + E[W_k^T X_k X_k^T W_k]$$

where

e_k = error signal (scalar)
W_k^T = weight values transposed (vector)
X_k = the tap values (vector)

When substituting the autocorrelation and cross-correlation functions, the definitions give

$$\text{MSE} = E[d_k^2] - 2r_{xd}W_k^T + W_k R_{xx}^T W_k$$

where

$R_{xx} = E[X_k X_k^T]$ = autocorrelation matrix
$r_{xd} = E[d_k X_k^T]$ = cross-correlation matrix
$X_k^T W_k = X_k W_k^T$

To minimize the MSE, the gradient (∇_w) is found and set to zero:

$$\nabla_w E[e_k^2] = 2R_{xx}W_k - 2r_{xd} = 0$$

Solving for W, which is the optimal weight value, gives

$$W_{opt} = R_{xx}^{-1} r_{xd}$$

The weight equation for the next weight value is

$$W_{k+1} = W_k - \mu \nabla_W E[e_k{}^2]$$

In the LMS algorithm, $e_k{}^2$ itself is the estimate of the MSE. Because the input signal is assumed to be ergodic, the expected value of the square error can be estimated by a time average. The weight equation then becomes

$$W_{k+1} = W_k - \mu \nabla_W (e_k{}^2) = W_k + 2\mu e_k X_k$$

This is known as the LMS algorithm and is used to update the weights or the filter response. The new weight value, W_{k+1}, is produced by summing the previous weight value to the product of the error (e_k) times the tap value (X_k) times a scale factor (μ), the latter determines the convergence rate and stability of the filter. It has been shown that under the assumption of ergodicity the weights converge to the optimal solution W_{opt}.

8.6 Digital/Analog ALE

In digital adaptive filters, it is assumed that the time for the error signal to be generated and processed to update the weight values is less than a clock cycle delay of the digital filter clock. However, since part of the feedback loop is analog, delay compensation is required. A quadrature method is used in the frequency conversion processes to allow ease of tuning across a very wide band of operation. This also provides a wider instantaneous bandwidth.

Unless the signal can be digitized at the RF, a digital/analog combination needs to be implemented. When ALEs are used at high frequencies, the RF signal must be down-converted using a local oscillator (LO) and a mixer before it can be processed and digitized by the digital filter (Figure 8-9).

A bandpass filter provides a coarse selection of the band for processing. The actual processing bandwidth of the signal is limited to the clock frequency of the digital filter due to

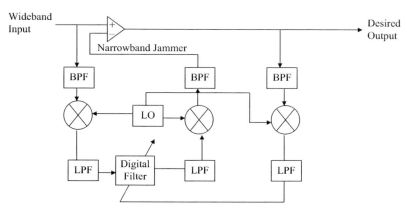

Figure 8-9   Analog/digital adaptive filter.

the Nyquist criteria to prevent aliasing. After the tone is generated by the digital filter, it is upconverted, using a mixer and the same LO, to the RF. The signal is fed through a bandpass filter to eliminate the unwanted sideband produced in the upconversion process and is then subtracted from the composite signal. The final output is split and used for the error feedback signal, which is downconverted in the same manner as the reference signal.

Large bandwidths can be achieved using a synthesizer to vary the LO frequency and to select a band where narrowband interference is a problem. This would not produce a large instantaneous bandwidth, but it would allow control over a large band of frequencies with relatively little hardware. However, either the sum term or the difference term needs to be filtered to process the signal correctly. One way to accomplish this is to use tunable filters, as shown in Figure 8-10.

The left part of the figure shows the standard conversion without filters. When using an LO to downconvert the signal, this process generates both the sum and difference terms.

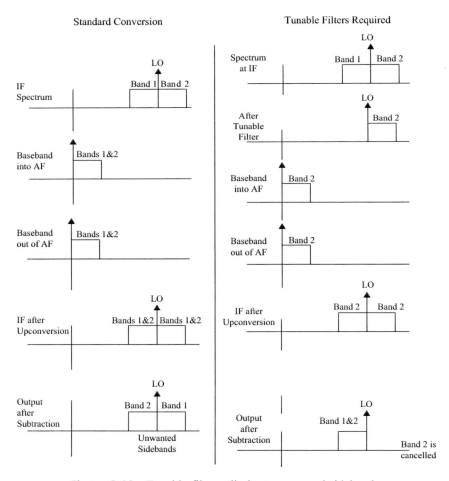

Figure 8-10   Tunable filters eliminate unwanted sidebands.

Therefore, the IF spectrum starts out with both sum and difference bands: band 1 is defined as the difference frequency band, and band 2 is defined as the sum frequency band. Without filtering, the second downconversion process places both bands at baseband, folded over on top of each other. They are passed through the adaptive filter (AF) and upconverted using an LO. The upconversion process generates the sum and difference bands, which contain both band 1 and band 2 (Figure 8-10). Therefore, when this composite signal is subtracted from the incoming IF spectrum at the top of the figure, only band 1 is eliminated in the difference band, and only band 2 is eliminated in the sum band. Thus, both band 1 and band 2 appear on the output, with band 2 in the lower band and band 1 in the upper band. Neither band is canceled out by the adaptive filter (Figure 8-10).

With the tunable filters, both sum and difference terms are generated in the down-conversion process, but for this case the unwanted difference band (band 1) is filtered out. Therefore, the second downconversion has only band 2 at the baseband. This sum band is fed through the adaptive filter process and upconverted using an LO. The upconversion generates the sum and difference bands, which contain only band 2 (Figure 8-10). When this composite signal is subtracted from the incoming IF spectrum at the top of the figure, the lower band has the results of the subtraction of band 1 and band 2. However, the important point is that the upper band contains no frequency because band 2 is subtracted from band 2, so band 2 is eliminated in that spectrum (Figure 8-10).

A quadrature upconversion/downconversion scheme can also be used to retrieve the desired signal, as shown in Figure 8-11. This configuration allows the LO to be positioned in the center of the band of interest and eliminates the need for tunable filters. The filter is similar to the system shown in Figure 8-9, except that in this configuration two digital filters and two channels in quadrature are required. A wide bandpass filter can be used in the IF section to select the entire band that the adaptive filter will tune across. The LO or synthesizer selects the band of interest, as shown in Figure 8-11. The in-phase (I) and quadrature (Q) signals are low-pass filtered. Both sidebands are downconverted to baseband with an in-phase signal from the oscillator for the I channel and a quadrature-phase signal from the oscillator for the Q channel. This method allows twice the processing bandwidth with a given clock frequency because it utilizes both sidebands in the process. A digital filter for both I and Q channels generates the respective tones. During the upconversion of the digital filter outputs, each channel produces both bands on each side of the carrier. However, the phase relationship of the bands provides cancellation of the unwanted side-bands when the signals are summed together so that only the desired sidebands are generated. These desired bands are subtracted from the input IF spectrum, and both bands are eliminated (Figure 8-11).

This behaves the same as a quadrature transmitter/receiver combination for sending two different signals and receiving them. Two signals are quadrature downconverted with an in-phase (I channel) and a 90° phase shift (Q channel) to give

$$\text{I channel} = \cos(\omega_1 t) + \cos(\omega_2 t)$$
$$\text{Q channel} = \cos(\omega_1 t - 90°) + \cos(\omega_2 t - 90°)$$

where

ω_1 = a frequency located in band 1
ω_2 = a frequency located in band 2

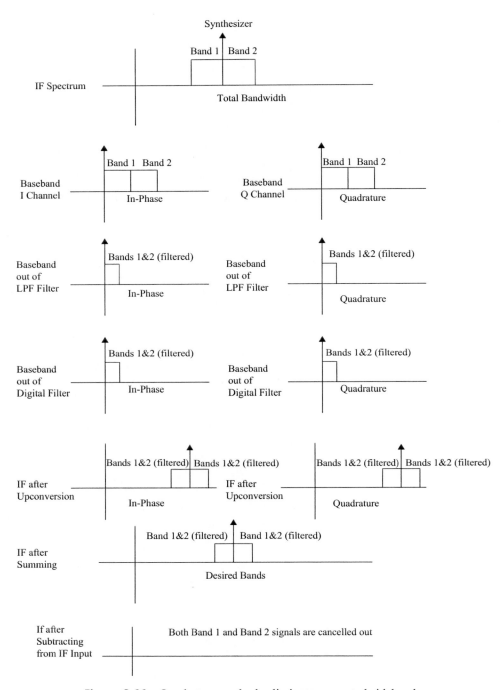

Figure 8-11   Quadrature methods eliminate unwanted sidebands.

The two digital filters generate these signals and deliver them to the upconversion process. The same LO is used for both the upconversion and downconversion. The in-phase LO is mixed with the I channel, and the 90° phase-shifted LO is mixed with the Q channel. The results are as follows and include the sum and difference terms:

$$\text{I channel} = \cos(\omega_0 + \omega_1)t + \cos(\omega_0 - \omega_1)t + \cos(\omega_0 + \omega_2)t + \cos(\omega_0 - \omega_2)t$$

$$\begin{aligned}\text{Q channel} &= \cos(\omega_0 t - 90° + \omega_1 t - 90°) + \cos(\omega_0 t - 90° - \omega_1 t + 90°)\\ &+ \cos(\omega_0 t - 90° + \omega_2 t - 90°) + \cos(\omega_0 t - 90° - \omega_2 t + 90°)\\ &= \cos[(\omega_0 + \omega_1)t - 180°] + \cos(\omega_0 - \omega_1)t\\ &+ \cos[(\omega_0 + \omega_2)t - 180°] + \cos(\omega_0 - \omega_2)t\end{aligned}$$

Amplitudes have been neglected to show phase response only. When summed together, the net result is

$$2\cos(\omega_0 - \omega_1)t + 2\cos(\omega_0 - \omega_2)t$$

The unwanted sidebands are eliminated because of the net 180° phase shift, as shown in Figure 8-10. If the desired frequency occurred on the high side of the LO, there would have been a sign change in the previous solution. Since the unwanted sidebands are eliminated because of the quadrature scheme, tunable filters are no longer required for the upconversion process.

Another problem to overcome in this implementation is LO bleed-through when using a nonideal mixer. LO bleed-through is the amount of LO signal appearing on the output of a mixer due to an imbalance of a double-balanced mixer. Since the idea is to eliminate jamming signals in the passband, the LO bleed-through will appear as another jamming signal, which is summed with the original signal. Designing a well-balanced mixer will help the problem, but there is a limit to how well this can be done. One way to reduce LO bleed-through is to couple the LO and phase shift the coupled signal so that it is 180° out of phase with the LO bleed-through and exactly the same amplitude and then to sum the coupled signal with the composite signal. However, since the synthesizer is changing in frequency and the group delay for the LO bleed-through is not constant, this becomes very difficult to accomplish. An alternative is to use an IF with a fixed LO to upconvert from baseband to the IF and then use the synthesizer to mix up to the desired frequency. Thus, the mixers at the lower frequency band provide better LO isolation, which in turn greatly reduces the fixed LO bleed-through. Further, the synthesizer bleed-through is out of band and can be easily filtered.

8.7 Wideband ALE Jammer Suppressor Filter

A wideband ALE jammer suppressor filter is shown in Figure 8-12. The adaptive filter is connected in parallel with the communication system, and the only components in line with the system are three couplers and one amplifier. The group delay through the amplifier is constant across the band of interest, with a deviation of less than 100 psec. The amplifier is placed in the system to isolate the reference channel from the narrowband channel. This prevents the narrowband signal from feeding back into the reference channel.

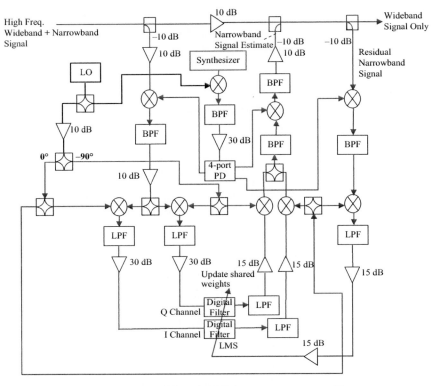

Figure 8-12   Wideband ALE jammer suppressor filter.

The wideband high-frequency composite signal is split using a −10 dB coupler, amplified and downconverted to an IF using an adjustable synthesizer. This provides the reference signal for the digital filter. The signal is filtered, amplified, and quadrature downconverted to baseband, where the quadrature signals are filtered and amplified and provide the reference signals to the digital filters. Elliptical low-pass filters are used to achieve fast roll-offs and relatively flat group delay in the passband. The digital filters produce the estimated narrowband signals. The outputs of the digital filters are low-pass filtered, amplified, and quadrature upconverted to the IF band and summed together to eliminate the unwanted sidebands. The signal is filtered and upconverted by the synthesizer to the desired frequency and then filtered, amplified, and subtracted from the composite signal to eliminate the narrowband signal. The output provides both the desired wideband signal and the error. This signal is split, and the in-phase error is downconverted in two stages to baseband to update the filter weights, as shown in Figure 8-12.

8.8 Digital Circuitry

The digital filter digitizes the reference input using analog-to-digital converters (ADCs) and feeds this signal through a decorrelation delay. This decorrelates the wideband signal and allows the narrowband signal, because of its periodicity, to remain correlated. The delayed signal is fed to a tapped delay line containing 16 different time-delayed signals or taps that are multiplied by the error feedback and then accumulated to update the weight values. The analog error signal is single-bit quantized and scaled by adjusting the μ value. The μ value determines the filter's sensitivity to the error feedback. The 16 taps are then multiplied by these new weight values and converted to analog signals, where they are summed together to form the predicted narrowband output. A selectable delay was incorporated in the design to provide adjustment for the time delay compensation. This was necessary because of the delay through the digital and analog portions of the filter. When the filter generates the predicted narrowband signal for cancellation, the error produced needs to update the portion of the signal that caused the error. If there is delay in this path, then the error will be updating a different portion of the signal from the part that generated the error. Since the analog portion produces a delay that is not quantized with regards to a certain number of clock cycles, a variable delay was designed to select small increments of a clock cycle. This allows for better resolution in selecting the correct compensation delay. A rotary switch, mounted on the board, is provided for adjusting the delay for each digital filter.

8.9 Simulation

A limited simulation of the ALE system was performed due to the complexity and large number of variables contained in the ALE system definition. Since simulation time is dependent on the sample rate of the computer clock, a baseband representation was used. The amount of computer time to do the simulation at high frequencies is impractical. Existing models were used to form the desired circuit. The frequency is swept across a 30 MHz band, and the output power is measured and displayed. Figure 8-13 shows the cancellation across the instantaneous bandwidth. The filter is unstable at the filter edges because of the phase distortion and aliasing effects. The spikes at the center frequency are a

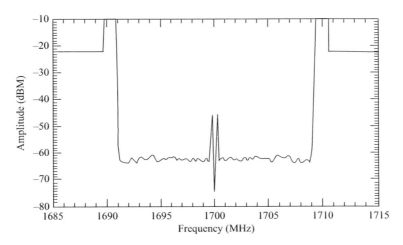

Figure 8-13   ALE simulation.

result of using only a 16-tap filter. The low-frequency components cannot be resolved in the tap delay line. The cancellation achieved in the simulation is greater than 40 dB, except around the center frequency.

8.10 Results

The adaptive filter cancellation bandwidth is similar to the results obtained in the simulation (Figure 8-14). This shows the response of the filter as a CW tone is swept across the selected band. The bandwidth is less than the simulation to suppress the phase distortion at the band edge. The center of the band is the response of the filter when it is mixed down to direct current (DC). Since the ALE is alternating current (AC)-coupled in the hardware, it cannot respond to DC or low frequencies, which results in no cancellation of the signal. The amount of cancellation across the band is approximately 10 dB, with up to 30 dB at certain frequencies. This can be compared to the 40 dB cancellation across the band achieved in the simulation.

Quadrature imbalance, which is difficult to measure at high frequencies using frequency conversion processes, is a major factor in the amount of suppression of the narrowband tone. Thus, the quadrature balance was estimated and the hardware tuned to achieve maximum performance. The synthesizer can be tuned to select a specified frequency to achieve maximum cancellation of that frequency across a given bandwidth. The amount of cancellation is dependent on many factors, such as I–Q balance for both phase and amplitude, phase linearity across the band, and stability and noise in the system.

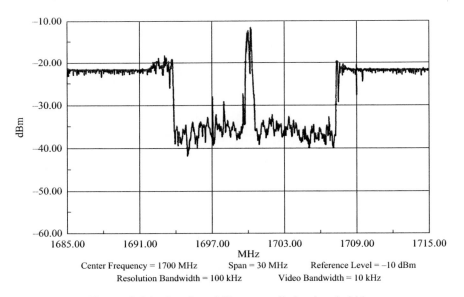

Figure 8-14   Results of filter cancellation bandwidth.

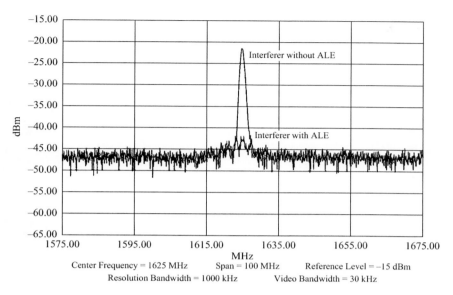

Figure 8-15 Cancellation for a single interferer.

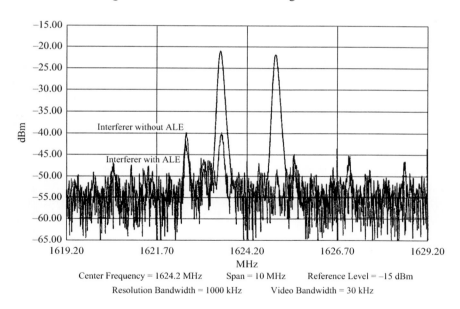

Figure 8-16 Cancellation for multiple interferers.

The cancellation for a single interferer is shown in Figure 8-15. The response shows a single interferer with and without the ALE in the system. The single tone is suppressed by approximately 20 dB. ALEs can suppress more than one interferer at a time, as shown in Figure 8-16. Generally there is degradation in performance in the amount of suppression with respect to the number of interferers. Also, more tones produce more spurious responses in the mixing processes in the system.

8.11 Amplitude and Phase Suppression Results

The main criterion for achieving the maximum cancellation of the interferer is to ensure that its replicated waveform is exactly equal in amplitude and exactly 180° out of phase. This provides perfect cancellation. If there is an error in the amplitude or an error in the phase, this will result in a nonperfect cancellation of the waveform. The cancellation performance degrades when the amplitude is not exact (Figure 8-17). In addition, if the phase of the cancellation waveform is not exactly 180°, the cancellation performance degrades (Figure 8-18).

Both these sources of errors need to be considered when analyzing the amount of cancellation that can be obtained. For analog systems, these parameters are very difficult to maintain over temperature, hardware variations, and vibrations. Digital systems are much better to use for cancellation, and a higher cancellation performance can be achieved. The sooner a system can digitize the signal and perform these types of functions in the digital domain, the better the performance that can generally be realized.

8.12 Gram-Schmidt Orthogonalizer

Signals on the x-axis and y-axis are by definition orthogonal and in this case are 90° out of phase (Figure 8-19). All signals can be represented by the sum of weighted orthonormal functions or the magnitudes of orthonormal functions. Orthonormal functions are normalized vectors with a magnitude of one. For example, if a signal $S_1(t)$ is placed on the X_1 axis, then its value is the orthonormal function times the magnitude a. If another signal $S_1(t)$ is

Suppression versus Amplitude

Amplitude ($B = 1$)	Suppression (dB)
1.005	52.1
1.01	46.1
1.02	40.1
1.04	34.2
1.06	30.7
1.08	28.3
1.10	26.4
1.12	24.9
1.14	23.7
1.16	22.6
1.18	21.7
1.20	20.8
1.22	20.1
1.24	19.4
1.26	18.8
1.28	18.2
1.30	17.7

Figure 8-17   Jammer cancellation performance with amplitude variations.

<u>Suppression versus Phase Error</u>

<u>Error (deg)</u>	<u>Suppression (dB)</u>
0.1	61.2
0.5	47.2
1.0	41.2
2.0	35.2
3.0	31.6
4.0	29.1
5.0	27.2
6.0	25.6
7.0	24.3
8.0	23.1
9.0	22.1
10.0	21.2
11.0	20.3
12.0	19.6
13.0	18.9
14.0	18.2
15.0	17.6

Figure 8-18   Jammer cancellation performance with phase variations.

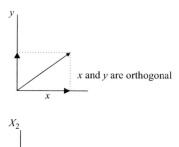

x and y are orthogonal

$S_2(t) = bX_1(t) + cX_2(t),$

$S_1(t) = aX_1(t)$

$X_1(t) = S_1(t)/a =$ first orthonormal function
$X_2(t) = [S_2(t) - bX_1(t)]/c =$ second orthonormal function

$a = \text{mag } S_1(t)$
$c = \text{mag } S_2(t) - bX_1(t)$

Figure 8-19   Orthonormal vectors.

somewhere between the X_1 axis and the X_2 axis, then it is the sum of the magnitude b of $X_1(t)$ and the magnitude c of $X_2(t)$ (Figure 8-19):

$$S_1(t) = aX_1(t)$$
$$S_2(t) = bX_1(t) + cX_2(t)$$

where $X_1(t)$, $X_2(t)$, ... are orthonormal functions, and a, b, c, ... are the weighting coefficients.

The constant a is simply the magnitude of $S_1(t)$, since the magnitude of $X_1(t)$, by definition, is unity. To solve for b, the second equation is multiplied by $X_1(t)$ and integrated:

$$\int S_2(t)X_1(t)dt = \int bX_1(t)X_1(t)dt + \int cX_2(t)X_1(t)dt$$

On the right side of this equation, the second term is equal to zero (the inner product of two orthonormal functions $= 0$) and the first term is equal to b (the inner product of an orthonormal function with itself $= 1$). Solving for b produces

$$b = \int S_2(t)X_1(t)dt$$

This is the projection of $S_2(t)$ on $X_1(t)$.

The constant (c) is determined by the same procedure except that $X_2(t)$ is used in place of $X_1(t)$ when multiplying the second equation. Therefore

$$\int S_2(t)X_2(t)dt = \int bX_1(t)X_2(t)dt + \int cX_2(t)X_2(t)dt$$

$$c = \int S_2(t)X_2(t)dt$$

This is the projection of $S_2(t)$ on $X_2(t)$. Note that $X_2(t)$ is not generally in the direction of $S_2(t)$.

The first orthonormal basis function is therefore defined as

$$X_1(t) = S_1(t)/a$$

where a is the magnitude of $S_1(t)$.

The second orthonormal basis function is derived by subtracting the projection of $S_2(t)$ on $X_1(t)$ from the signal $S_2(t)$ and dividing by the total magnitude:

$$X_2(t) = [S_2(t) - bX_1(t)]/c$$

where c is the magnitude of the resultant vector $S_2(t) - bX_1(t)$. Thus, the magnitudes are equal to

$$a = S_1(t)/X_1(t) = abs(S_1)$$

$$b = \int S_2(t)X_1(t)dt \leq S_2, \; X_1 \geq \text{inner product}$$

$$c = \int S_2(t)X_2(t)dt \leq S_2, \; X_2 \geq \text{inner product}$$

A phasor diagram is provided in Figure 8-19 to show the projections of the vectors.

8.13 Basic GSO

A basic GSO system is shown in Figure 8-20. The weight (w_1) is chosen so that the two outputs, V_o and W_o, are orthogonal; that is, the inner product $<V_o,W_o> = 0$. This gives the result:

$$<J + S_1, J + S_2 - w_1(J + S_1)> = 0$$
$$<J,J>(1 - w_1) - w_1<S_1,S_1> = 0$$
$$w_1 = \frac{|J|^2}{|J|^2 + |S_1|^2} = \frac{1}{1+p}$$
$$p = \frac{1-w_1}{w_1} = \frac{|S_1|^2}{|J|^2}$$

Note the following assumptions:

1. Same J in both inputs.
2. J, S_1, S_2 are orthogonal.
3. $|J|^2, |S_1|^2, |S_2|^2$ are known.
4. $p \ll 1$. This means the jammer is much larger than the signal S_1.

The outputs are

$$V_o = J + S_1$$
$$W_o = J + S_2 - w_1(J + S_1) = J(1 - w_1) + S_2 - w_1S_1 = J(p/(1+p)) + S_2 - S_1(1/(1+p))$$

Since $p \ll 1$, then

$$W_o = J_p + S_2 - S_1 = |S_1 - S_2|^2$$

This shows that the jammer signal has been attenuated in the W_o output.

Suppose only a jammer exists in one of the inputs, S_1, and the jammer plus signal is in S_2, as shown in Figure 8-20. Taking the projection of S_2 on the orthonormal function ($Q_1 = S_1/|S_1|$) provides the amount of jammer present in S_2:

$$b = <S_2, \frac{S_1}{|S_1|}>$$

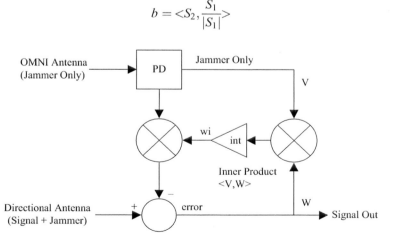

Figure 8-20 Basic GSO used for jammer suppression.

This scalar quantity multiplied by Q_1 produces the jammer vector bQ_1. Subtracting the jammer vector from S_2 gives the amount of signal present (S_p) in S_2:

$$S_p = S_2 - bQ_1 = S_2 - <S_2, \frac{S_1}{|S_1|}> \frac{S_1}{|S_1|}$$

Thus, the jammer is eliminated. In the implementation of these systems, the $|S_1|$ values are combined in a scale factor k where

$$k = 1/|S_1|^2$$

The final result is then

$$S_p = S_2 - k <S_2, S_1> S_1$$

The constant (k) is incorporated in the specification of the weight value or in the integration process during the generation of the weights in an adaptive system.

8.14 Adaptive GSO Implementation

If a system uses an omnidirectional antenna and assumes jammer only for S_1 (since the jammer is much larger in amplitude than the desired signal) and a directional antenna for signal plus jammer for S_2 (since the antenna will be pointed toward the desired signal), the previous example will apply. The technique starts with updating weights to force the outputs to be orthogonal so that the inner product $<v,w> = 0$. The weights are updated by the inner product of the outputs.

Assume the only jammer signal is $S_1(t)$. The magnitude of the jammer is equal to $a = abs(S_1)$. Signal $S_2(t)$ is made up of a signal vector and a jammer vector. Therefore, the magnitude of the signal vector $S_2(t)$ that contains no jammer is equal to

$$S_v = cx_2(t) = S_2(t) - bx_1(t)$$
$$x_1(t) = S_1(t)/abs(S_1)$$
$$b = <S_2,x_1> = <S_2, S_1/abs(S_1)>$$
$$S_v = S_2 - [<S_2, S_1/abs(S_1)> S_1/abs(S_1)]$$
$$S_v = S_2 - [<S_2, S_1> S_1/abs(S_1)^2] = S_2 - S_1[<S_2, S_1>/abs(S_1)^2]$$

Weight value $= w = <S_2, S_1>/abs(S_1)^2$
$$S_v = S_2 - wS_1 = S_{dir} - w_1 S_{omni}$$

An adaptive filter configured as an adaptive noise canceller can be used as a GSO jammer suppressor (Figure 8-21).

This shows a quadrature system with separate I and Q outputs and separate weight generators. The error signal is produced by subtracting the weighted reference input signal (the received signal from the omnidirectional antenna) from the signal received from the directional antenna:

$$e = S_{dir} - w_1 S_{omni}$$

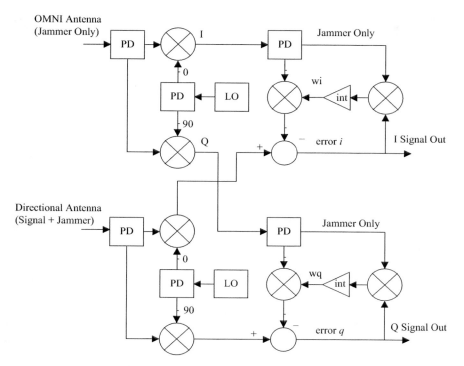

Figure 8-21   Quadrature adaptive system.

The square error is therefore

$$e^2 = S_{dir}{}^2 - 2w_1 S_{dir} S_{omni} + w_1{}^2 S_{omni}{}^2$$

The mean square error is

$$\text{MSE} = \overline{e^2} = w_1{}^2 S_{omni}{}^2 - 2w_1 S_{dir} S_{omni} + S_{dir}{}^2$$

Taking the gradient of the MSE and setting this equal to zero, the optimum weight value can be solved:

$$2w_1 S_{omni}{}^2 - 2S_{dir} S_{omni} = 0$$
$$w_{1(opt)} = S_{dir} S_{omni} / S_{omni}{}^2$$

Since the error output is the desired signal output, then

$$e = S_{dir} - (S_{dir} S_{omni} / S_{omni}{}^2) S_{omni}$$

The inner product is defined as

$$<X_1, X_1> = \int X_1^2(t) dt = E[X_1^2(t)] = \overline{X_1^2(t)} = |X_1(t)|^2$$

$$<X_1, X_2> = \int X_1(t) X_2(t) dt = E[X_1(t) X_2(t)] = \overline{X_1(t) X_2(t)}$$

Therefore the error can be expressed as

$$e = S_{dir} - \left(\frac{<S_{dir}, S_{omni}>}{|S_{omni}|^2} \right) S_{omni}$$

The second term on the right side of this equation is the projection of S_{dir} on the orthonormal function $(S_{omni}/|S_{omni}|)$ times the orthonormal function. This determines the amount of jammer present in S_{dir}. This result is then subtracted from S_{dir} to achieve the desired signal (e) and eliminate the jammer. The LMS algorithm assumes that the gradient of the MSE can be estimated by the gradient of the square error, which turns out to be twice the error times the reference.

8.15 Intercept Receiver Comparison

Some receivers, known as intercept receivers, are designed to intercept the transmissions of an unknown transmitter. Electronic countermeasure (ECM) receivers, also a type of intercept receiver, are designed to listen to broadcasts from other sources (Figure 8-22). Table 8-1 lists the most common types along with their disadvantages and advantages. This is not intended to be a comprehensive list but is provided to give a general idea of what type of receivers can detect the desired signal.

Using intercept receivers to determine the type of jamming signal can help tremendously in deciding what type of anti-jam technique to use. For example, if all of the jammers are broadband, then an adaptive filter might not be the best type of anti-jammer to use.

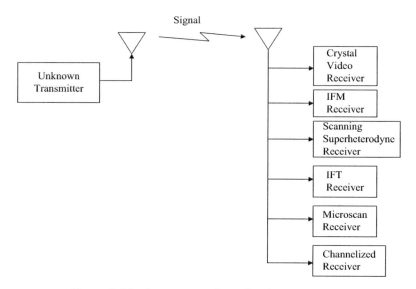

Figure 8-22 Intercept receivers for signal detection.

Table 8-1 Common intercept receivers and their advantages and disadvantages

Type of Receiver	Description	Advantages	Disadvantages
Crystal video receiver	Uses a crystal video detector circuit that searches to see if a signal is present. Generally for narrowband detection (to limit the noise into the detector) unless tunable filters are employed.	Small in size, simple to design, and less expensive compared with other types of intercept receivers.	Does not have a broad frequency detection ability unless tunable filters, which are complex and expensive, are used in the front end.
Instantaneous frequency measurement (IFM) (delay line discriminator receiver)	Usually consists of a delay line discriminator design to detect one single frequency.	High frequency resolution detection capability and the ability to detect wide frequency bandwidths.	Cannot handle multiple frequencies at the same time.
Scanning superheterodyne receiver	Uses a standard superheterodyne receiver with a tuning oscillator to cover the wide frequency bandwidth. Not instantaneous and dependent on the sweep time of the oscillator. Sometimes referred to as a spectrum analyzer receiver.	Good frequency resolution detection capability and coverage of a wide frequency dynamic range.	Does not cover the band instantaneously; dependent on scan time, and the spurs that are generated in the scanning process result in false detections of the signal.
Instantaneous Fourier transform (IFT)	Uses an acousto-optic device such as a Bragg cell. Employs optical rays that project on a surface at different points dependent on the incoming frequency, which can then be determined by the location of the ray.	Good frequency resolution detection capability, multiple frequency handling, and instantaneous wideband coverage.	Small dynamic range (20–30 dB) and the size of the receiver.
Microscan receiver	Often referred to as a chirped receiver. Uses a surface acoustic wave (SAW) device that is excited by an impulse input signal. Since all frequencies are contained in an impulse, these frequencies are mixed with the incoming signal and delayed by means of acoustic waves, and the output is time dependent on the resultant frequencies. Basically an extremely fast scanning receiver.	Wideband instantaneous frequency coverage, good frequency sensitivity, and very small size.	Limited dynamic range capability (30–40 dB) and the pulse information is lost. The dynamic range reduction is due to the sidelobes that are generated in the process.
Channelized receiver	Uses multiple frequency channels that provide true instantaneous frequency processing. Basically several receivers fused together, with each receiver containing bandwidths to cover the total desired broad bandwidth.	Good frequency resolution detection capability, detectability of multiple frequencies, and a wide frequency IDR.	Size and cost.

8.16 Summary

Burst jammers are effective at capturing the AGC time constant to maximize the ability to jam a receiver with minimum average power. The burst clamp is used to mitigate this type of jammer by preventing the AGC from being captured.

Adaptive filters can be configured as ALEs to suppress undesired narrowband signals. When these filters are used at high frequencies and over large bandwidths, modifications need to be made. The time delay through portions of the system need to be compensated for and is accomplished in the digital filter by using a tap value to generate the signal and a delayed tap value to update the weights. Also, for small variations in delay, the clock is modified in the digital filter. Once the delay is set, constant delay over the band of operation is important for proper operation of the filter. A variable synthesizer is used in the design to achieve a wide operational bandwidth for the ALE. This allows the filter to be positioned across a given bandwidth with minimal hardware. A quadrature scheme is used to eliminate filtering constraints and provide twice the processing bandwidth for the ALE. However, the performance of the system relies on the quadrature channels being balanced in both phase and amplitude. The LO bleed-through problem is reduced by using a double downconversion scheme, since the isolation of the LO signal is greater for lower frequencies. The isolation for the conversion process is established in the first low-frequency mixer because the LO signal for the second mixer lies outside of the passband and can be filtered.

Gram-Schmidt orthogonalizers can be used to reduce the effects of jamming signals. One of the assumptions in this approach is that the jammer signal level is much higher than the desired signal level. The basic GSO has two inputs, with one containing more signal than jammer. This applies to having two antennas with one of the antennas directed toward the signal providing higher signal power. The error signal for feedback in updating the weight value is produced by subtracting the weighted reference input signal from the received signal, which contains the higher level of desired signal. When the weight has converged, then the jamming signal is suppressed.

Spread spectrum systems can reduce a signal's detectability, and the research is on to design better intercept receivers.

References

Bishop, F. A., R. W. Harris, and M. C. Austin. "Interference Rejection Using Adaptive Filters." *Electronics for National Security Conference Proceedings*, pp. 10–15, September 1983.

Bullock, S. R. "High Frequency Adaptive Filter." *Microwave Journal*, September 1990.

Haykin, S. *Adaptive Filter Theory*, 4th ed. Prentice-Hall, NJ, 2001.

Haykin, S. *Communication Systems*, 5th ed. Wiley, NJ, 2009.

Widrow, B., Glover Jr, J. R., McCool, J. M., Kaunitz, J., Williams, C. S., Hearn, R. H., "Adaptive Noise Cancelling: Principles and Applications." *Proceedings IEEE*, 1975.

Widrow, B., and S. D. Stearns. *Adaptive Signal Processing*, Prentice-Hall, NJ, 1985, pp. 354–361.

Problems

1. What is meant by capturing the AGC of a system?
2. What would be a good pulse frequency for a burst jammer given an AGC response time of 1 μsec?
3. What is the main difference between a digital FIR filter and an adaptive filter?
4. Why is either filtering or quadrature method required to operate the adaptive filter in the RF world?
5. What does the μ value in the LMS algorithm represent?
6. What is the result of increasing the μ value on convergence time, stability, and steady-state accuracy?
7. Why is the assumption that the jammer is the only signal present in the omnidirectional antenna of a GSO jammer suppression filter a good assumption?
8. When might the assumption that the jammer is the only signal present in the omnidirectional antenna of a GSO jammer suppression filter be a bad assumption?
9. What is the best detection receiver if cost, size, and complexity are not issues, and why?
10. If a narrowband signal is present in the desire passband, what technique can be utilized to reduce its effect on the desired signal?

CHAPTER 9

Cognitive Systems

Operating communication systems in a changing environment requires the need to develop a cognitive system to mitigate the effects the environment has on communications or data links. A simple definition is as follows:

> *Cognition is the ability for a system or systems to monitor, record, sample, test, and be cognitive or aware of the surrounding environments and then to adapt, modify, or change the system to improve the quality of service, including learning from past experiences.*

In other words, the basic concept is to develop a system, radio, antenna, network, and then to use the available resources to monitor the environment and make an optimal change to the system to improve the wireless link's quality of service (QoS).

9.1 The Environment

Many elements of the environment require the need to incorporate cognitive capabilities for a communications/data link system (Figure 9-1). The following sections list the major factors, but these are not exhaustive. Many other hindrances can be present, and additional cognitive abilities can become necessary as new developments evolve and as the environment changes.

9.1.1 Jammers

One of the major environmental hindrances to any communications is jamming. This can be either unfriendly or friendly jammers. Unfriendly jammers purposely jam the data link to disable the system. They are unwanted users in the volume space. They can be very sophisticated and focused primarily on jamming the signal. They can also use cognitive methods to optimize the jamming effects on the desired data link. Friendly jammers are users in the proximity of the desired data link and do not intentionally jam the desired signal. There can be other equipment and radios on the platform operating simultaneously with the data link (Figure 9-1).

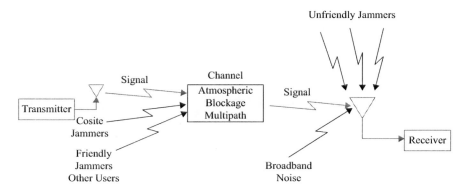

Figure 9-1 Environmental conditions affect the desired signal.

Unfriendly jammers. There are several types of unfriendly jammers, including barrage, pulsed, and continuous wave (CW). The type of jammer is selected to cause the greatest harm to the desired data link. Barrage jammers send out a barrage of frequencies to ensure that they jam the frequency of the desired data link. A pulsed jammer produces a very large pulse or spike that can be very detrimental to the data link and still only using minimum average power. In addition, if the pulse jammer can determine the automatic gain control (AGC) time constant of the desired receiver, it can maximize its effect as a jammer. This is accomplished by sending a large jamming pulse that causes the receiver's AGC to reduce its gain. When the pulse is turned off, the receiver's signal is lost because the gain is set too low, so the AGC takes time to adjust the gain until it can retrieve the signal. Since the jammer knows the response time of the AGC, it will know when to send the next jamming pulse to start the process over again (see Chapter 8). CW jammers can increase the power and jam the desired signal by overpowering the received signal.

For a cognitive jammer, it has the ability to assess the environment of the desired data link and select the type of jammers, the pulse duration, the amount of power required, and other parameters to improve its effectiveness. Unfriendly cognitive jammers generally pose the biggest threat to the communications link since they can adapt to the changes the receiver makes to mitigate the jammer. For example, if the receiver realizes that a pulse jammer has captured the AGC, it can change the time constant. The cognitive jammer monitors the receiver's AGC's response time and changes the pulse timing to maintain maximum jammer efficiency. So no matter what anti-jam process the receiver uses to mitigate the interference, the cognitive jammer adapts to the changes. In addition, a cognitive jammer can learn from the past receiver's mode of operation and then can make more intelligent changes in the future (Figure 9-1).

Friendly jammers. There are many types of friendly jammers—basically anything that is not an intentional jammer. Mainly this includes the equipment either on or in proximity to the platform that causes unintentional interference to the data link. It can also involve multiple users that are trying to communicate at the same time. This equipment can operate at the same frequency, or it can have harmonics or spurious signals that jam the data link. It also can be powerful enough to jam the data link by saturating the front end or increasing the noise floor level, which reduces the sensitivity of the receiver. These types of jammers are often referred to as cosite or co-located jammers, since they exist together at the site or platform of the data link. This is generally on-board equipment or radios that may have other purposes or missions (Figure 9-1).

9.1.2 Channel Degradation

The channel or path of the data link can be degraded by various factors such as atmospheric changes and blockage from obstacles like hills, buildings, and multipath (Figure 9-1). All of these elements can reduce the desired signal level or increase the noise, which in turn can degrade the signal level or quality of service (QoS).

In addition, broadband noise caused by adjacent equipment can degrade the data link by raising the noise floor, which causes the data link to have an insufficient signal-to-noise ratio (SNR).

9.2 Basic Cognitive Techniques

Several cognitive techniques can mitigate the effects of jammers and channel degradation to improve the data link's QoS, including dynamic spectrum access (DSA); adaptive power gain control; modification of the waveform including the order or type of modulation; spread spectrum and error correction; adaptive filters; cosite radio frequency (RF) tunable filters; dynamic antenna techniques using active electronically scanned arrays (AESAs) such as multiple-in, multiple-out (MIMO) antenna systems; and network reconfigurations such as multi-hop, ad hoc meshed networks, self-forming, and self-healing.

9.2.1 Dynamic Spectrum Access

The cognitive capability of DSA is focused on the frequency of operation of the data link. This is generally included in the cognitive radio (CR), but the software resource manager controller can also handle it using a frequency synthesizer in the front end before the radio. The system is capable of scanning the frequency spectrum (miniature spectrum analyzer) to be cognitive of the frequencies and use of the spectrum to make an intelligent decision of what frequency is optimal for use by the data link. The cognitive system continuously monitors or pulses the environment between transmissions to evaluate the spectrum and uses that information to determine the best frequencies with minimal noise and jamming for the data link (Figure 9-2).

This requires system and network coordination to switch frequencies between two users or multiple nodes in a network.

A simple example of DSA between two users is as follows:

1. The data link exhibits a poor QoS or link loss.
2. The base informs the remote of the desired frequency if possible.
3. The base switches to an unused frequency with minimal noise and jamming, or "white space."
4. The base stays on that frequency for a specified time looking for the remote's signal.
5. The remote searches for the base's new frequency using a random search of possible frequencies.

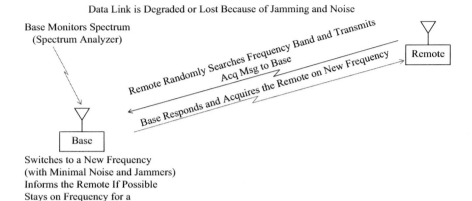

Figure 9-2 Dynamic spectrum allocation.

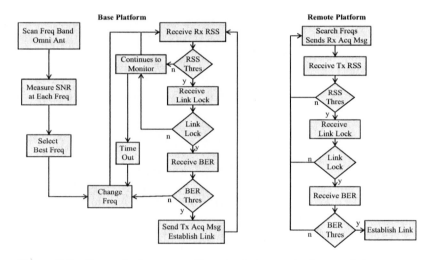

Figure 9-3 Dynamic frequency allocation between the base and the remote.

6. The remote sends a short acquisition sequence on each frequency and listens for a response.
7. The base responds with a short acquisition sequence upon detection of the remote's signal.
8. The process repeats until acquisition is successful.

The example is shown in Figure 9-2. In addition, a flow diagram shows the dynamic spectrum allocation process between a base and a remote (Figure 9-3).

Alternative to Implementing DSA Using Random Frequency Switching. Another alternative to DSA is when the base and the remote have a known random switching frequency sequence. This is known and stored into memory of both systems and can be changed or reprogrammed into the devices before the mission. During communications between the systems, this random frequency sequence can be changed or adapted to the environment.

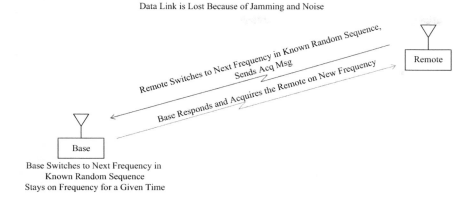

Figure 9-4   Dynamic spectrum allocation alternative.

The process of events of the alternative DSA is as follows:

1. The data link exhibits a poor QoS or link loss.
2. The base switches to the next frequency in the known pseudo-random sequence and stays there for a specified period of time.
3. The base informs the remote of the next frequency if possible.
4. The remote switches to the next frequency in the known random sequence and sends short acknowledgment sequence back to the base.
5. The base responds with a short acknowledgment sequence upon detection of the remote's signal and acquires the remote.
6. The process is repeated until acquisition is successful.
7. Over a period of time or number of unsuccessful tries, the system reverts to the previous DSA process.

This method is shown in Figure 9-4. The disadvantage of this process is that it does not use the cognitive capability and actually can switch to a worse frequency until it finds a good frequency. In addition, there is a probability that the systems will become out of sync and will have to revert to the first DSA random search process. However, the advantage of this DSA process is that the search time for the remote is significantly reduced.

9.2.2 Adaptive Power/Gain Control

There are three basic methods for implementing adaptive power/gain control:

1. Adapt the power/gain based on the received signal strength indication (RSSI) and bit error rate (BER)/LinkLock.
2. Adapt the power/gain level based on the range using the navigation data from either an external source or an internal source to the data link.
3. Use an integrated solution that combines both of these approaches using the range for coarse control and the RSSI/BER/LinkLock for the fine control.

Adaptive Power/Gain Control Using RSSI and BER/LinkLock. This method uses the measured RSSI to determine the amount of gain control in a closed loop system with both the base station and the remote user. The BER or LinkLock indicator is used to verify that the RSSI is the desired remote user and not a jammer or other unwanted signal measurement.

The base station monitors the RSSI that has been verified, and if the power is lower than a threshold the base station increases the power/gain of its output signal to the remote.

The remote signal measures the RSSI and compares it with the previous RSSI from the base to see if there is an increase or decrease in signal level. If the signal level increases, then the remote station increases its signal level by raising power/gain of its output signal to the base station. The remote looks at the delta RSSI received from the base for power/gain control (Figure 9-5).

If the base station receives an RSSI greater than the threshold it has set, it lowers its power/gain to decrease the signal sent to the remote. Upon reception of the lower power from the base station, the remote station lowers its power/gain to send a lower power signal to the base station (Figure 9-5).

The base station sets the power control for each of the remote users using a threshold. This can be a fixed threshold for each of the users or a dynamic threshold for each of the users if the environments including noise and jammer levels are constantly changing. A summary of the tasks for both the base and the remote stations are as follows:

Base Platform:	**Remote Platform:**
Receives RSSI from the remote	Receives RSSI from the base
Receives BER/LinkLock	Receives BER/LinkLock
High RSS: Reduces power output	Delta lower RSS: Reduces power output
Low RSS: Increases power output	Delta higher RSS: Increases power output
Sets AGC threshold level	Adjusts on delta RSS level
Standard power control/gain adjust	

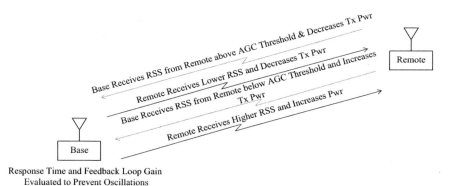

Response Time and Feedback Loop Gain
Evaluated to Prevent Oscillations

Figure 9-5 Adaptive power control between the base user and the remote user.

Control Theory Design for a Stable System. Using this method requires basic control theory since it is a feedback closed loop system and needs to be designed to prevent oscillations. The loop incorporates both the base and the remote stations.

The closed loop feedback system is shown in Figure 9-6. The system uses time constants for each of the blocks and includes both a loop filter to establish a pole and a zero for the root locus to ensure stability and also an integrator for producing a zero steady-state error for a step response. See Chapter 4 for detailed analysis of control systems for AGCs.

Figure 9-7 shows an example of a closed loop system with the time constants for each of the blocks. The example includes the integrator, loop filter, and a fixed threshold

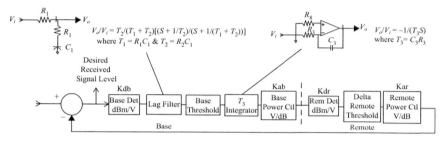

Base and remote second order AGC closed loop response.

Open Loop Transfer Function = $Tsol = G(s) = Kdb*Kab*Kdr*Kar*$
where: $K = Kdb*Kab*Kdr*Kar*(T_2/[T_1 + T_2])*(1/T_3)$

$$\frac{T_2}{(T_1+T_2)}\frac{\left(S+\dfrac{1}{T_2}\right)}{\left(S+\dfrac{1}{T_1+T_2}\right)}*\frac{1}{T_3 S} = \frac{K\left(S+\dfrac{1}{T_2}\right)}{S\left(S+\dfrac{1}{T_1+T_2}\right)}$$

Closed Loop Transfer Function = $Tscl = G(s) / [1 + G(s)H(s)] =$

$$\frac{K\left(S+\dfrac{1}{T_2}\right)}{S^2+\left(\dfrac{1}{T_1+T_2}+K\right)S+\dfrac{K}{T_2}}$$

Figure 9-6 Closed loop analysis using control theory to ensure stability.

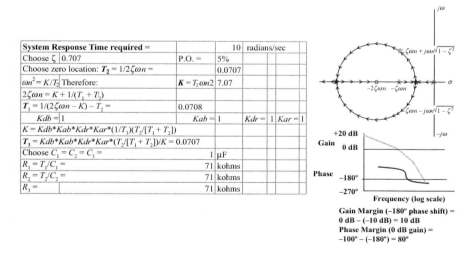

System Response Time required =		10	radians/sec
Choose ζ	0.707	P.O. =	5%
Choose zero location: $T_2 = 1/2\zeta\omega n =$			0.0707
$\omega n^2 = K/T_2$ Therefore:	$K = T_2\omega n2$	7.07	
$2\zeta\omega n = K + 1/(T_1 + T_2)$			
$T_1 = 1/(2\zeta\omega n - K) - T_2 =$		0.0708	
$Kdb = 1$		$Kab = 1$	$Kdr = 1$ $Kar = 1$
$K = Kdb*Kab*Kdr*Kar*(1/T_3)(T_2/[T_1+T_2])$			
$T_3 = Kdb*Kab*Kdr*Kar*(T_2/[T_1+T_2])/K = 0.0707$			
Choose $C_1 = C_2 = C_3 =$		1	µF
$R_1 = T_1/C_1 =$		71	kohms
$R_2 = T_2/C_2 =$		71	kohms
$R_3 =$		71	kohms

Gain Margin (–180° phase shift) =
0 dB – (–10 dB) = 10 dB
Phase Margin (0 dB gain) =
–100° – (–180°) = 80°

Figure 9-7 Detailed example of using control theory to ensure stability and prevent oscillations.

(can be variable for more agile system). In a practical case, multiple RSSIs are averaged for a given output value and are verified with a valid LinkLock indicator. The overall response time required needs to be slower than dynamics of the averaged RSSI.

Description of the Closed Loop Analysis for a Base and Remote Power Control. As the RSSI is received at the base, it is detected to produce a voltage level responding to the power of the RSSI. This voltage is fed through a lag filter to generate poles and zeros to stabilize the control loop. The base threshold sets the voltage level threshold, responding to the desired received signal level that is optimal for the detection of the signal. The integrator is used to ensure that there is zero steady-state error and that it holds the desired receive signal exactly at the correct level. If there is an error from the threshold, at steady state, it will be amplified by theoretical infinity to ensure no errors when the input level is not changing. The voltage-controlled amplifier converts the voltage level to a gain level, which is subtracted from the maximum gain to a value that amplifies the RSSI signal to the desired signal strength set by the threshold.

With the addition of the remote in the feedback loop, the gain of the detector and the gain of the power control are added to the forward path of the feedback loop. The delta threshold adjusts the level depending on the previous level.

For the base, a large signal input causes a large signal gain, which is subtracted from the maximum gain, so a large signal reduces the gain. Adding the remote function, this large gain is further amplified by the gains in the remote, which further reduces the gain from the maximum gain.

So in summary, a large signal into the base reduces the signal level output, and this delta-reduced signal into the remote reduces the signal further, which is an extra gain toward a smaller signal or gain.

The loop shows that the larger the RSSI, the larger the gain for both the base and remote and therefore the smaller the overall gain since it is subtracted from maximum gain in the base. If it was just the base, then the reduction and feedback loop would end at the dotted line (Figure 9-6). With the remote as part of the feedback loop, it adds to the gain increase, which reduces the overall gain response faster than just the base.

In the actual performance, when the signal is high at the base the base reduces the signal level and the remote detects this reduction of signal level and further lowers the signal level, which is accomplished in the total feedback loop, as shown in Figure 9-6.

Control theory feedback analysis determines the loop filter parameters to prevent oscillations. ζ is chosen for a 5% overshoot, which is a trade-off between response time and settling time of the control. The overall response time is chosen at 10 radians/sec, which provides ample time for power control. This allows the response time to be fast enough for high aircraft maneuvers and also provides the time to complete the power control in the total loop response. The root locus plot shows the stability of the system (Figure 9-7). For further information regarding control theory, the reader is referred to Chapter 4.

A Bode plot analysis is also included to verify a stable closed loop system (Figure 9-7). The Bode plot determines the gain and phase margin of the system as an indicator of stability.

Adaptive Power Using Navigation Data (External or Internal) for Range. The range can be calculated using the available navigation data, such as an input from an external source like a global positioning system (GPS), or it can be included in the data link itself, such as the internal data message of the common data link (CDL). The process is shown in Figure 9-8.

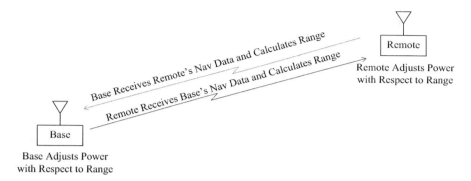

Figure 9-8   Adaptive power control between a base and a remote using navigational data for range.

For the range solution, a simple stepped power control can be used. Here is an example of a simple stepped range solution using range thresholds and power steps.

Simple Stepped Power Control for Range Data:
If range is greater than 20 nmi, set to full power.
If range is less than 20 nmi, reduce power by 10 dB.
If range is less than 5 nmi, reduce power by another 10 dB.
Add hysteresis (1–5 nmi) to prevent continual switching at the transition.

Note: Finer step sizes can be used depending on the dynamics of the system.

A summary of the tasks for both the base and the remote stations are as follows:

Base Platform:
Calculates range from navigation data
Adjusts power with respect to range
Sends the power level to the remote
If range is not available:
Use closed loop analysis

Remote Platform:
Calculates range from the navigation data
Adjusts power with respect to range
Sends the power level to the remote
If range is not available:
Use closed loop analysis

Integrated Solution of Closed Loop and Range. This solution combines the closed loop RSSI and BER/LinkLock with the range solution using navigation data. The navigation data for range is used for a coarse power control (e.g., 10 dB steps), while the closed loop RSSI and BER/LinkLock method is used for fine power control (continuous adjustments) (Figure 9-9).

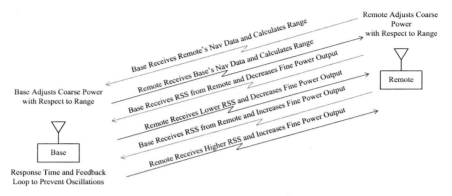

Figure 9-9 Integrated solution for adaptive power control.

A summary of the tasks for both the base and the remote stations are as follows:

Base Platform: Calculates range from navigation data Adjusts coarse power output with respect to range Receives RSSI from the remote Receives BER/LinkLock High RSSI: Reduces power output Low RSSI: Increases power output Fine power output adjustment Sets AGC threshold level Standard power control/gain adjust	**Remote Platform:** Calculates range from navigation data Adjusts coarse power output with range Receives RSSI from the base Receives BER/LinkLock Delta lower RSSI: Reduces power output Delta higher RSSI: Increases power output Fine power output adjustment Adjusts on delta RSSI level

A flow diagram shows the adaptive power control process between a base station and a remote station (Figures 9-10 and 9-11).

Another alternative in cellular communications is to allow the base to send out power control information to users, telling them what power they should use at the measured range. This assumes that the link is available and that it has the means of getting this message to the remote stations. This same technique could be used in this application but is not as highly dynamic as the closed loop process.

9.2.3 Cognitive Techniques Using Modulation Waveforms

Trade-offs between modulation types can be adapted to the changing environments using software-defined radio (SDR) techniques. The ability to change modulations in real time as the environment changes provides a unique cognitive solution. The modulation can be changed based on the SNR of the received signal. In a clean environment, where the SNR is high, a higher order modulation can be used to increase the data throughput. If the environment changes to a noisy jamming environment, which reduces the SNR, then a more

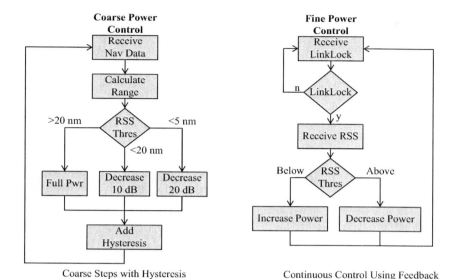

Figure 9-10   Base station adaptive power control flow diagram.

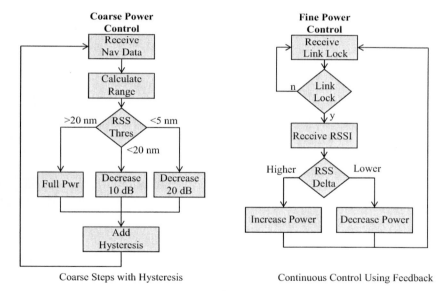

Figure 9-11   Remote station adaptive power control flow diagram.

robust, lower order modulation type can be used. The noise or jammer immunity for two different types of modulations is shown in Figure 9-12.

In a clean environment, the modulation with the maximum data rate is used, or 16-state quadrature amplitude modulation (16-QAM). In noisy and jamming environments, the system adapts by switching to lower order modulation with better noise and jammer immunity

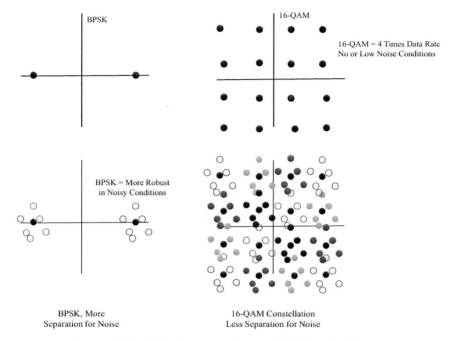

Figure 9-12 Noise immunity between BPSK and 16-QAM.

at the expense of lower data rates, or binary phase-shift keying (BPSK). This is a trade-off between data rate and anti-jam capability: with higher data rate, there is less anti-jam; with lower data rate, there is more anti-jam. The threshold for the SNR to switch the system to a different modulation is either set or can be an adaptive threshold, which has higher performance at the cost of complexity.

The 16-QAM modulation provides four times the data rate compared with BPSK. However, it is affected more from a noisy environment and is much harder to detect without errors. The BPSK modulation is much more robust in a noisy environment and is easily detected but has a lower data rate (Figure 9-12). So the cognitive solution adjusts the modulation according to the conditions of the environment. With an SDR, both modulations are available, and the type of modulation is selected very quickly. Other modulations can be made available for selection as needed. The required E_b/N_o for several types of modulations is shown in Figure 9-13.

9.2.4 Spread Spectrum for Increased Process Gain against Jammers

Another method that can be used as a compromise between data rate and anti-jam is the amount of spread spectrum used. The more spread spectrum that is used, the more anti-jam at the expense of a lower data throughput. This technique can also be employed in a cognitive sense. If the data link is operating in a harsh environment with jammers, more spread spectrum can be applied. If the data link is operating in an environment that is free from jammers and interference, then minimal or no spread spectrum is required and the data rate is improved.

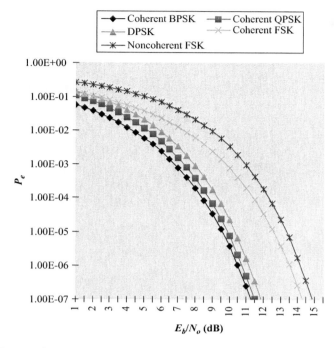

Figure 9-13   Required E_b/N_o for a given probability of error for different modulations.

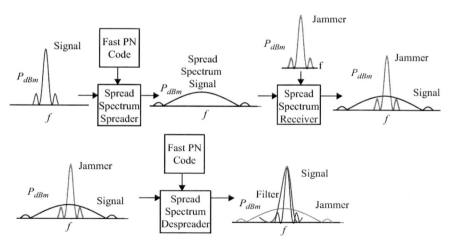

Figure 9-14   Spread spectrum techniques to provide anti-jam capabilities.

The main two types of spread spectrum are direct sequence and frequency hop. Spread spectrum systems are discussed in Chapters 2 and 6 for further details. Either of these can be changed rapidly with an SDR for use in a cognitive system. The system determines the amount of jamming by monitoring the spectrum, using the signal-to-jammer ratio, or BER. An example of spread spectrum is shown in Figure 9-14. The narrowband desired signal is spread via a fast pseudo-noise (PN) code and sent out of the transmitter as a spread spectrum signal. Both the spread spectrum signal and the narrowband jammer are then delivered to the receiver. The receiver contains the same fast PN code as the transmitter, and when the codes are aligned the design signal is despread to the narrowband desired signal. At the same time, the fast PN code in the receiver spreads out the jammer so that the signal is now much higher than the jammer and can be easily filtered and detected (Figure 9-14).

9.2.5 Adaptive Error Correction

Error correction can also be altered depending on the jamming environment. The more error correction, the better the anti-jam performance at the expense of a decreased data throughput, since it requires additional bits to provide error correction. If the environment is clear, with low jamming levels, then the system can operate with less to no error correction, which increases the data throughput. If the environment is noisy and being jammed, then error correction is increased at the cost of reducing the data rate. This also can be cognitive with the determination of the jamming environment similar to the spread spectrum scenario.

9.2.6 Adaptive Filter for Jammer Mitigation

The adaptive filter can reduce narrowband jammers and changes its response as the jammer changes frequencies. This is discussed in detail in Chapter 8. By its nature, the filter is adaptive in frequency. If the environment contains a high level of narrowband interference, the adaptive filter can be incorporated into the receiver to eliminate these narrowband jammers. If the narrowband interference is not present, the adaptive filter can be disabled to improve the processing time of the received signal. In addition, the number of weights or the length of the tapped delay line can also be adaptive dependent on the type of interference. If the jammers are CW, then the number of weights and length of the tapped delay line can be reduced to improve process time. If the jammers are broader band, then more weights and longer tapped delay lines are required.

A digital adaptive filter automatically adjusts to a changing narrowband jammer. It uses an adaptive line enhancer to obtain the narrowband jammer, which in turn is used to cancel the unwanted jammer (Figure 9-15).

The de-correlation delay is the key to the operation of the adaptive filter since it separates the narrowband jammer from the wideband signal. The adaptive filter has the ability to cancel jammers in both the digital domain and also at RF or intermediate frequency (IF) if needed with degraded performance.

Other adaptive methods for reducing unwanted jamming signals include the Graham-Schmidt orthogonalizer (GSO; see Chapter 8) and cosite RF filtering.

9.2.7 Dynamic Antenna Techniques Using AESAs

Several techniques can be included in the cognitive system with AESAs. Many of these can be employed with mechanically steered antennas, but the AESA provides a much improved means to accomplish these dynamic capabilities.

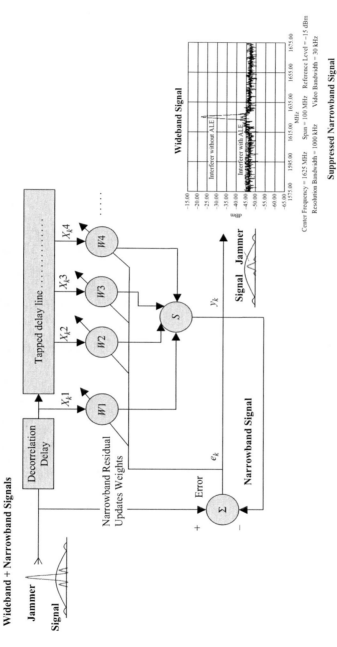

Figure 9-15 Digital adaptive filter reduces the narrowband jammers in the environment.

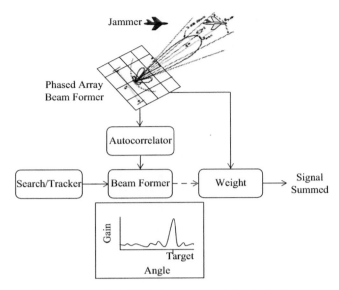

Figure 9-16   Beam former steers the AESA beam toward the desired target and away from the jammer.

Beam Steering. Beam forming and steering can be accomplished quickly using an AESA. The technique is to adapt the pointing angle of the AESA to steer away from the jammer and point it in the direction of the desired user (Figure 9-16). This is dependent on the beam width and how far spatially the jammer is from the desired user. The system determines the position of the jammer and the position of the user and then establishes where to point the beam with the best signal from the desired user and the minimum signal from the jammer.

Null Steering. AESAs are versatile and can create a null at the location of the jammer. The system determines where the jammer is spatially located and the AESA steers a null in the direction of the jammer to reduce the jammer level into the receiver (Figure 9-17).

Beam Spoiling and Narrowing. Beam spoiling involves widening the beamwidth. Beam narrowing involves narrowing the beamwidth. Both of these techniques can be easily accomplished using an AESA (Figure 9-18). The narrow beam is better for power and improved spatial performance over jammers. The wide beam is better for wider coverage and better volume search performance. The cognitive AESA adapts to the desired beamwidth for coverage versus jamming performance. Beam spoiling reduces search time but lowers the link margin and increases vulnerability to jammers. The system reviews the jamming environment and system requirements and then determines whether a narrow beam is required for interference rejection or a wider beam can be used for improved coverage performance.

9.2.8 Multipath Communications

Multipath is generally the unwanted path that interferes with the desired direct path of the signal. However, it can be used as the desired path if the direct path is either blocked or

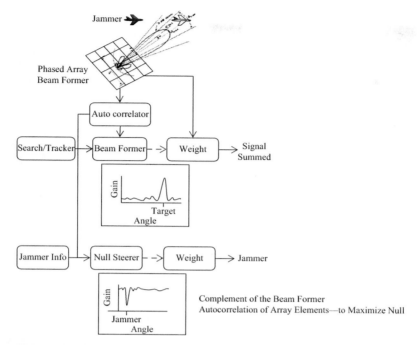

Figure 9-17   Null steerer creates and steers a null of the AESA toward the jammer.

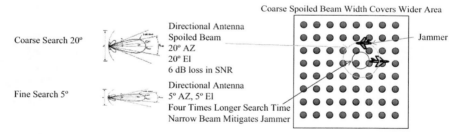

Figure 9-18   Adaptive beam spoiling and beam narrowing provide coverage and anti-jam trade-offs.

jammed. It may be the only feasible path to the target (Figure 9-19). This technique can be useful to transmit around obstructions or to point the antenna away from the source of jamming. It uses the alternate multipath of the signal for the communications path (Figure 9-19). The system can determine if there is blockage or jammers in the main path and automatically steer the AESA to a multipath position to receive the desired signal.

9.2.9 Multiple Antennas

Multiple antennas can be used on the base platform, the remote user, or both. The types of multiple antennas are shown in Figure 9-20.

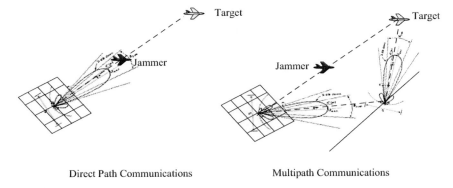

Direct Path Communications   Multipath Communications

Figure 9-19   Multipath used as the desired path for communications when the direct path is blocked or jammed.

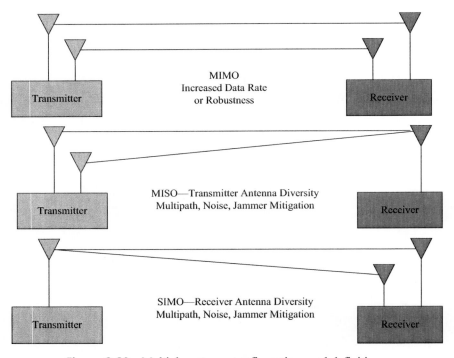

Figure 9-20   Multiple antenna configurations and definitions.

MIMO defines multiple antennas at both the transmitter and the receiver for increased data rates or robustness to the harsh environment including anti-jam, multipath, and noise performance. Multiple-in, single-out (MISO) identifies multiple antennas at the transmitter and a single antenna at the receiver for transmitter antenna diversity. The single-in, multiple-out (SIMO) defines a single antenna at the transmitter and multiple antennas at the receiver, which is used for receiver antenna diversity that is primarily used to mitigate multipath.

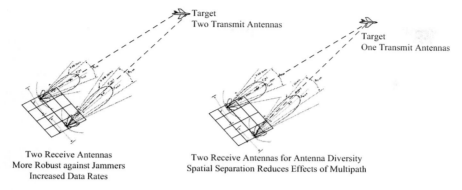

Figure 9-21   MIMO trade-offs between data rate, anti-jam, and multipath.

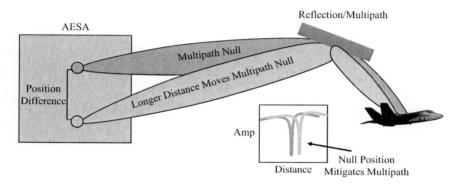

Figure 9-22   Antenna diversity mitigates multipath signals.

MIMO is used to increase data rates or improve anti-jam. The trade-off between data rate and anti-jam can be adaptive due to the environmental conditions. If the jammer and noise levels are high, MIMO is used for improved anti-jam performance by sending the same data through multiple antennas using multiple paths. If one of the paths is unreliable due to jamming or multipath, the other paths contain reliable data. If the jammer and noise levels are low, then MIMO is used for improved data rates by sending different data in each of the paths and then combining these data paths in the receiver.

In addition, antenna diversity can be used to mitigate multipath (Figure 9-21). This includes using multiple antennas at the receiver end to cancel the unwanted multipath. Since the path lengths are different, if one of the antennas is in a multipath null, then the other one is not in the null since the null changes according to position and distance to the transmitter (Figure 9-22).

9.2.10 Network Configurations

Networks can be adaptively configured to provide the optimal solution for a cognitive system. One of the aspects of a network is the ability to do multi-hop. Multi-hop networks can communicate to one node of the network through another node, or relay node. For example, if a direct path to a specific remote user is blocked or being jammed, the cognitive network can sense the problem and perform a multi-hop to the remote user via another user or node (Figure 9-23). If the system contained power control only for the adaptive process, the power would continue to increase to overcome the obstacle making the users vulnerable for detection and also causing unwanted jamming of other nearby network users or other systems. In the case of blockage, DSA would not be successful since the blockage would attenuate all frequencies. Multipath communications may be an alternative, but it may be difficult to find a clear multipath direction and is actually more complicated than simply using the network configuration to solve the problem. Implementing a cognitive system approach that can evaluate and process the information and select the optimal solution from several options and capabilities would produce the best approach. In this case, the system would evaluate all of the possible solutions and choose the network multi-hop solution for the optimal performance.

9.2.11 Cognitive MANETs

Cognitive networks are used to analyze and reconfigure the network according to the changing environment. Many network capabilities can be monitored and made adaptive. Mesh and mobile ad hoc networks (MANETs) have the ability to interconnect multiple nodes either directly or using multi-hop techniques. The latter uses a node in the network as a relay to send communications to another node, which extends the network's range and also allows the network to adapt to a changing environment. For example, in Figure 9-23, two nodes desire to communicate with each other, but the path or channel is being blocked, which prevents a stable link. The network adapts using another node in the network as a relay to overcome this environmental blockage so that communication is restored between the two nodes. This scenario could also be used if a jammer in the communication path can be overcome through a multi-hop network.

Since MANETs are constantly changing due to the mobility of the network, they can use adaptive techniques as they self-form and self-heal the ever-changing network and environment. In addition, a cognitive network possesses the ability to learn from the consequences of the cognitive action that was utilized to account for the changing environment. This is important in determining how well the cognitive change affects the overall system or network and using that knowledge to provide a better overall solution. This introduces a smart learning process that can continually improve the QoS of the network and system.

Figure 9-23   Network multi-hop solution overcomes blockage or jamming of the direct link.

MANETs can monitor and adapt to many changing environmental capabilities. Some of these are listed in Table 9-1.

The list in Table 9-1 is extensive and constantly growing with new technology and increased network demands. Many techniques, including gaming theory, are used to determine the best approach for adapting networks for optimal performance. The network decisions are modeled using all of the capabilities of the individual components to provide a complete cognitive solution. These approaches attempt to provide the optimal solutions given many variables that exist in a MANET. However, the basic concepts are common to all adaptive techniques and cognitive networks.

Another application in developing a cognitive network or system is to prevent interference to other operators of the band including primary users. When the industrial, scientific, and medical band became available for digital cellular telephones, strict requirements were necessary to avoid interference with its primary users (i.e., operators in those industries). Spread spectrum techniques were used to prevent jamming.

Cognitive networks and systems can also prevent interference to the primary users and other operators by sensing the bands of operation and techniques incorporated by the

Table 9-1 Cognitive possibilities for an adaptive MANET

Capability	Description
Traffic load and flow (congestion from multiple users)	Schedules operation times, priority users, data rate throughput.
Usage with respect to time of day or missions	Schedules users at different times of the day, and according to mission parameters.
Nodes coming in and out of the network	Acquires and raise awareness of nodes, more or less nodes, more or less bandwidth.
Media access control (MAC) layer organization	Adapts MAC layer for improved networking performance.
Optimization of network for minimum hops, best path in the network	Minimizes multi-hops unless needed, provide the shortest best path for communications.
Network topology including multi-hops	Extends the range of the network and prevents blockage and jamming effects with multiple hops.
Network conditions	Monitors network conditions, jammers, noise, and blockage and configures the network for optimal performance.
Number of hops permitted	Adapts to the environment to determine the number of hops permitted, depends on bandwidth and timing.
Link quality between nodes	Adapts to link quality by changing a function, frequency, modulation, power, or multi-hops.
Priority messaging and hierarchy	Provides adaptability for priority users to ensure the required bandwidth for them.
Connectivity versus coverage	Makes trade-offs between how well the network is connected versus how far the network can extend.
Time division multiple access (TDMA) structure	Adapts the timing and slot assignment for optimal network performance.
Carrier sense multiple access (CSMA) and collision avoidance	Senses the carrier to prevent users from jamming each other.
Types of communications such as broadcast, multicast, unicast	Adapts to the type of communications that is needed in the network given environment and mission.

primary users and rapidly change their cognitive capabilities to prevent interference. Cognitive networks can also learn users' patterns and adapt this information to further enhance their cognitive capability. For a simple example, if a user is operating the band only at a specific time, then the cognitive network can learn that characteristic and not use the band at that time to prevent interference. The cognitive techniques listed in Table 9-1 can be used to prevent interference in accordance with the learned information of the cognitive network.

9.3 Cognitive System Solution

The main cognitive focus of this chapter has been on DSA, power control, and networking. A complete cognitive system solution would include these and also would monitor all of the environmental parameters and control all of the available hardware, software, and networking capabilities.

A cognitive system therefore evaluates all of the hardware and software modules that make up the system. The modem contains many aspects of a cognitive system, including DSA, power control, modulation type, error correction, adaptive filters, and spread spectrum. The antenna system also has cognitive capabilities such as beam steering, null steering, MIMO, antenna diversity, beam spoiling, and beam narrowing. AESAs provide the means to accomplish many of these cognitive capabilities. Other hardware such as the RF front ends, transmitter, and receiver can provide cosite mitigation, multipath mitigation, power control, and DSA. Software and system configurations such as networking topology, trade-offs and optimization, and software monitor and control comprise the rest of the cognition capabilities.

The cognitive system discovers what capabilities are available to the system, evaluates all possible mitigation techniques, determines which solutions are optimal given the known parameters, and implements and controls the hardware and software modules (Figure 9-24). The cognitive system determines the best solution for time, effort, vulnerability, optimization, range, and QoS. These decisions are evaluated with respect to a price or cost to the system, which determines the impact to the system's performance before the decision is made. This allows for the optimal solution with the least impact to the system's performance.

An example of a cognitive system is shown in Figure 9-25. The system uses a cognitive monitor-and-control (M&C) system that is central to all of the hardware and software modules. This block determines the optimal cognitive solution for the system by monitoring the environment and then controlling the available hardware and software modules.

A flow diagram shows the process of the cognitive system (Figure 9-26). The incoming energy is monitored and detected and passed on to the next process, which performs the trade-offs. The next process determines the optimal solution and selects the device used (e.g., AESA, power control module), and the final process changes the selected device to mitigate the problem.

The QoS of any data link system can be reduced by a low level signal where the SNR is not large enough to prevent a high BER in reference to the noise floor, or it can have a sufficient power level that is being jammed, which produces a low signal-to-jammer ratio (SJR). Each of these scenarios needs to be addressed in a slightly different manner by the cognitive solution. An example using some of the capabilities of a cognitive system addressing both these needs is shown in Figure 9-27. If the signal is being jammed by a narrowband jammer, an adaptive filter could be enabled to mitigate the jammer. In addition, it could implement further

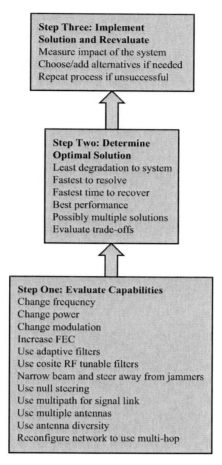

Figure 9-24 Cognitive system processes to provide the optimal solution for the system.

mitigation techniques such as changing the frequency, increasing the power, or changing the modulation, beam steering, or beam nulling. The broadband signal does not utilize the adaptive filter but has other techniques at its disposal. In addition, both narrowband and broadband jammers can be mitigated by reconfiguring the network or using multipath communications. If the system does not have some of these capabilities, they can be skipped, and if they have additional capabilities they can be employed. When the degradation is due to a low signal level and not jamming, many of the same techniques can be used, but without the adaptive filter or beam steering and nulling. MIMO and antenna diversity would apply if the signal is in a noisy environment or in a multipath null (Figure 9-27).

Since there are many capabilities and system solution options for each of the scenarios, it is not practical to describe them all. The optimal solution will depend on the application. One approach to combining several capabilities is shown in Figure 9-28 for the base platform and in Figure 9-29 for the remote platform. The decision blocks with $C = n$ denote the number of times the change has occurred. For example, if the path shows to change the frequency, if the

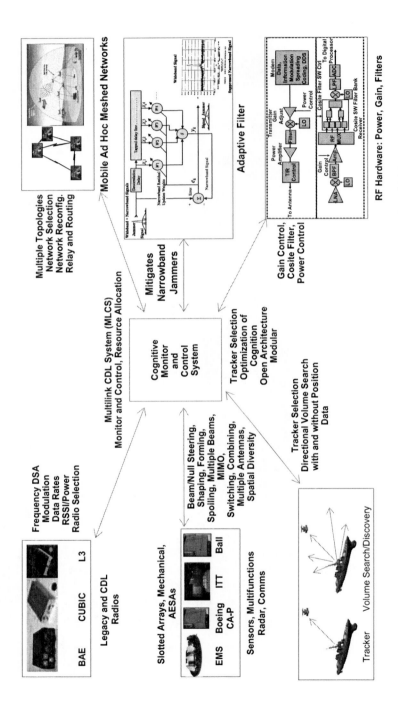

Figure 9-25 Cognitive system example provides an optimal solution using all available resources.

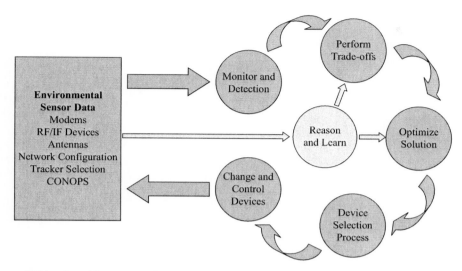

Figure 9-26   Cognitive system flow diagram showing the process for cognitive implementation.

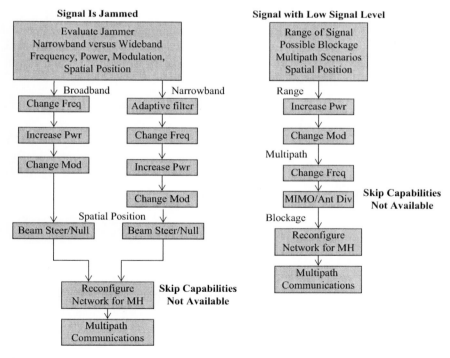

Figure 9-27   Cognitive system addresses both the low SJR and the SNR separately.

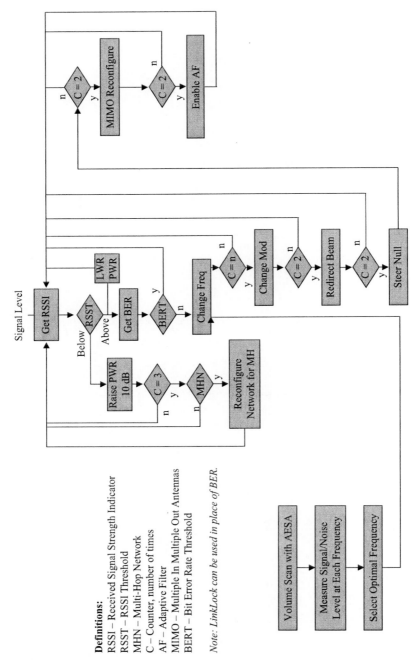

Definitions:
RSSI – Received Signal Strength Indicator
RSST – RSSI Threshold
MHN – Multi-Hop Network
C – Counter, number of times
AF – Adaptive Filter
MIMO – Multiple In Multiple Out Antennas
BERT – Bit Error Rate Threshold

Note: LinkLock can be used in place of BER.

Figure 9-28 Cognitive system of the base platform using multiple cognitive capabilities.

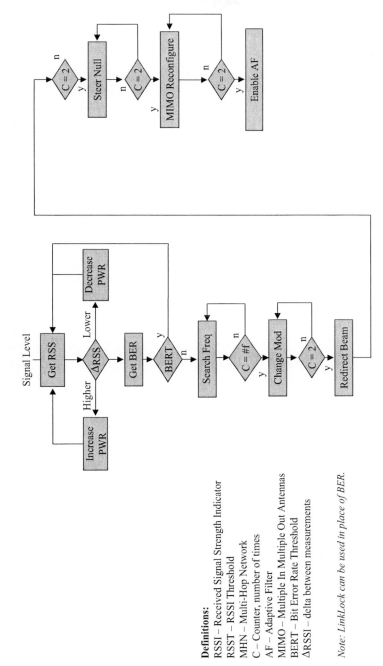

Definitions:
RSSI – Received Signal Strength Indicator
RSST – RSSI Threshold
MHN – Multi-Hop Network
C – Counter, number of times
AF – Adaptive Filter
MIMO – Multiple In Multiple Out Antennas
BERT – Bit Error Rate Threshold
ΔRSSI – delta between measurements

Note: LinkLock can be used in place of BER.

Figure 9-29 Cognitive system of the remote platform with multiple cognitive capabilities.

number of times the frequency has changed in the flow equals to C, then the flow continues to the next change, in this case the change in modulation.

The base platform approach starts with the power control capability by comparing the received RSS to a threshold and raising or lowering the power level, which is verified by the BER or LinkLock. The latter is generally sufficient to determine if it is the correct signal. The numbers in the decision blocks indicate the number of times it changes. It then branches off to either a multi-hop network or frequency control. If this is not successful, then it changes the modulation type. Beam forming and null steering are next, and if that is unsuccessful then MIMO reconfiguration and enabling the adaptive filter are utilized (Figure 9-28).

The remote platform approach starts the power control capability by receiving the RSS level and the BER/LinkLock verifier. If the BER/LinkLock is good, then the received RSS level is compared with the previously received RSS level. If it is lower then the remote decreases its power output, and if it is higher then the remote increases its power output. If the BER/LinkLock is bad, then it branches off to a frequency search to find the correct base frequency. If that is unsuccessful a given number of times, then the modulation is changed, the beam and null steering are enabled, and finally MIMO is used and the adaptive filter is enabled (Figure 9-29).

Cognitive devices have progressed from traditional relatively static designs to variable capabilities (Figure 9-30). However, for a system to provide a complete, optimal, cognitive solution by evaluating all of the variable devices, the system cognition uses the information of the environment and available devices and makes the optimal cognitive system solution, as shown in Figure 9-30. If the individual devices are cognitive without the system solution, the individual controls may not be optimum, that is, the aforementioned multi-hop network solution. The M&C system is used for all the capabilities in the system and can make the optimal system cognitive choice.

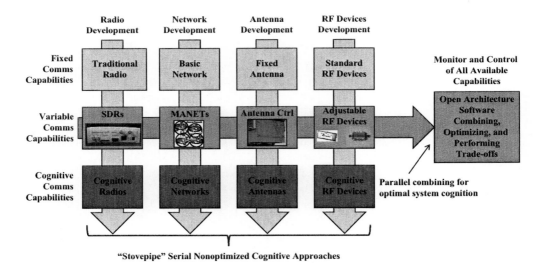

Figure 9-30 Serial versus parallel evolution of cognitive capabilities.

9.4 Summary

Cognitive systems are used to adapt the operating system to the changing environment. This can be accomplished by many methods suggested in this chapter, and many additional methods will be available in the future. The optimal cognitive system solution evaluates a system's available capabilities as well as all of the available knowledge about the changing environment, and then it calculates, makes trade-offs, and determines the best solution for the system to adapt to these environmental changes with minimal impact to performance. In addition, the cognitive system contains a learning capability that uses past experiences and impact/results of changes and applies this information to make smart decisions in the future.

References

Hossain, Ekram, and Vijay Bhargava. *Cognitive Wireless Communication Networks*. Springer, NY, 2007.

Doyle, Linda E. *Essentials of Cognitive Radio*. Cambridge University Press, MA, 2009.

Fett, Bruce A. *Cognitive Radio Technology*, 2nd ed. Academic Press, MA, 2009.

Wyglinksi, Alexander M., Maziar Nekovee, and Y. Thomas Hou. *Cognitive Radio Communications and Networks: Principles and Practice*. Elsevier, NY, 2009.

Rabbachin, Alberto, Tony Q.S. Quek, Hyundong Shin, and Moe Z. Win, "Cognitive Network Interference," *IEEE Journal*, 2011, **29**(2), pp. 480–493.

Thomas, Ryan W. "Cognitive Networks," PhD diss., Virginia Tech, 2007.

Problems

1. List at least three environmental factors that drive the demand for cognitive systems.
2. What are seven type cognitive techniques that could be used to mitigate problems?
3. What does DSA stand for, and how is it used to mitigate problems?
4. Describe two methods that can be used for switching frequencies.
5. Describe two methods that can be used for power control.
6. What technique is used to prevent oscillations in one type of power control?
7. Do one or both users establish a gain control threshold?
8. What technique is used to prevent constant power switching in a changing environment?
9. What is used to ensure that power threshold level is maintained in the AGC?
10. What technique is used to eliminate narrowband jammers with a broadband signal?
11. How does an AESA reduce the effects of a jammer?
12. What are the two advantages of using MIMO antennas?
13. What is antenna diversity, and how can it reduce multipath?
14. How can multipath be used as an advantage when the direct path is unavailable?
15. What technique does a network use to extend the range and eliminate direct path jamming?
16. What advantages do integrated system cognition have over just a cognitive radio?

CHAPTER 10

Broadband Communications and Networking

Broadband refers to technology that distributes high-speed data, voice, and video. It is used for broadband communications, networking, and Internet access and distribution and provides a means of connecting multiple communication devices. Several generations of wireless communication products—designated as 1G (first generation) through 4G (fourth generation)—have evolved over the years and will continue to advance in the future.

Broadband is also used in the home to connect to the outside world without having to run new wires. In this application, information is brought to the home in various ways including power lines, phone lines, wireless radio frequencies (RFs), fiber-optic cable, coaxial cable, and satellite links. This information generally comes into the home at a single location, so it is necessary to have a way to distribute it throughout. Along with the distribution of information, networking plays an important role in the connection and interaction of different devices in the home.

The military is investigating several networking techniques such as the Joint Tactical Radio System (JTRS) and Link 16 to allow multiple users in a battle scenario for communications, command, control, and weapon systems.

10.1 Mobile Users

Wireless communications have progressed from simple analog cellular telephones to high-speed voice, data, music, and video using digital communications techniques. 1G devices provide wireless analog voice service using advanced mobile phone service (AMPS) but no data services or Internet connection.

2G devices use digital communications with both voice and data capability. The data rate for 2G is from 9.6 to 14.4 kbps. Several techniques are used to provide 2G communications, including code division multiple access (CDMA), time division multiple access (TDMA), global system for mobile communications (GSM), and personal digital cellular (PDC) for one-way data transmission. This type of connection does not have always-on data connection, which is important for Internet user applications.

3G is an enhanced version of the 2G digital communications with an increased data rate of 114 kbps to 2 Mbps; always-on data for Internet applications; broadband services such as

Table 10-1 Generations of mobile radios

Generation	Data Rate	Modulation Access	Additional Features
1G	Analog	AMPS	Voice, no data or Internet services
2G	9.6–14.4 kbps	GSM, CMDA, TDMA, PDC	Voice, data, no always-on for Internet
3G	114 kbps-2 Mbps	CDMA (W-CDMA), CDMA-2000, TD-SCDMA	Always-on for Internet, video, music, multimedia, voice quality, enhanced roaming
4G	1 Gbps	SDRs, OFDM, MIMO, IFDMA, Turbo codes, Cognitive radios	Adaptation for interoperability and real-time modulation changes

video, music, and multimedia; superior voice quality; and enhanced roaming. The three basic forms of 3G are wideband CDMA (W-CDMA), CDMA-2000, and a time division scheme called time division synchronous code division multiple access (TD-SCDMA).

4G provides a means of adapting the communication link so that it is interoperable with multiple communication techniques and higher data rates up to 1 Gbps. This is accomplished using software-defined radios (SDRs) for real-time modulation changes and cognitive radio techniques that sense the environment and adjust the modulation and other parameters according to the signals present. A summary of the four generations is shown in Table 10-1.

10.1.1 Personal Communications Services

The cellular telephone used analog techniques for many years before the creation of personal communications services (PCS). This technique uses digital communication modulation to send digital data wirelessly. To enable multiple users in the same frequency band, CDMA, TDMA, and frequency division multiple access (FDMA)—or a combination of any of these multiple access schemes—were implemented (Figure 10-1). The Federal Communications Commission (FCC) allowed companies to develop their own multiple access solution and chose not to impose a standard for the development of PCS. This caused an issue with interoperability. To combat this issue, two main solutions emerged. The first, CDMA, was developed and adopted early on by several companies as a standard for sending and receiving digital communications. The second, GSM, was patterned after the standard in many European countries. Most PCS hardware accommodates one of these two approaches. Many PCS telephones are capable of using more than one technique. Also, other companies competing for the interoperability "standard" used TDMA and a combination of CDMA/TDMA.

Some systems have both PCS and cellular modes to allow the end user a choice and to provide additional coverage and versatility. Since the onset of PCS hardware, the cellular industry also has developed digital communication techniques to enhance its performance and to compete with PCS. Some systems incorporate PCS, digital, and analog cellular all in one handset, but the trend now is to eliminate the analog and old cellular devices, which are obsolete, and to focus on the PCS technology and beyond.

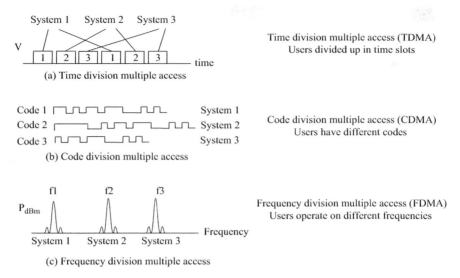

Figure 10-1   Multiple access methods for PCS communications.

GSM. The GSM standard originated in Europe and has been used for years, which has allowed for many improvements. The standards in Europe include GSM-900 and DCS-1800. In addition, the Digital European Cordless Telecommunications (DECT) in Europe set a popular standard that provides digital wireless communications and the DECT wireless telephone.

The United States followed the standard set in Europe by using GSM technology, but at a slightly different frequency. The standard for GSM in the United States is called PS-1900, which operates at 1900 MHz band.

Modulation for GSM. Gaussian minimum shift keying (GMSK; see Chapter 2) is used to modulate GSM system because of its reduced sidelobes and out-of-band transmissions, making it a spectrally efficient waveform with its ability to send data at a high rate for digital communications. The bandwidth specified to send voice or data over the air is 200 kHz. This allows multiple users to be on the band at the same time with minimal interference. During this process, known as FDMA, the users occupy a portion of the band that is free from interference. GSM also incorporates frequency-hopping (FH) techniques, controlled by the base to further prevent multiple users from interfering with each other.

The specification for GSM makes it possible to send control bits to set power output, which also helps to prevent interference from other users since they should use only the minimum power required for a reliable and quality connection. The power control helps the near–far problem, which occurs in a handset that is close to the base. If the handset is allowed to transmit at full power, then it will jam most of the other handsets. If power control is used, then the handset's power is reduced to the necessary power needed. This prevents jamming and saturation of the base and allows for handsets that are farther away from the base unit to turn up the power and reach the base. The optimum system would be to have the power level close to the same level regardless of the distance from the base station. The specification for GSM also allows adjustment of the duty cycle.

The data rate for GSM is approximately 277 kbps, with a 13 kbps vocoder. Two types of GSM are specified: Reflex 25, which uses two frequency channels; and Reflex 50, which uses four frequencies for twice the data speeds. The rate of change from one frequency to another is 2.4 kHz. Digital signal processors (DSPs) reduce the distortion in the overall system and detect the information reliably.

10.1.2 Cellular Telephone

Cellular telephones have been around for several decades. The first ones used an analog frequency modulation system that operated in the 800 MHz band under AMPS. With the arrival of PCS digital telephones, cellular needed to transition to digital modulation techniques to compete in the marketplace. Cellular had the advantage of an already established and well-proven infrastructure and allocated antenna sites. The PCS industry had to start from scratch. Cellular companies developed a digital modulation system and combined it with analog technology to create what they called the dual mode AMPS. They also implemented TDMA IS-54/IS-136 and CDMA techniques as digital cellular operation.

10.1.3 Industrial, Scientific, and Medical Bands

The FCC designated frequency bands to be used solely by industrial, scientific, and medical (ISM) organizations (Table 10-2). These ISM bands were chosen for the PCS community as secondary users with strict requirements to not interfere with the existing primary users of the bands. This included the use of spread spectrum techniques to minimize interference.

With the development of PCS telephones, the FCC opened up the ISM bands for digital PCS communications with requirements that need to be met to use these bands for wireless telephony communications. These requirements also apply to other applications such as broadband communications and home networking, which are now using the ISM bands for RF communications. The following ISM bands are now popular for use with RF solutions:

902–928 MHz
2.4–2.5 GHz
5.725–5.875 GHZ

Table 10-2 Industrial, scientific, and medical bands used for communications

ISM Bands	
Freq Range	**Center Freq**
13.533–13.567 MHz	13.560 MHz
26.957–27.283 MHz	27.120 MHz
40.66–40.70 MHz	40.68 MHz
902–928 MHz	915 MHz
2.4–2.5 GHz	2.45 GHz
5.725–5.875 GHz	5.8 GHz
24–24.25 GHz	24.125 GHz

Also, ISM higher frequency bands (24 to 24.5 GHz) are currently being considered for residential use in fixed wireless systems.

10.2 Types of Distribution Methods for the Home

Three types of distribution methods or mediums are used to distribute high-speed voice, data, and video throughout the home: transmission over power lines; transmission over phone lines; and transmission through the air using RF communications.

10.2.1 Power Line Communications

A power line communications (PLC) system modulates voice, data, and video signals and sends them over the existing alternating current (AC) power lines in the home. This method uses the infrastructure of the existing power lines inside the home and provides very broad coverage without adding additional wires. Also, since the hardware requires coupling to the power lines but no antennas or high-frequency components, this method is relatively low cost.

The power line medium itself is fairly noisy. In addition, the different lengths of wiring and different terminations lead to several impedance mismatches, which cause peaks and nulls in the amplitude response of the spectrum (Figure 10-2). Thus, most systems use sophisticated digital transmission design methods to mitigate the problems with medium and frequency agility to prevent operation in amplitude nulls.

Orthogonal Frequency Division Multiplexing. One of the more common techniques used for PLC is orthogonal frequency division multiplexing (OFDM; see Chapter 2). This method uses multiple frequencies, each of which is orthogonal, to send the data. This allows independent detection of all the frequency channels used, with minimum interference between channels. OFDM is used in conjunction with digital modulation schemes such as binary phase-shift keying (BPSK), quadrature phase-shift keying (QPSK), and quadrature amplitude modulation (QAM). This allows higher throughput of the data utilizing multiple channels in parallel and allows for a very robust system to drop frequency channels that are being jammed or are unusable due to the amplitude spectral response.

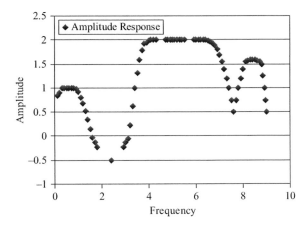

Figure 10-2   Spectrum of the amplitude response of the power line.

10.2.2 Home Phoneline Networking Alliance

The Home Phoneline Networking Alliance (PNA) transmits signals over the existing telephone line infrastructure. Therefore, wherever there is a telephone line present, voice and data can be transmitted without running additional wires. Home PNA has two standards: 1.0 and 2.0. The 1.0 standard is capable of 1 Mbps and was released in fall 1998. The Home PNA is used for shared Internet access, printer and file sharing, and network gaming. The 2.0 standard increased the speed to 10 Mbps and must be backward-compatible with the 1 Mbps 1.0 standard. The Home PNA is essentially an Ethernet network that uses the telephone lines for infrastructure. Home PNA developers now provide USB and Ethernet adapters to the Home PNA network, and have increased the data rates greater than 100 Mbps.

The data and voice are coupled to the telephone line and transmitted to other devices via the telephone lines. This system is ideal if the existing home has multiple telephone lines. In many homes, however, the telephone line infrastructure does not provide a means to distribute signals into every room in the house. This is a major limitation to Home PNA. The power line and RF methods provide a much better solution for full coverage throughout the home.

10.2.3 Radio Frequency Communications

Radio frequency is a method for distributing voice, data, and video throughout the home. It provides the required bandwidth to extend high-definition television (HDTV) and other broadband signals and offers connectivity to external wireless devices. However, RF can be degraded by multipath and interference from other RF signals.

Radio frequency technologies exhibit problems with multipath, blockage, and band saturation from other users, causing nulls in the amplitude spectrum and distorted signals. Spread spectrum schemes and antenna diversity can help to mitigate these issues. Networking standards including the IEEE 802.11 standard and Bluetooth are established and verified. Table 10-3 summarizes the different home networking methods.

Table 10-3   Home networking systems

Home Networking System	Implementation	Advantages	Disadvantages
Power Line	Apply modulation on A/C power lines Utilizes OFDM	Power lines in all rooms and garage No new wires Low cost	Power line noise Variable impedances and multiple terminations
Phone Line	Apply modulation on the telephone lines	Phone lines existing	Not all rooms have telephone lines Limited coverage
RF	RF modulation GMSK, ISM bands 802.11, Bluetooth, WiMAX, LTE, others	Good coverage Adaptable to external wireless devices	Multipath, RF interference

10.2.4 IEEE 802.11

The IEEE 802.11 standard specifies both the physical (PHY) layer and the medium access control (MAC) layer and has been adopted as a standard for home RF networking. The PHY layer can use both RF and infrared (IR). RF uses spread spectrum techniques, including either frequency hopping or direct sequence, and operates in the 2.4 GHz ISM band. The 802.11 standard specifies initial data rates of either 1 or 2 Mbps and has increase the throughput well over 100 Mbps. Further developments include operation in the 5 GHz band with much higher data rates.

The MAC layer uses the carrier sense multiple access with collision avoidance (CSMA/CA) protocol. This permits the network to have multiple users and reduces the chance of transmittal at the same time. Each node checks to see if the channel is being used, and if not it transmits data. If the channel is busy, it randomly selects the amount of time that it waits before transmitting again. Since there are multiple time selections, the probability is very low that two nodes trying to enter the network will select the same random time slot, which further reduces the chance of collisions.

To ensure that a connection is made between two nodes, the transmitting node sends out a ready-to-send (RTS) packet, and the receiving node, upon successful reception of the RTS packet, sends a response of clear-to-send (CTS), which lets the transmitting node know that it is communicating with that particular node and not with another one. Then the transmitting node sends the data, and the receiving node acknowledges reception by returning an acknowledge (ACK) to the transmitter, which is verified by a cyclic redundancy check (CRC).

Wi-Fi networks are short-range, high-bandwidth networks primarily developed for data. Their quality of service (QoS) is similar to fixed Ethernet, where packets can receive different priorities based on their tags. For example, Voice over Internet Protocol (VoIP) traffic may be given priority over Web browsing.

Wi-Fi runs on the MAC's CSMA/CA protocol, which is connectionless and contention based. The 802.11 specification defines peer-to-peer (P2P) and ad hoc networks, where an end user communicates to users or servers on another local area network (LAN) using its access point or base station. Wi-Fi uses the unlicensed spectrum, generally 2.4 GHz, using direct sequence spread spectrum (DSSS), multicarrier OFDM to provide access to a network. Subscriber stations that wish to pass data through a wireless access point (AP) are competing for the AP's attention on a random interrupt basis, which causes distant subscriber stations from the AP to be repeatedly disturbed by closer stations, greatly reducing their throughput.

Wi-Fi is supported by most personal computer operating systems, many game consoles, laptops, smartphones, printers, and other peripherals. If a Wi-Fi–enabled device such as a PC or mobile phone is in range, it connects to the Internet. In addition, both fixed and mobile computers can network to each other and to the Internet. The interconnected access points, called hotspots, are used to set up the mesh networks. Routers that incorporate DSLor cable modems and a Wi-Fi access point give Internet access to all devices connected (wirelessly or by cable) to them. They can also be connected in an ad hoc mode for client-to-client connections without a router. Wi-Fi enables wireless voice applications—Voice over Wireless LAN (VoWLAN) or Wireless VoIP (WVoIP)—and provides a secure computer networking gateway, firewall, Dynamic Host Configuration Protocol (DHCP) server, and intrusion detection system. Any standard Wi-Fi device will work anywhere in the world. The current version of Wi-Fi Protected Access encryption (WPA2) is not easily defeated.

QoS is more suitable for latency-sensitive applications (voice and video), and power-saving mechanisms (WMM Power Save) improve battery operation.

Wi-Fi's distance is limited to typically 32 m (120 ft.) indoors and 95 m (300 ft.) outdoors. The 2.4 GHz band provides slightly better range than 5 GHz. In addition, directional antennas can increase the distance to several kilometers. Zigbee and Bluetooth are in the range of 10m but continue to increase.

Wireless network security is a concern for Wi-Fi users. Attackers gain access by monitoring Domain Name Service (DNS) requests and responding with a spoofed answer before the queried server can reply.

The security methods for Wi-Fi include the following:

1. Suppress AP's Service Set Identifier (SSID) broadcast. SSID is broadcast in the clear in response to client SSID query. However, since it is also transmitted for other signals, this is not a valid security method.
2. Allow only computers with known MAC addresses to join the network. Unfortunately, MAC addresses are easily spoofed.
3. Use Wired Equivalent Privacy (WEP) encryption standard. This can be easily broken.
4. Use Wired Protected Access (WPA) with TKIP. However, more secure is no longer recommended.
5. WPA2 encryption standards are still considered secure and are used presently.
6. Future encryption standards will be forthcoming due to the sophisticated attacks on security.

Wi-Fi APs default with no encryption and provide open access to other users. Security needs to be configured by the user via graphical user interface (GUI). Other security methods include virtual private networks (VPNs) or a secure Web page.

Unless it is configured differently, Wi-Fi uses 2.4 GHz access points that default to the same channel on startup. This causes interference if there are multiple users. In addition, high-density areas and other devices at 2.4 GHz can cause interference with many Wi-Fi access points. An alternative is to use the 5 GHz band, but the interference will increase as more and more users operate in that band.

Wi-Fi uses an AP that connects a group of wireless devices to an adjacent wired LAN for Internet connection or networking. The access point relays the data between wireless devices and a single wired device using an Ethernet hub or switch. In this way, wireless devices can communicate with other wired devices. Wireless adapters are used to permit devices to connect to a wireless network. The connection to the devices is accomplished using external or internal interconnects such as a PCI, USB, ExpressCard, Cardbus, and PC card. Wireless routers integrate WPA, Ethernet switch, and internal router firmware applications to provide Internet Protocol (IP) routing, network address translation (NAT), and DNS forwarding through an integrated WAN interface. Wired and wireless Ethernet LAN devices connect to a single WAN device such as a cable or DSL modem. This allows all three devices (mainly the access point and router) to be configured through one central utility or integrated Web server that provides Web pages to wired and wireless LAN clients and to WAN clients.

10.2.5 Bluetooth

Bluetooth (IEEE 802.15) is a standard that operates at 2.4 GHz in the ISM band (2.4–2.4835 GHz). The band allows 79 different channels with a channel spacing of 1 MHz. It starts with

2.402 GHz, which allows for a guard band of 2 MHz on the lower band edge, and ends with 2.480 GHz, which allows for a guard band of 3.5 MHz on the upper band edge. These band allocations and guard bands are the U.S. standards. Other countries may operate at slightly different frequencies and with different guard bands.

The following three types of classes deal mainly with power output and range:

- Class 1: +20 dBm
- Class 2: +4 dBm
- Class 3: 0 dBm

The higher the power output, the greater the range of the system. However, both Class 1 and 2 require power control to allow only the required power to be used for operation. Class 1 operation is required to reduce the power to less than 4 dBm maximum when the higher power output is not required. The power control uses step sizes from 2 dB minimum to 8 dB maximum. The standard method of power control uses a received signal strength indication (RSSI). Depending on the RSSI in the receiver, this information is sent back to the transmitter to adjust the transmitted power output.

The modulation scheme specified in the Bluetooth standard uses differential Gaussian frequency-shift keying (DGFSK) with a time bandwidth (BT) of 0.5 and a modulation index between 0.28 and 0.35, with "+1" being a positive frequency deviation and "−1" being a negative frequency deviation. The frequency deviation is greater than 115 kHz. The symbol rate is 1 Msymbols/sec. Full duplex operation is accomplished using packets and time division multiplexing (TDM; see Chapter 2), and for this application it is referred to as time division duplexing (TDD).

Frequency hopping is used for every packet, with a frequency hop rate of 1600 hops/sec, providing one hop for every packet. The packets can cover from one to five time slots, depending on the packet size. Each time slot is 625 µsec long, which corresponds to the hop rate. However, since the frequency hop rate is dependent on the packet size, if the packet is longer than the slot size, then the hop rate will decrease according to the packet size. For example, if a packet covers five time slots, then the hop rate would be 1/5 of the 1600 hops/sec, or 320 hops/sec, which equals 3.125 msec dwell time, which equals 5 × 625 µsec.

The standard packet is made up of three basic sections: access code, header, and payload (Figure 10-3). The access code is 72 bits long and is used for synchronization, direct current (DC) offset compensation, and identification. The detection method used is a sliding correlator. As the code slides in time against the received signal, when the code matches up to the received signal the correlation of the two produce a peak, and when that peak is greater than a set threshold it is used for a trigger in the receiver timing.

Bluetooth forms networks that are either point to point (PTP) or point to multipoint (PMP). PMP is called a piconet and consists of a master unit and up to seven slave units, which can be in multiple piconets, called scatternets, using TDM. However, the piconets cannot be time or frequency synchronized.

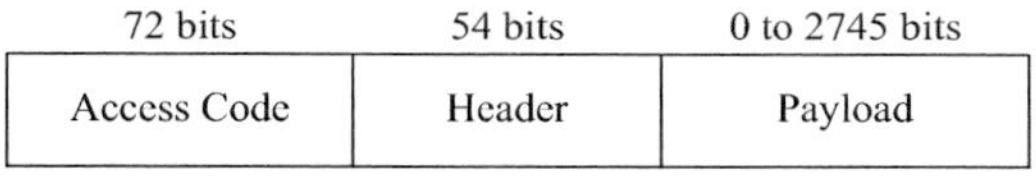

72 bits	54 bits	0 to 2745 bits
Access Code	Header	Payload

Figure 10-3   Basic structure of a standard packet define in the Bluetooth specification.

There are three types of access codes: channel access code, device access code, and inquiry access code. The channel access code identifies the piconet. The device access code is used for applications such as paging. The inquiry access code is for general inquiry access code (GIAC) and checks to see what Bluetooth devices are within range. The dedicated inquiry access code (DIAC) is used for designated Bluetooth units in a given range. The header is 54 bits long and determines what type of data (e.g., link control) is being sent. A wide variety of information types are specified in the Bluetooth specification. The payload is the actual data or information that is sent. Depending on the amount of data, this can range from 0 to 2745 bits in length.

Three error correction schemes are used with Bluetooth: $\frac{1}{3}$-rate forward error correction (FEC); $\frac{2}{3}$-rate FEC; and automatic repeat request (ARQ). Error detection is accomplished using CRCs. The standard also specifies a scrambler or a data whitener to prevent long constant data streams (all "1"s for a period of time or all "0"s for a length of time), which cause problems in carrier and tracking loops.

The capacity of the Bluetooth standard can support an asynchronous data channel, asynchronous connectionless link (ACL), three point-to-point simultaneous synchronous voice channels, synchronous connection-oriented (SCO) links, or an asynchronous data channel and a synchronous voice channel. The synchronous voice channel is 64 kbps for quality voice in both directions. The asymmetrical speeds for Bluetooth are up to 723.2 kbps and 57.6 kbps on the return channel or 433.9 kbps if both links are symmetric.

The Bluetooth system is required to have a minimum dynamic range of 90 dB. The receiver for Bluetooth is required to have a minimum sensitivity of -70 dBm and needs to operate with signals up to $+20$ dBm.

The Bluetooth (802.15), Wi-Fi (802.11) and Worldwide Interoperability for Microwave Access (WiMAX) (802.16) and LTE specifications are summarized in Table 10-4. The initial Bluetooth specification was noted for its short range and was used mainly for device control applications. With the increase in power output, up to $+20$ dBm, its range has increased so that it can now be used for multiple applications. However, Bluetooth was not designed for high-speed data rates and will eventually need to be redesigned or replaced by devices that are focused on high-speed data rates and broadband communications.

10.2.6 WiMAX

WiMAX is an infrastructure that uses PMP links to provide mobile Internet users a connection to the Internet Service Provider (ISP), for example, a wireless laptop computer. Using a PTP connection to the Internet, users gain access via cell phones to a fixed location.

WiMAX provides P2P and ad hoc networks, which give end users the ability to communicate with other users or servers on another LAN using its access point or base station. WiMAX is the IEEE 802.16 standard for broadband wireless access with the following options:

802.16d : fixed WiMAX
802.16e : mobile WiMAX
802.16m : WiMAX II for 4G

A WiMAX Forum was established in June 2001 to ensure conformity and interoperability.

Table 10-4 RF commercial standards

RF System	Performance	Waveform/Security	Modulation	Data Rate	Other
IEEE 802.11 WiFi 802.11b/g – mobile	RF or infrared (IR) Ubiquitous 100 m Fixed Range APs provide Internet-access – "Hot Spots"	ISM band 2.4 GHz Protected Access encryption (WPA2) Attacks via DNS	Spread Spectrum Direct Sequence Frequency Hop	1 or 2 Mbps Future 11 Mbps CCK	CSMA/CA protocol. Compete for access point AP random interrupt basis Near/Far Problem
Bluetooth IEEE 802.15	FEC, ARQ, CRC Sensitivity: –70 dBm DR = 90 dB Saturation: +20 dBm	ISM band 2.4 GHz 79 channels 1 MHz spacing PTP, PMP, TDM FEC, Scrambler	Class 1–3: +20 dBm +4 dBm, 3: 0 dBm DGFSK BT=.5 Mod Index .28-.35 FH1600 h/sec, 1 h/pktet Full Duplex – TDM/TDD	Symbol Rate –1 Msps Synch – 64 kbps Asynch – 723.2 kbps	Full Duplex TDD ACL, SCO PTP PMP
WiMAX- Worldwide Interoperability for Microwave Access 802.16d – fixed 802.16e – mobile 802.16m – WiMAX II for 4G	ISP to end user Voice, 4G, TDD, FDD, TRANSEC, FIPS, DIA CAP, JITC 4G mobile broadband	Multiple Bands PMP & PTP BWs: 1.4 MHz–20 MHz 200 active users in every 5 MHz cell	BPSK, QPSK 16 & 64 QAM TDD, TD-SCDMA GSM/GPRS, CDMA, W-CDMA (5 MHz) UMTS 3GPP2 networks – cdmaOne & CDMA2000	2-15 Mbps (original) Downlink 100 Mbit/s, Uplink 50 Mbit/s	OFDMA MIMO Time slots assigned Resolves Near/Far
LTE – Long Term Evolution Voice, 4G, GSM/ GPRS, CDMA, W-CDMA (5 MHz) UMTS	TDD, FDD, Security: 3% overhead, TRANSEC, FIPS, DIA CAP, JITC 4G mobile broadband	Multiple Bands Flexible bandwidths – 1.4 MHz to 20 MHz 200 active users in every 5 MHz cell	QPSK, 16 & 64 QAM TDD, TD-SCDMA GSM/GPRS, CDMA, W-CDMA (5 MHz) UMTS 3GPP2 networks – cdmaOne & CDMA2000	Downlink 100 Mbit/s, Uplink 50 Mbit/s	MIMO, OFDM SC-FDMA Turbo codes Intereaving

The bandwidth and range of WiMAX can provide the following functions:

1. Connection to Wi-Fi hotspots to the Internet.
2. The "Last mile" connection to replace or enhance cable and DSL.
3. Both data (for computers) and telecommunications (wireless telephone) services.
4. Internet connectivity backup to fixed services such as cable and DSL.
5. Broadband access to the Internet.

WiMAX is part of the Radio Communications Sector of International Telecommunication Union (ITU-R) or IMT-2000. It provides international interoperability for 3G mobile communications. Current mobile communications that use WiMAX include handsets (e.g., cellular smartphones, Android handset from Google, 3G EV-DO devices, Qualcomm's IPR); PC peripherals (e.g., PC cards or USB dongles); and laptops and electronics devices (e.g., game terminals, MP3 players).

Using WiMAX, cable companies can gain access to wireless networks such as the mobile virtual network operator, which has the ability to move between the new Clearwire and Sprint 3G networks. It replaces GSM and CDMA and the wireless backhaul for 2G, 3G, and 4G networks.

The North America Backhaul for urban cellular consists of a copper T1 line, and for remote places a satellite is used. In most other regions, urban and rural backhaul is provided by microwave links such as Local Multipoint Distribution Service (LMDS) or Multichannel Multipoint Distribution Service (MMDS). Since copper wire is not adequate for broadband communications, the microwave backhaul provides 34 Mbps and 1 Gbps speeds with latencies in the order of 1ms.

The WiMAX MAC/data link layer consists of a base station that assigns the subscriber an access slot upon entry that can be increased or decreased depending on demand or priorities. The scheduling algorithm allows users to compete for the time slot, but once they are in a permanent access slot is allocated to them. This provides stability under overload conditions and is much more bandwidth efficient than other methods. The base station controls the QoS by balancing time slot assignments with application needs of subscriber stations. The connection is based on specific scheduling algorithms.

The WiMAX physical layer is certified, which gives vendors interoperability. The frequency range is 10 to 66 GHz, with some of standard frequencies being 2.3 GHz, 2.5 GHz, and 3.5 GHz (Table 10-4). The backhaul generally uses frequencies in the 5 GHz unlicensed band. In addition, the analog TV bands (700 MHz) are being used for WiMAX applications. Details of the WiMAX data link are shown in Table 10-4.

The available bandwidth is shared between users, so with more users the individual performance can deteriorate. Most users can expect around 2–3 Mbps. Multiple-in, multiple-out (MIMO) techniques are used for increased coverage.

WiMAX allows for self-installation, power consumption, frequency reuse, and bandwidth efficiency. Fast Fourier transforms (FFTs) are used to provide scaling to the channel bandwidth for constant carrier spacing of 1.25 MHz, 5 MHz, 10 MHz, or 20 MHz. WiMAX provides higher spectrum efficiency in wide channels and cost reduction in narrow channels using the following multiple techniques:

● Intelligent dynamic frequency selection (I-DFS)
● Advanced antenna diversity
● Hybrid automatic repeat-request (HARQ)

- Adaptive antenna systems (AAS) and MIMO technology
- Turbo coding and low-density parity check (LDPC)
- Downlink subchannelization for coverage versus capacity
- Extra QoS class for VoIP applications
- Stationary and mobile (4G)
- Dynamic burst algorithm, which adjusts burst profile for signal-to-noise ratio (SNR) at the same time
- Modulation agility, which has more bits per OFDM/scalable orthogonal frequency division multiple access (SOFDMA) symbol for high SNR, less bits for low SNR, more robust burst profile
- Integration with an IP-based network, which gives a defined connection with an IP-based core network at the ISP and a base station that works with other types of architectures
- Packet switched mobile networks

WiMAX maintains a flexible architecture to allow remote and mobile stations of varying scale and functionality and to provide base stations of varying size (e.g., femto, pico, and mini as well as macro).

10.2.7 LTE

LTE is a next generation of communication and networking. LTE is based on 4G radio technology, including TDD and frequency division duplex (FDD). Security for LTE requires approximately 3% overhead and contains the following: Transmission Security (TRANSEC), *Federal Information Processing Standards* (FIPS), Department of Defense Information Assurance Certification and Accreditation Process (DIACAP), and Joint Interoperability Test Command (JITC). This technology uses 4G mobile broadband and all-IP flat architecture with data rates (peak) of 100 Mbps for the downlink and 50 Mbps for the uplink. The latency is less than 10 ms for round-trip time and less than 5 ms latency for small IP packets. If MIMO is used, the peak download rates are 326.4 Mbps for 4×4 antennas and 172.8 Mbps for 2×2 antenna for every 20 MHz of bandwidth. The peak upload rate is 86.4 Mbps for every 20 MHz.

LTE provides flexible bandwidths of 1.4 MHz to 20 MHz with low operating expenditures (OPEX). TDD in LTE is aligned with TD-SCDMA as well, which provides coexistence and seamless connection with GSM/General Packet Radio Service (GPRS), code division multiple access (CDMA), Wideband - W-CDMA, Universal Mobile Telecommunications System UMTS, and 3rd Generation Partnership Project 3GPP2 networks including cdmaOne and CDMA2000 (IS-95 CDMA).

LTE supports multicast broadcast single frequency network (MBSFN) for mobile TV. It uses an evolved packet system (EPS) and evolved UMTS Terrestrial Radio Access Network (UTRAN) (Evolved E-UTRAN) on the access side and evolved packet core (EPC) and system architecture evolution (SAE) on the core side.

This technology uses five different terminal classes, from voice-centric to high-end terminal, and 200 active users in every 5 MHz cell (200 active data clients). It also supports W-CDMA at 5 MHz and has an optimal cell size of 5 km with 30 km and 100 km cell sizes being supported.

The 3GPP, which uses 3G technology, proposed the Transmission Control Protocol/Internet Protocol (TCP/IP) with studies into all IP networks (AIPN). The higher level protocols such as LTE SAE are flat network architectures. They are efficient supports of mass-market usage of any IP-based service and are the evolution of the existing GSM/WCDMA core network. The Release 8 E-UTRA for UMTS operators is geared to be used over any IP network, WiMAX, Wi-Fi, and other wired networks.

3GPP uses orthogonal FDMA (OFDMA) for the downlink (tower to handset) and single carrier FDMA (SC-FDMA) for the uplink. It also incorporates MIMO with up to four antennas per station. Turbo coding is used along with a contention-free quadratic permutation polynomial (QPP) turbo code internal interleaver. OFDM is used for downlink.

The time domain data structure is as follows:

Radio frame = 10 ms
10 subframes of 1 ms each
Subframe contains two slots
Each slot = 0.5 ms
Subcarrier spacing = 15 kHz

Twelve of these subcarriers together (per slot) are called a resource block:

One resource block = 180 kHz
6 resource blocks fit in a carrier of 1.4 MHz
100 resource blocks fit in a carrier of 20 MHz

The Physical Downlink Shared Channel (PDSCH) includes all the transmitted data. It uses various types of modulations including QPSK, 16-state quadrature amplitude modulation (16-QAM) and 64-QAM, depending on the required data throughput and the available SNR. The Physical Multicast Channel (PMCH) is the broadcast transmission using a single frequency network (SFN). The Physical Broadcast Channel (PBCH) sends the most important system information within the cell.

Single-user MIMO is used to increase the user's data throughput. Multi-user MIMO is used for increasing the cell throughput.

The uplink is a precoded OFDM called an SC-FDMA. This is to compensate for high peak-to-average power ratio (PAPR) for normal OFDM requiring high linearity. SC-FDMA groups resource blocks to reduce linearity requirements. There are two physical channels in the uplink signal: Physical Random Access Channel (PRACH) for initial access and when the UE is not uplink synchronized; and Physical Uplink Shared Channel (PUSCH) for all data using quadrature phase-shift keying (QPSK), 16-QAM, and 64-QAM. The MIMO provides spatial division multiple access (SDMA) and increases the data rate depending on the number of antennas that are used. This in turn ensures that more than one mobile unit may reuse the same resources.

The following has been demonstrated with the results shown:

HDTV streaming > 30 Mbps, video supervision, and mobile IP-based handover between LTE and High-Speed Downlink Packet Access (HSDPA)
Bit rates up to 144 Mbps
Data rates of 200 Mbps with < 100 mW
HD video streaming with peak rates of 96 Mbps downlink, 86 Mbps uplink, and VoIP
Data rates Mbps traveling 110 km/h
M700 mobile platform peak data rates 100 Mbps downlink and 50 Mbps uplink
100 Mbps uplink transfer speeds

Most carriers currently supporting GSM or HSPA networks will eventually upgrade to LTE network. Companies are upgrading to LTE as their 4G technology but will provide an interim solution using HSUPA and HSPA+. Upgrade paths for CDMA networks will also migrate to LTE.

10.3 LMDS

LMDS is a PMP distribution service that allocates communications signals with a relatively short RF range to multiple end users. In this multipoint system, the base station or hub transmits signals in a PMP method that is basically a broadcast mode. The return path from the subscriber to the base station or hub is accomplished via a PTP link.

The LMDS system is basically an end-to-end microwave radio link that enables VoIP, local and long distance telephony services, high-speed Internet access and data transmissions, and video conferencing and TV. It uses digital wireless transmission systems operating in the 28 GHz range for systems in the United States and 24–40 GHz for overseas systems, depending on the area and country. LMDS can be employed in either an asymmetric or symmetric configuration. The asymmetric system has a data rate that is different depending on the direction of the transmission; for example, the downstream has a higher data rate than the upstream, as is the case with asynchronous digital subscriber loop (ADSL). The symmetric configuration has the same data rate regardless of which direction the data are sent.

The FCC has assigned frequency as blocks to the end user. Block A users are assigned 1150 MHz of bandwidth in different frequency bands:

Frequency	Bandwidth
27.5–28.35 GHz	850 MHz
29.1–29.25 GHZ	150 MHz
31.075–31.225 GHz	150 MHz
Total bandwidth	1150 MHz

Block B users are left with the remainder of the bandwidth, which equals 150 MHz total. This total bandwidth allocation of 1300 MHz for LMDS represents the largest FCC bandwidth allocation of the wireless data communications systems including PCS, cellular, direct broadcast satellite (DBS), MMDS, digital audio radio services, wireless communications service, and interactive and video data and over two times the bandwidth of AM/FM radio, very high frequency and ultra high frequency TV, and cellular telephone combined. Most of the above systems are allocated a fraction of the LMDS allocation, for example the ISM band for PCS at the 900 MHz band is only 26 MHz. The standard intermediate frequency (IF) for typical LMDS systems ranges from 950 to 2050 MHz.

Many LMDS systems use the concepts of bandwidth-on-demand and shared resources by integrating their systems with asynchronous transfer mode (ATM). IP transport methodology is also used to send data and VoIP. LMDS is used to span the "last mile" to the user's facilities. These systems transmit very high data rates up to 500 Mbps each way. However, the distance is limited to about 2 to 4 miles. One of the reasons for this short range can be attributed to the rain fade at these frequencies of up to 30 dB of attenuation in the link analysis. Using the bandwidth allocated to LMDS, these systems can provides up to 85 Mbps to the residential users and up to 155 Mbps to the commercial users. A typical LMDS can provide up to 155 Mbps downstream and a return link of 1.544 Mbps. These high-speed data rates are compatible with the T1 speed requirements up to and including OC-3 connections. Using just the first part of the Block A bandwidth of 850 MHz and using QPSK modulation, this system can provide 100 simultaneous T1 lines. With the additional bandwidth and higher order modulation schemes, this capacity can be increased.

All communication systems involved with providing multiple users access require a method of multiplexing these users so that they can all access the same system. The basic multiplexing schemes used are CDM/CDMA, TDM/TDMA, and FDM/FDMA. The difference in the multiplexing methods and the multiple access methods is that the multiplexing method assigns the code, time slot, or frequency to a user regardless of whether the user is using the system, whereas with multiple access schemes the user is assigned a code, time slot, or frequency as needed when the user accesses the system. This assignment is only for that period of use, and then that code, time slot, or frequency is released and made available for the next user after the original user exits the system.

Most LMDS systems today use either FDM/FDMA or TDM/TDMA to provide multiple access for a given system. FDMA works well for applications where users are connected on a more continuous basis, and TDMA is better for when users are more periodic. If two users require the link on a continuous basis, TDMA would not be suitable because they would have to share the time allocation, but with FDMA the time is always allocated to each of the users and the access is separated in frequency. If the two users are periodic in their usage and do not require continuous access to the system, then TDMA might be the method to use because it takes advantage of when the subscriber is not using the services to increase either the number of users on the system or the overall data rate of the connection. These factors and others are considered when deciding on which type of multiplexing is optimum. Many times both of these methods are used in a system to provide better access, more users, and higher speeds for each user with minimal interference between them.

FDMA may be the best solution to large customers since they are using the Internet on a fairly continuous basis. However, for small end users, the requirement is more periodic. They require high usage for downstream applications while they are downloading large files, but generally the usage and speed requirements are lower during upstream use. Therefore, to provide an optimum use system, when the demand is high, and when the demand is low, a TDM/TDMA scheme is generally used.

Orthogonal methods and spatial antenna separation also make it possible to have multiple users. Some examples of orthogonal techniques include OFDM, orthogonal phase systems like QPSK, and orthogonal antenna polarizations, using vertical and horizontal polarizations for different users. Satellite communications currently use this technology.

Spatial antenna separation, or space sectoring, is another multiple access scheme that allows multiple subscribers to use the same communications system without jamming each other by having directional antennas pointed to different space segments of the intended users. The number of users and spatial separation depends on the beam width and the amount of isolation between beams.

To optimize a communications link, various combinations of orthogonal techniques, multiple access schemes, and spatial antenna separation are used to provide the optimum solution and the maximum amount of end users per system.

LMDS uses different types of digital modulation schemes depending on complexity and bandwidth efficiency. Some of the more popular modulation schemes include BPSK, QPSK, O-QPSK, 8PSK, using both standard and differential on any of the modulation types, and various QAM systems including 16-QAM and 64-QAM. The higher the order of modulation scheme, the higher the data rates for a given bandwidth. However, these systems are generally more complex, require higher signal levels (or have a reduce range), and are costlier to implement.

Modulation techniques provide a higher data rate for bandwidth efficiency, and multiple access techniques allow multiple users to access the system. The combination of these methods is used to maximize the capacity of a given system by providing the maximum users per base station or site. The higher capacity that can be achieved for a given base station provides more coverage or less base stations per given coverage. Therefore, the modulation and multiple access schemes are carefully designed for each system installed.

Since LMDS systems have a relatively short RF range, the base stations or hubs are spaced a few kilometers apart and are linked together to provide service up to several thousand end users. The main reason for reduced coverage in LMDS systems is atmospheric conditions, mainly rain. Their advantages over other wireless systems is that they are line-of-sight (LOS), that their antennas are fixed at the sight, that they are usually mounted at a high elevation (rooftops), and that often the antennas are directional so multipath and blockage is generally not a problem after the antennas are mounted and operational. However, the atmosphere is constantly changing, which affects the range of the system. Antenna position and mounting are important factors along with the type of modulation used to provide the range required for coverage and capacity.

Four elements make up an LMDS system: the customer premises equipment (CPE), which is the end user of the system; the base station or hub, which services multiple end users; the fiber-based infrastructure, which is the wired connection to the central office (CO); and the network operations center (NOC), which provides the networking and can operate with or without a CO.

Examples of CPEs include LANs, telephones, faxes, Internet, video, television, and set-top boxes (STB). The interfaces to these devices are digital signal, DS-0, DS-1 structured and unstructured T1/E1, T3/E3, DS-3, OC-1, OC-3/STS-3 fiber optics, ATM and video communications, plain old telephone (POTS), frame relay, Ethernet 10BaseT/100BaseT, and others. Both scalable and nonscalable network interface units (NIUs) at the CPE are used to go between the incoming LMDS signals and the devices being used at the CPE. The scalable NIU is used in large business and commercial uses. It is a flexible system that can be configured for the application, and it is chassis based. Therefore, the same chassis can be used for many types of systems and applications.

The main elements of the NIU are the modem and the data processor that supports the different type of external connections or interfaces. The nonscalable unit is for small and medium-sized users and is for a fixed application and interfaces. This unit is designed specifically for certain types of interfaces: T1/E1, T3/E3, POTS, 10BaseT, video, frame relay, and ATM.

The network-node equipment (NNE) connects the wireline functions to the LMDS wireless link and contains the processing, multiplexing/demultiplexing, compression/decompression, modem, error detection and correction, and ATM.

The base station or hub provides the interface between fiber infrastructure and wireless infrastructure. The elements of a standard base station are antenna system and microwave equipment for both transmitting and receiving, downconverter/upconverters, modulators/demodulators, and interface to the fiber. The base station antennas are mounted on rooftops and other high places to prevent blockage from structures since the LMDS frequencies are LOS.

Local switching in the base station can permit communications between users operating in the same network without going through the wired infrastructure if a NOC is provided. The fiber-based infrastructure is the wired connections mainly to the CO. This provides the connections from the base stations or NOC to the local CO.

The NOC consists of the network management system (NMS) equipment consisting of optical network (SONET) optical carrier OC-12 and OC-3, DS-3 links, central office CO equipment, ATM and IP switching systems, and interconnections with the Internet and public switched telephone networks (PSTNs). Some implementation schemes do not use a NOC, and the information from the base station is connected through fiber to ATM switches or CO equipment at the CO so that all communications must go through the CO. By using a NOC, if users on the same network desire to communication they can do so directly through the NOC bypassing the CO.

The current applications for LMDS are for both the suburban and rural communities throughout the world where it becomes difficult and impractical to run copper wire. In large cities the buildings and roads are already established, and there is no practical way of running copper wire except to tear up roads and alter buildings, which becomes very disruptive, costly, and impractical. This is the case for many large cities in the United States and also many worldwide cities. Also, in many of the smaller cites this can pose a major problem, requiring the same disruptions and demolitions. Since LMDS is a wireless solution, it can be installed on rooftops of buildings without having to run wires or tear up roads and will have a minimum impact to the community while simultaneously providing high-speed communications through its extended bandwidth.

For rural communities where the population is scarce, it becomes impractical and costly to provide services to the few that live in rural communities. In addition, they are spread out so that the infrastructure of a wired solution becomes more costly running miles of copper wire and often times is not made available to these customers. The LMDS offers a solution by providing a wireless link to these remote communities at a lower cost and less time to install.

Throughout the world in many remote, underdeveloped countries there does not currently exist a wired infrastructure for use with telephone, Internet, video applications. LMDS can provide a solution for these countries by setting up a wireless infrastructure that is less costly and requires less time to install. Often in these countries it takes years after a request is made to provide a hard-wired line to these end users, and sometimes the request is never fulfilled. LMDS not only can provide a immediate solution to the problem but also can extend the coverage that the wired lines provide. This enhances the coverage of the hard-wired fiber and coax lines to these remote places.

Presently, most of the applications are corporate because the cost to implement them residentially is too high. However, over the next several years, they will become more viable for home use.

10.4 MMDS

MMDS is a wireless service that operates from 2.2 to 2.4 GHz. It has a range of approximately 30 miles, which is LOS at these frequencies. It was originally designed to provide one-way service for bringing cable TV to subscribers in remote areas or in locations that are difficult to install cable. Other systems use bandwidths of 200 MHz in the band just above 2.5 GHz and also in the Ka-band at 24 GHz. The power output allowed is up to 30 W, and OFDM is used to enhance the number of users and increase the speeds. This provides up to 10 Mbps during peak use and can provide speeds up to 37.5 Mbps to a single user.

MMDS has the capacity to support up to 33 analog channels and more than 100 digital channels of cable television. Although MMDS was mainly designed for this cable TV service where wiring was impractical and costly to run, in 1998 FCC passed rules to allow MMDS to provide data and Internet services to subscribers. MMDS is used as a short-range inexpensive solution.

The IEEE 802.16 working group for broadband wireless access networks sets the requirements for both the MMDS and LMDS and other wireless "last mile" type technologies. Also, OFDM and VOFDM technology has been proposed for use in the MMDS system solution. The lower frequency bands, 2.5 and 5.0 GHz, which fall into the MMDS bands, are more appealing at the present time due to cost, availability, and quickness to install. LMDS may provide a long-term future solution to the "last mile" connection, but it will take more time to get the infrastructure in place for this LOS technology. The International Telecommunications Union (ITU) and the European Telecommunications Standards Institute (ETSI) are also involved in setting standards for LMDS and MMDS.

10.5 Universal Mobile Telecommunications System

UMTS is a 3G mobile networking that provides wide-area wireless cellular voice telephone, video calls, and broadband wireless data with data rates up to 14.4 Mbps on the downlink and 5.8 Mbps on the uplink using W-CDMA and GSM. It supports up to 21 Mbps data transfer rates with HSDPA and up to 42 Mbps HSDPA+. UMTS, using 4G technology, operates 100 Mbps downlink and 50 Mbps uplink using OFDM. In addition, UMTS combines W-CDMA, TD-CDMA, or TD-SCDMA, GSM's Mobile Application Part (MAP) core, and the GSM family of speech codecs. The modulation is 16-QAM operation at a frequency of 1710–1755 MHz and 2110–2155 MHz. Users access the 3G broadband services using a cellular router, Personal Computer Memory Card International Association (PCMCIA), or USB card.

A 3G phone can be used as a gateway or router to provide connection of Bluetooth-capable laptops to the Internet.

10.6 4G

4G provides wireless communications, networking, and connection to the Internet for voice, data, multimedia messaging service (MMS), video chat, mobile TV, HDTV, and digital video broadcasting (DVB). 4G is a spectrally efficient system that gives high network capacity and high throughput data rates: 100 Mbps for on the move (OTM) and 1 Gbps for fixed location. It also provides a smooth handoff across heterogeneous networks, seamless connectivity, and global roaming across multiple networks. 4G contains high QoS for next-generation multimedia support of real-time audio, high-speed data, HDTV video content, and mobile TV. It has been designed to be interoperable with existing wireless standards, which include an all-IP, packet switched network. 4G systems dynamically share and utilize network resources to meet the minimal requirements. They employ many techniques to provide QoS such as OFDM, MIMO, turbo codes, and adaptive radio interfaces as well as fixed relay networks (FRNs) and the cooperative relaying concept of multimode protocol. The packet-based (all-IP) structure creates low latency data transmission.

4G technology is incorporated into LTE and the higher speed version of WiMAX (WiMAX IEEE 802.16 m) for 3GPP, which uses orthogonal FDMA (OFDMA), SC-FDMA, and MC-CDMA.

Interleaved frequency division multiple access (IFDMA) is being considered for the uplink on the 4G system since OFDMA contributes more to the PAPR. IFDMA provides less power fluctuation and thus avoids amplifier issues. It needs less complexity for equalization at the receiver, which is an advantage with MIMO and spatial multiplexing. It is capable of high-level modulations such as 64-QAM for LTE.

IPv6 support is essential when a large number of wireless-enabled devices is present because it increases the number of IP addresses, which in turn provides better multicast, security, and route optimization capabilities.

Smart or intelligent antennas are used for high rate, high reliability, and long range communications. MIMOs provide spatial multiplexing for bandwidth conservation, power efficiency, better reception in fading channels, and increased data rates.

10.7 Mobile Broadband Wireless Access IEEE 802.20

The Mobile Broadband Wireless Access (MBWA) standard covers mobile units moving at 75 to 220 mph (120 to 350 km/h). High-speed dynamic modulation and similar scalable OFDMA capabilities are utilized to handle these speeds and provide fast hand-off, FEC, and cell edge enhancements. The mobile range is approximately 18 miles (30 km) with an extended range of 34 miles (55 km). The data rate goals for WiMAX2 and LTE going forward are 100 Mbps for mobile applications and 1 Gbps for fixed applications.

10.8 MISO Communications

MISO used in cellular network implementations contains multiple antennas at the base station and a single antenna on the mobile device, which is often referred to as space time code (STC). This improves reception, range, data rate, and spectral efficiency. Auto-negotiation is used dynamically between each individual base station and mobile station to adjust specific parameters such as transmission rates. In addition, multiple mobile stations can be supported with different MISO capabilities as needed, which maximizes the sector throughput by leveraging the different capabilities of a diverse set of vendor mobile stations.

This technique uses different data bit constellations (phase/amplitude) that are transmitted on two different antennas during the same symbol. The conjugate or inverse of the same two constellations are transferred again on the same antennas during the next symbol. The data rate with STC remains the same with the received signal being more robust using this transmission redundancy. Similar performance can be achieved using two receive antennas and one transmitter antenna, which is generally referred to antenna diversity and relies on the fact that while one of the antennas is blocked or in a multipath null the other antenna receives a good signal.

10.9 MIMO Communications

MIMO is used to provide spatial multiplexing (SMX) using two antennas at both ends. One antenna transmits one data bit, and another antenna transmits another bit simultaneously; this makes up one symbol. The receiver contains two antennas to separate the signals and receives the two bits and the same time. The receiver then multiplexes these bit streams into a data stream that provides a data rate at twice the original data rate transmitted out of each antenna with no increase in bandwidth. Therefore, the data rate is twice as fast compared with using STC with only one receive antenna.

MIMO can be used to improve the link and robustness. Using MIMO for a more robust link, a single data stream is replicated and transmitted over multiple antennas. The redundant data streams are each encoded using a mathematical algorithm known as space time block codes. With such coding, each transmitted signal is orthogonal to the rest, which reduces self-interference and improves the capability of the receiver to distinguish between the multiple signals. With the multiple transmissions of the coded data stream, there is increased opportunity for the receiver to identify a strong signal that is less adversely affected by the physical path. MIMO is fundamentally used to enhance system coverage (Matrix A).

MIMO can also be used to increase data throughput. The signal to be transmitted is split into multiple data streams and each data stream is transmitted from a different base station transmit antenna operating in the same time-frequency resource allocated for the receiver. In the presence of a multipath environment, the multiple signals will arrive at the receiver antenna array with sufficiently different spatial signatures allowing the receiver to readily discern the multiple data streams. Spatial multiplexing provides a very capable means for increasing the channel capacity (Matrix B).

This technique can be used to increase the data rate with adding the number of antennas. For example, WiMAX uses four antennas at both ends with either a four times increase in data rate, improved signal quality and twice the data rate, or additional improvement in signal quality with the same date, which reduces error rates caused by blockage, multipath, or jamming.

MIMO or MISO can be configured for multiple users that are spatially separated. This allows users to be on the same frequency with the base using multiple directional antennas, each of them pointed in the direction of each of the users. In addition, synchronizing the users can provide further improvement using TDMA and the multiple antennas.

Active electronically steered arrays (AESAs) can be used to provide adaptive antenna steering (AAS) for multiple beams in a single antenna or aperture. In addition, beamforming can be used to shape the beam for the best reception. Another way to improve performance is via cyclic delay diversity (CDD), in which multiple signals are delayed before transmission. At the receiver, these signals are combined using the specified delays for a more robust received signal. The closer the signal can get toward a flat channel at a certain power level, the higher the throughput.

10.10 QoS

Quality of Service (QoS) is a measure of the guaranteed level of performance regardless of capacity. It uses different priorities for different applications, users, or data rates. It monitors the level of performance and dynamically controls the scheduling priorities in the network to guarantee the performance by this scheduling, or it can use the reserve capacity if needed. The required bit rate, delay, jitter, packet dropping probability, and bit error rate are part of the QoS. A possible alternative to QoS is to provide margin to accommodate expected peak use.

QoS has requirements for many parameters such as service response time, signal loss, signal-to-noise ratio, cross-talk, echo, interrupts, frequency response, and loudness levels. The grade of service (GoS) is used in the telephony world and relates to capacity and coverage such as maximum blocking probability and outage probability.

QoS can be the cumulative effect on subscriber satisfaction. Generally, four types of service bits and three precedence bits are provided in each message that are redefined as DiffServ Code Points (DSCP) for the modern Internet. For packet-switched networks, QoS is affected by human factors such as stability of service, availability of service, delays, and user information and also technical factors such as reliability, scalability, effectiveness, maintainability, and GoS.

Some of the problems that occur in transmissions include the following:

1. Dropped packets. Can prevent communications and possible retransmission of the packets causing delays.
2. Delay. Excessive delay due to long queues or a less direct route can cause problems with real-time communications such as VoIP.
3. Jitter. Variation in delay of each packet can affect the quality of streaming audio or video.
4. Out-of-order delivery. Different delays of the packets can cause the packets to arrive in a different order causing problems in video and VoIP.
5. Error. Distorted or corrupted packets causing errors in the data, needs error detection or ARQ.

Many applications require QoS, including streaming multimedia, IPTV, IP telephony or voice over IP (VOIP), and video teleconferencing (VTC).

There are basically two types of QoS: inelastic, which requires a minimum bandwidth and maximum latency; and elastic, which is versatile and can adjust to the available bandwidth. Bulk file transfer applications that rely on TCP are generally elastic. A service-level agreement (SLA) is used between the customer and the provider to guarantee performance, throughput, and latency by prioritizing traffic. Resources are reserved at each step on the network for the call as it is set up; this is known as the Resource Reservation Protocol (RSVP). Commercial VoIP services are competitive with traditional telephone service in terms of call quality without QoS mechanisms. This requires margin above the guaranteed performance to replace QoS, and it depends on the number of users and the demands.

Integrated services (IntServ), which reserve the network resources using RSVP to request and reserve resources through a network, are not realistic because of the increase in users and the demand for more bandwidth. Differentiated services (DiffServ) use packets marked according to the type of service needed. This service uses routers for supporting multiple queues for packets awaiting transmission. The packets requiring low jitter (VoIP or VTC) are given priority over packets in the other queues. Typically, some bandwidth is allocated by default to the network control packets. Additional bandwidth management mechanisms include traffic shaping or rate limiting.

The TCP rate control artificially adjusts the TCP window size and controls the rate of ACKs being returned to the sender. Congestion avoidance is used to lessen the possibility of port queue buffer tail drops, which also lowers the likelihood of TCP global synchronization. QoS deals with these issues concerning the Internet. The Internet relies on congestion-avoidance protocols, as built into TCP, to reduce traffic load under conditions that would otherwise lead to Internet meltdown. Both VoIP and IPTV necessitate large constant bitrates and low latency, which cannot use TCP. QoS can enforce traffic shaping that can prevent it from becoming overloaded. This is a critical part of the Internet's ability to handle a mix of real-time and non–real-time traffic without imploding. ATM network protocol can be used since it has shorter data units and built-in QoS for video on demand and VoIP. The QoS priority levels for traffic type are listed as follows:

0 = Best Effort
1 = Background
2 = Standard (Spare)
3 = Excellent Load (Business Critical)
4 = Controlled Load (Streaming Multimedia)
5 = Voice and Video (Interactive Media and Voice) < 100 ms latency and jitter
6 = Layer 3 Network Control Reserved Traffic < 10 ms latency and jitter
7 = Layer 2 Network Control Reserved Traffic = Lowest latency and jitter

ITU-T G.hn standard provides QoS via contention-free transmission opportunities (CFTXOPs), which allocates the flow and negotiates a contract with the network controller and supports non-QoS operation of contention-based time slots.

Multi Service Access Everywhere (MUSE) defines a QoS concept along with Platforms for Networked Service (PLANETS) to define discrete jitter values including best effort with four QoS classes: two elastic and two inelastic. This concept includes end-to-end delay where the packet loss rate can be predicted. It is easy to implement with simple scheduler and queue length, nodes can be easily verified for compliance, and end users notice the difference in quality. It is primarily based on the usage of traffic classes, selective call admission control (CAC) concept, and appropriate network dimensioning.

The Internet is a series of exchange points interconnecting private networks, with its core being owned and managed by a number of different Network Service Providers (NSPs). The two approaches to QoS in modern packet-switched networks are parameterized and differentiated.

The parameterized system is based on the exchange of application requirements within the network using IntServ, RSVP, and a prioritized system where each packet identifies a desired service level to the network. DiffServ implements the prioritized model using marked packets according to the type of service they need. At the IP layer, DiffServ code point (DSCP) markings use the first 6 bits in the TOS field of the IP packet header. At the MAC layer, VLAN IEEE 802.1q and IEEE 802.1D do the same. Cisco IOS NetFlow and the Cisco Class Based QoS (CBQoS) Management Information Base (MIB) is leveraged in the Cisco network device to obtain visibility into QoS policies and their effectiveness on network traffic. Strong cryptography network protocols such as Secure Sockets Layer, I2P, and VPNs are used to obscure the data being transferred. All electronic commerce on the Internet requires cryptography protocols. This affects the performance of encrypted traffic, which creates an unacceptable hazard for customers. The encrypted traffic is otherwise unable to undergo deep packet inspection for QoS. QoS protocols are not needed in the core network if the margin is high enough to prevent delay. Newer routers are capable of following QoS protocols with no loss of performance. Network providers deliberately erode the quality of best effort traffic to push for higher-priced QoS services. Network studies have shown that adding more bandwidth was the most effective way to provide for QoS.

10.11 Military Radios and Data Links

10.11.1 The Joint Tactical Radio System

The Joint Tactical Radio System (JTRS), sometimes referred to as "Jitters," is a military data link system utilizing SDRs that provide interoperability on the battlefield for voice, data, and video communications. In addition, the system deliver command, control, communications, computers, and intelligence (C4I), including missile guidance command, control, and communications for air-to-air, air-to-ground, and ground-to-ground systems. The JTRS affords the ability to handle multiband, multimode, and multiple channel radios. It was initially designed to cover an operating spectrum of 2 to 2000 MHz, but this has been expanded to beyond 2000 MHz to allow higher frequency operation and provide more bandwidths for multiple users and increased data rates.

The main principle behind JTRS is to provide a software programmable radio that is interoperable with all existing legacy radios. In addition, the system needs to provide wideband networking software for mobile ad hoc networks and security or encryption to prevent unwanted users and jammers from entering the network. There are several challenges facing the JTRS.

● Interoperability. There is a need to provide interoperability with different modulation and frequency waveforms for all legacy and future radios on the battlefield. This is a challenge to be compatible with so much diversity in the current communication data links. The RF sections are the most challenging, with components such as filters, antennas, and amplifiers not being able to cover the entire frequency bandwidth. RF integrated circuits (RFICs) are being developed to help mitigate these problems. Tunable filters can be used, but the cost generally increases with complexity. In addition, work is being done to design broadband antennas and antenna switching arrays to cover the wide band of frequencies. The modulation waveform for generation and detection is designed with SDRs that can be software and firmware programmed for different modulation/demodulation schemes.

- Anti-jam. To provide interoperability with multiple radios and data links, the system is required to handle a very wide bandwidth. The wider the bandwidth for operation, the more vulnerable the system is for jammers to affect the signal integrity and performance. Unless these different types of frequencies and modulation schemes are switched from one to another, including their bandwidths, jamming could be a problem.
- Security. Encryption is required to prevent unauthorized users from entering the network. This adds complexity and cost but is necessary to provide the security needed for military operations. One of the major problems is the lack of encryption for each of the legacy radios in the network. The system must be able to handle different encryption schemes or provide a universal encryption system.
- Cost. For several applications, especially for simple controls such as missile guidance, the size and cost of the data link are critical to the customer. With enhanced data link capability, this cost increases, and it may not be practical to use a full capability data link and network protocol. The complexity of the data links used for these types of applications needs to be minimized to avoid size and cost increases.
- Networking. The network must provide interoperability for multiple communication links with different types of radios. Increases in interoperability require networking to avoid collisions of the multiple users. The network protocols have to be designed to handle different legacy radios and future radios and possibly to provide a gateway that can be used for translating different legacy radios. SDRs can be reprogrammed to provide different modulation/demodulation methods, protocols, signal processing, and operational frequencies. They can also be used to provide a network-centric gateway to translate these differences between different radios. In addition, critical networking aspects such as self-forming and self-healing in mesh type networks are utilized to provide highly versatile, high-performance networks. Due to rapid movement on the battlefield coupled with the joint aspect of operations with multiple users from the military, the operational network needs to be dynamic with a constant addition and subtraction of nodes. This network management needs to be real time and versatile to handle the demands and scenarios that are required by the users.

10.11.2 SDRs

Software-defined radios are software programmable radios that can handle different waveforms and multiple modulation/demodulation techniques and can be modified using real-time loaded software in the field or programmed before operation. To extend the use of SDRs, a system with the ability to sense the current waveforms and automatically adjust and adapt to the radios and environment has been developed known as cognitive radios (CRs).

SDRs have the capability to program the current waveforms to be interoperable with multiple systems. This is accomplished using DSPs or field-programmable gate arrays (FPGAs) and reprogramming these devices for the waveform and modulation/demodulation processes for the desired radio or data link.

A program that combines legacy and future radios for interoperability is called the Programmable Modular Communications System (PMCS). It uses SDRs as a key component in allowing software programmable waveform modulation and demodulation, encryption, signal processing, and frequency selection.

10.11.3 Software Communications Architecture

Software communications architecture (SCA) is essential to the JTRS program and provides a basis for software waveforms. It is a nonproprietary, open systems architecture framework and is required to ensure operability between users. SCA compliance governs the structure and operation of the JTRS using programmable radios to load waveforms, to provide the communication and data link functions, and to allow for networking of multiple users. The SCA standard provides for interoperability among various types of radios by using the same waveform software that can be loaded in multiple users, providing network-centric capabilities. Security is required and becomes a challenge in the overall develop of the radios. Type 1 security is generally required for development of the JTRS radios. Software upgradability is also important in the development of the JTRS radios, which allows new waveforms in the future to be compatible with the JTRS system.

10.11.4 JTRS Radios (Clusters)

The JTRS is designed to be interoperable with legacy data link communication systems. In addition, future growth and capabilities are incorporated to extend the ability of the system to provide new modulation and higher data rate capacities. JTRS development is broken down into five types of radios, each for a particular use and requirements:

- Ground Mobile Radio (GMR) program (Cluster 1): This effort is involved with developing radios for Army and Marine Corps ground vehicles, Air Force Tactical Air Control Parties, and Army rotary-wing applications. This also includes developing a wideband network waveform (WNW), which is the next-generation IP-based waveform to provide ad hoc mobile networking. Recently Cluster 1 has been revised for ground vehicles only.
- JTRS Enhanced Multi-Band Inter/Intra Team Radio (JEM) (Cluster 2): This effort is involved with upgrading an existing handheld radio by adding JTRS capability to an existing enhanced multiband inter/intrateam handheld radio (MBITR). This is directed by the Special Operations Command (SOCOM). In addition, this radio is SCA compliant. A further development is a JTRS enhanced MBITR (JEM), which includes rigorous government evaluations and tests.
- Airborne, Maritime, Fixed Site (AMF) program (Clusters 3–4): Clusters 3 and 4 have been combined to form this new program. This effort will incorporate the Army, Air Force, and Navy support. Cluster 3 was the maritime/fixed terminal development for the Navy. Cluster 4 was led by the Air Force and provided Air Force and Naval aviation radios for rotary and fixed-wing aircraft.
- Handheld, Manpack, Small Form Fit (HMS) (Cluster 5): This effort includes handheld radios, manpack radios, and man-portable radios that are small form fit radios for the Army's future combat systems and other platforms such as unattended ground sensors, unattended ground vehicles, unmanned aerial vehicles (UAVs), robotic vehicles, intelligent munitions, and weapon and missile systems. These radios are under development with the Soldier Level Integrated Communications Environment (SLICE) program, which utilizes the soldier radio waveform (SRW) and operates in different bands at frequencies of 450 MHz to 1000 MHz and 350 MHz to 2700 MHz. Burst data rates are from 450 kbps to 1.2 Mbps and 2 kbps to 23.4 kbps for low probability of intercept (LPI).

The Mobile User Objective System (MUOS) is a cellular network that provides support for the HMS radios. This will help in ultra high frequency Demand Assigned Multiple Access (DAMA) protocols for satellite communications.

A summary of the different JTRS radios is shown in Table 10-5.

10.11.5 Waveforms

Originally planned to span a frequency range of 2 MHz to 2 GHz, the JTRS has been now been expanded to frequencies above this top range to satisfy space communications requirements. A selection of legacy radios for possible inclusion in the JTRS system is listed in Table 10-6a and Table 10-6b.

10.11.6 JTRS Network Challenge

The initial concept for the JTRS was to have all of the waveforms and networking protocols available at each JTRS radio. To do this requires high-performance DSPs, FPGAs, and a lot of memory. Generally the main obstacles are antennas and RF circuits to handle the vast differences in frequencies and bandwidth. This requires size, cost, complexity, and power. This may be acceptable for some applications, but small size, weight, and power requirements for portable radios, man-pack radios, and weapon and missile systems are critical.

Table 10-5   JTRS radio development

JTRS Radios	Users	Characteristics	Features
GMR – Ground Mobile Radio program (Cluster 1)	Army and Marine Corps ground vehicles, Air Force Tactical Air Control Parties, and Army rotary-wing	Next generation IP-Based waveform to provide ad-hoc mobile networking	Wideband network waveform WNW Revised for ground vehicles only: GMR.
JEM – JTRS Enhanced Multi-Band Inter/Intra Team Radio (Cluster 2)	Army, Navy, Air force Special Operations Command (SOCOM)	Upgrade handheld radio by adding JTRS capability to an existing MBITR 30–512 MHz	JTRS enhanced MBITR or JEM includes rigorous government evaluations and tests
AMF – Airborne, Maritime, Fixed Site Program (Cluster 3&4)	Air Force and Navy	For rotary and fixed wing Remote and maritime vessels	Airborne Maritime Fixed-Station JTRS AMF
HMS – Handheld, Man-pack, Small Form Fit (Cluster 5)	Army Soldier Level Integrated Communications Environment (SLICE)	SRW 450–1000 MHz 350–2700 MHz Burst data rates 450 kbps–1.2 Mbps LPI 2 kbps–23.4 kbps MUOS, DAMA	JTRS Man-pack, JTRS HMS, unattended ground sensors and vehicles, UAVs, robotic vehicles, intelligent munitions, weapon and missile systems.

Table 10-6a Types of legacy radios

Communication Link	Frequency of operation	Modulation	Data Rate	Features
SINCGARS – Single Channel Ground Air Radio System	30–87.975 MHz	FM, CPFSK	16 kbps	Single channel (8 possible) or FH (6 hop sets, 2320 freq/hop set)
EPLRS: – Enhanced Position Location Reporting System	425.75–446.75 MHz	CP-PSK/TDMA		BW–24 MHz. Burst 825-1060 mS, 2 mS time slot. FH, 83 MHz channels, 57 kbps VHSIC SIP and 228 kbps VECP
HAVE QUICK II – military aircraft radio UHF used by the Air Force.	225–400 MHz	AM/FM/PSK	16 kbps digital voice	FH
UHF SATCOM	225–400 MHz	SOQPSK, BPSK, FSK, CPM	75 bps – 56 kbps	MIL-STD-188-181, -182, -183 and -184 protocols, DAMA
WNW – Wideband Networking Waveform	2 MHZ to 2 GHZ	OFDM, DQPSK	5 Mbps	
Soldier Radio and Wireless Local Area Network WLAN	1.755–1.850 GHz, and 2.450– 2483.5 GHz	TBD	16 kbps voice, 1 Mbps data	IEEE 802.11b, 802.11e and 802.11g.
Link 4A	225–400 MHz		5 kbps	TADIL C
Link 11	2–30 MHz and 225–400 MHz	QPSK	1364 and 2250 bps	TADIL A
Link 11B	225–400 MHz	FSK	600, 1200, & 2400 bps	TADIL B
Link 16	960–1215 MHz	MSK	2.4, 16 kbps & data 28.8 kbps to 1.137 Mbps	TADIL J, Ant Diversity, FH, TDMA
Link 22	3–30 MHz, 225–400 MHz	MSK	Compatible with Link 16	FH, TDMA
VHF-AM civilian Air Traffic Control	108–137 MHz, 118–137 MHz	AM	Analog	BW – 25 kHz US, 8.33 kHz European

Table 10-6b Types of legacy radios

Communication Link	Frequency of operation	Modulation	Data Rate	Features
High Frequency (HF) ATC Data Link	1.5–30 MHz	NA	300, 600, 1200 and 1800 bps.	Independent Side Band (ISB) with Automatic Link Establish (ALE), and HF ATC
VHF/UHF-FM Land Mobile Radio (LMR)	Various, see Features	NA	16 kbps.	25–54, 72–76, 136–175, 216–225, 380–512, 764–869, 686–960 MHz
SATURN: Second generation Anti-jam Tactical UHF Radio	225–400 MHz	PSK	NA	
IFF – Identification Friend or Foe. IFF/ADS/ TCAS will support data at	1030 & 1090 MHz	NA	689.7 bps.	
DWTS – Digital Wideband Transmission System Shipboard system	1350–1850 MHz	NA	Multiple from 144 to 2304 kbps.	
Cellular Radio and PCS	824–894, 890–960 & 850–1990 MHZ	Various, see Features	10 kbps, 144/ 384 kbps and 2 Mbps	TR-45.1 AMPS, IS-54 TDMA, IS-95b CDMA, IS-136HS TDMA and GSM and 3GSM, 2.5 G, 3G, WCDMA and CDMA-2000.
Mobile Satellite Service (MSS)	1.61–2 [2.5] GHz.	NA	2.4 to 9.6 kbps.	includes VHF and UHF MSS bands, LEOs & MEOs, Iridium, Globalstar, and others.
Integrated Broadcast Service Module (IBS-M)	225–400 MHz	NA	2.4, 4.8, 9.6 and 19.2 kbps	
BOWMAN – UK Tri-Service HF, VHF and UHF tactical communications system	NA	NA	16 kbps digital voice	
AN/PRC-117G© – man-portable radio platform	30 MHz–2 GHz	NA	Up to 5 Mbps	embedded, programmable, INFOSEC capabilities.
VHF FM	30–88 MHz	FM	16 kbps	
VHF AM ATC (Extended)	108–156 MHz	AM	16 kbps	
VHF ATC Data Link (NEXCOM)	118–137 MHz	D8PSK	4.8 kbps Voice 31.5 kbps data	
UHF AM/FM PSK	225–400 MHz & 225–450 MHz	AM/FM/, PSK	16 kbps	
MUOS	240–320 MHz	NA	2.4, 9.6, 16, 32 & 64 kbps	support the Common Air Interface

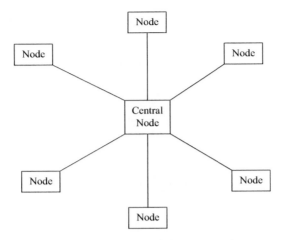

Figure 10-4 Star topology network.

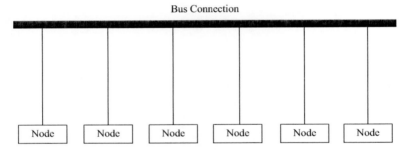

Figure 10-5 Bus topology network.

10.11.7 Gateway and Network Configurations

A possible alternative to this concept is to use a gateway, which provides all of this capability and can be used to translate different signals for different radio users. The gateway has the network control, using a network system that connects all of the different users to a single point, known as a star topology, and also employs TCP and IP (Figure 10-4).

Another type of network, called a bus topology network, allows users to communicate directly with each other without requiring a gateway to relay the information (Figure 10-5). All of the users are connected to a bus, and according to various protocols to avoid sending data at the same time they are able to communicate with everyone else who is connected. In addition, the bus topology can include multiple gateways, linked together in the bus network, and the users of the bus can send and receive their data to and from the gateway. The gateway then translates and sends the data to other users connected to the bus.

A ring network connects all of the users in a ring (Figure 10-6), and the data are passed around it, going from user to user. If the protocol is set up to receive the sent message, then the data are received. Here again, the ring network can include multiple gateways, and the

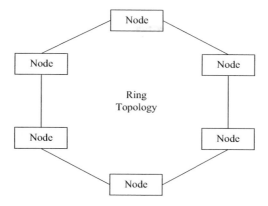

Figure 10-6   Ring topology network.

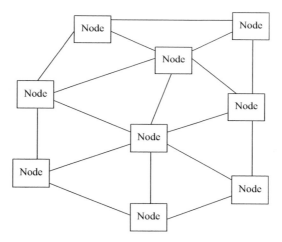

Figure 10-7   Wireless mesh network.

users on the ring can send and receive their data to and from the gateway, and the gateway would translate and send the data to other users connected to the ring.

A high-performance network used in many applications is known as a mesh network. It is used in wireless applications for both commercial and military networks. The main advantage of mesh networks is their ability to establish and maintain network continuity and route signals and links in the presence of interference and node dropouts in the network. They are able to maintain network integrity and perform real-time repair of the network during a failure without shutting down. This is known as self-forming and self-healing. They do not contain a central hub that controls each of the users; instead, the data links are established directly between users, which reduces single point failures (Figure 10-7).

Wireless mesh networks require minimum system administration, maintenance, and support. They are very useful in establishing networks for broadband communications. They overcome inherent connectivity limitations and provide automatic configuration and routing.

Whatever type of network is used, the main purpose is provide an overall system to communicate with many different radios and links into one big communication network. One of the steps to take in the near future is to develop interoperability—for example, making the SINCGARs radio interoperable with the HAVEQUICK II radio. From there, the system must expand and grow by adding other legacy radios, and as some of the legacy radios become obsolete they can be replaced with SDRs.

LAN standards. Standards have established to help set the direction and interoperability of different manufacturer's hardware, and they are continually being updated and new standards established to keep up with the new technologies, speed, and applications. The Institute of Electrical and Electronic Engineers (IEEE) established a standards committee known as the Local Area Network Standards Committee 802. This group created the 802.3 standard for bus topologies using CSMA/CD and the 802.5 standard dealing with token rings.

Ethernet. Ethernet uses a CSMA/CD scheme, in which a node senses if the bus is clear, then sends out a transmission, and monitors if there is another node trying to transmit. If it detects another transmission, then it sends out a signal telling all nodes to cease transmission. The other nodes cease for a random period of time and then try again if the channel is clear. This helps to prevent an endless loop of collisions between nodes trying to use the bus. Propagation delay can also increase the probability of collision. Someone may be using the system, and the other node may not have received his transmission due to the propagation delay so the node attempts to use the system.

There are four common standards using Ethernet: 10BASE-5, 10BASE-2, 10BASE-T, 100BASE-T, and 1000BASE-T. The designation tells the user the physical characteristics. The 10 describes the operating speeds (10 Mbps), the BASE stands for baseband, the 5 is distance (500 m without a repeater), and the T stands for unshielded twisted pair. 100 BASE-T provides 100 Mbps data rates under the IEEE 802.3u standard using CSMA/CD technology. The 100BASE-TX is a version that provides 100 Mbps over two pair of CAT-5 unshielded twisted pair (UTP) or two pair of Type 1 shielded twisted pair (STP). If other types of cables are used, the designation changes. For example, fiber cable uses X, 1000BASE-X; coax cable uses CX, 1000BASE-CX. Other terms for 1000 Mbps or 1 Gbps Ethernet are GbE or 1GigE.

Network layers. In the discussion of networking, several layers of the design have been established to help in designing interoperable systems and to use different layers with other designs. The standard for network layers is the Open System Interconnection (OSI) Seven-Layer Model. Some of these layers may not be included in the design of a network system, but the following gives an overview of the different types and basically what they do. More detailed description can be found in other references:

1. Physical layer. The lowest level network layer and is mainly concerned with voltages, currents, impedances, power, signal waveforms, and also connections or wiring such as the RS-232 serial interface standard.
2. Data link layer. Used to communicate between the primary and secondary nodes of a network. Involved with activating, maintaining, and deactivating the data link. Also concerned with framing the data and error correction and detection methods.
3. Network layer. Specifies the network configuration and defines the mechanism where messages are divided into data packets.

4. Transport layer. Controls end-to-end integrity, interface or dividing line between the technological aspects and the applications aspects.
5. Session layer. Network availability, log-on, log-off, access, buffer storage, determines the type of dialog, simplex, duplex.
6. Presentation layer. Coding, encryption, compression.
7. Application layer. Communicates with the users program. Controls the sequence of activities, sequence of events, general manager.

A chart showing the different layers is given in Figure 10-8.

10.11.8 Link 16

Link 16 (TADIL J) uses J messages and has been looked at to provide interoperability and networking in a complete battlefield system. This discussion addresses Link 16 Class 2, Tactical Data Link for C3I.

The Joint Tactical Information Distribution System (JTIDS) uses Link 16 as its core. The Multifunctional Information Distribution System (MIDS), with its reduced size, is used in the F/A-18 aircraft. The MIDS provides interoperability between the MIDS and JTIDS. The Army is using JTIDS with a rate of less than 8 kbps and employs a combinational interface of the Position Location Reporting System (PLRS) and JTIDS hybrid interface.

In addition, the MIDS radio has been upgraded to be compatible and interoperatable with JTRS requirements and is referred to as MIDS JTRS.

Link 16 is a multiuser system with low duty cycle, pseudo-randomly distributed in the frequency code domain on a slot-by-slot basis. The range is 300 nm nominal, with the ability to extend the range to 500 nm. Any Link 16 terminal can operate as a relay to further increase the range of the system. The power output is 200 W low power and 1000 W high power if a high-power amplifier (HPA) is added to the system. The power can be reduced by eliminating both the voice channels and tactical air navigation (TACAN).

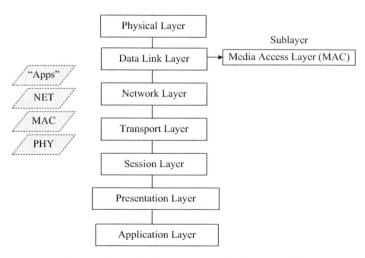

Figure 10-8   OSI seven network layer model.

The system operates at least two antennas, for antenna diversity, to prevent multipath and jamming. The antennas used are broadband vertical antennas in the ultra high frequency (L-band) 960 MHz to 1215 MHz operating frequency range. Special filtering is added to prevent interference with the TACAN, distance measuring equipment (DME), and identification friend or foe (IFF) by using a dual-notch bandpass filter, which eliminates IFF at 1030 MHz and 1090 MHz. The frequency of operation of the Link 16 system is in two bands: ultra high frequency, 300–1000 MHz; and L-band, allocated 960–1215 MHz (969–1206 MHz) with an IF of 75 MHz.

10.11.9 Link 16 Modulation

The modulation of Link 16 is minimum shift keying (MSK). A single message consists of 5 data bits. which are spread at a 5 MHz rate using 32-bit cyclic code shift keying (CCSK) to produce an approximately 8 dB processing gain: $10\log(32/5)$. Link 16 also uses frequency hopping, with each message pulse on a different frequency per pseudo-noise (PN) code, with up to 20 different hop PN codes to prevent mutual interference. Therefore, each hop frequency contains one message, which contains 5 data bits spread by 32 MSK chips. The message pulse is 6.4 μsec on and 6.6 μsec off, for a total of 13 μsec.

The data rates are variable, with rates of 28.8 kbps, 31.6 kbps, 57.6 kbps, and 115.2 kbps, with a capacity up to 238 kbps, which can be expanded to 2 Mbps. In addition to data, the Link 16 provides two digital voice channels with rates of 2.4 kbps and 16 kbps.

In addition, Link 16 provides FEC. This is a Reed-Solomon (RS) FEC: (31,15). The data word is 15 5-bit characters (70 bits plus 5 bits parity) and is expanded to 31 RS characters (155 bits). If there is a need to receive more than half bits error free, then the entire message can be regenerated. FEC and interleaving across hops have the ability to correct channel bits that are lost, but the voice channel contains no FEC.

An important part of Link 16 is encryption. The system uses COMSEC/TRANSEC, encrypted by KGV-8B. In addition, it is encrypted using frequency hopping and jitter (the random time period the terminal waits to enter system).

10.11.10 TDMA

Hopped MSK pulses are assigned a time slot for TDMA, up to 60 hops/time slot at a rate of 76.923 kHps, or 1/13 μsec. The data link hops both the data and the sync pulse. The frame is 12 sec, with 1536 time slots/frame. A total of 24 hours is divided into 112.5 epochs, which is equal to 12.8 minutes each. The epochs are divided into 98,304 time slots, which equals 7.8125 msec/time slot. Access to Link 16 rotates among all users every 12.8 minutes (98,304 slots). Each cycle is an epoch, and assignments are made against a 12 sec repeating frame.

10.11.11 "Stacked" Nets

Stacked nets are networks that allow multiple users to occupy the same time slots because the users have different PN codes (20 total) selected for their hop sequence and some process gain due to MSK.

10.11.12 Time Slot Reallocation

Time slot reallocation (TSR) was developed for the Navy and is similar to a DAMA system but without a central control station. To avoid collisions and contention access, the user waits for the next time slot if one time slot is busy.

10.11.13 Bit/Message Structure

The first pulse is the start of the time slot, called jitter, followed by the synch pulse, then the header timing pulses containing the following information:

ID
Address
Classification
FEC instructions
RS FEC for the header is (16,7) and interleaved.

The message can be single or double pulse. The standard is double pulse, with the same data in each pulse. There are additional variations to this standard. The time allocated for each part of the message is as follows:

Time slot jitter pulse = 2.418 msec
Time propagation allowed in the design = 2.040 msec
Time for the data plus header = 3.354 msec
In extended slot, jitter time is used for data; total data time = 5.772 msec

A summary of the Link 16 performance specifications is shown in Table 10-7.

Link 16 has been the preferred method of communication in the military arena. The challenge is for industry to generate a network design centered on this technology and determine if it is feasible to use this data link for all applications, or perhaps a subset, or possibly a totally new method for accomplishing interoperability for C4I.

Table 10-7   Link 16 performance specifications

RF System	Frequency	Modulation	Data Rate	Other Networking
Link 16	300–1000 MHz 960–1215 MHz (969–1206 MHz)	Spread Spectrum MSK 32 chips/5 bits FH 1 hop/message LPI, anti-jam TDMA FEC Encryption Double pulse	31.6, 57.6, and 115.2 kbps 238 kbps capacity, expandable to 2 Mbps 200 W Low power 1000 W High power with HPA	Stacked nets Time slot reallocation (TSR) 20 different PN codes

10.12 Summary

Broadband and home networking will shape the future. New standards are being reviewed as new technologies are developed and as the data rates increase. With several incoming signals to a home, such as voice, data, and video, there is a need to provide optimal distribution throughout to allow for easy access.

Networking is becoming important on the military battlefield, and JTRS and Link 16 play important roles in the interoperability of communication devices. The development of these and other new technologies will provide the military with a network for all communication devices in the future.

References

Lough, Daniel L., T. Keith Blankenship, and Kevin J. Krizman. "A Short Tutorial on Wireless LANs and IEEE 802.11." http://decweb.bournemouth.ac.uk/staff/pchatzimisios/research/IEEE%20802_11% 20Short%20Tutorial.htm

Core Specification of the Bluetooth System Specification Volume 1, Version 1.1 February 22 2001

Profiles Specification of the Bluetooth System Specification Volume 2, Version 1.1 February 22 2001

Problems

1. What are the three mediums used in broadband distribution in the home?
2. What is a common technique is used in power line communications to maximize the frequency band with multiple users?
3. What criteria are used to determine if two signals are orthogonal?
4. Describe how the desired signal can be retrieved among other signals in an OFDM system.
5. What is a major drawback of Home PNA?
6. What are the three main challenges that face the JTRS?
7. How do SDRs help to solve the interoperability problem?
8. What are the four standard network topologies?

CHAPTER 11

Satellite Communications

Satellite communications are becoming a viable means of providing a wide range of applications for both the commercial and military sectors. The infrastructure for distributing signals covers the widest range of communications methods; even the most remote places on Earth like the South Pole can have communications via satellite. Satellite's bandwidth, field of view and availability, along with combining this technology with other types of communications systems, produce a ubiquitous infrastructure that provides communications worldwide.

11.1 Communications Satellites

Communications satellites are sent into space in a geostationary orbit and are made up of a space platform and the payload. The payload is the equipment and devices mounted on the space platform. One satellite may contain multiple payloads for multiple applications, including different data links operating with their own frequencies and antennas, which are all mounted to the space platform and sent into space.

Each satellite is equipped with attitude control to ensure that the antennas, which are attached on the spacecraft, are all pointed toward the earth and that the antenna beam is focused on the intended area on the earth (Figure 11-1).

Several types of disturbances change both the attitude of the geostationary satellite and the orbit of the satellite by a slight amount. The satellite's attitude stabilizers compensate for the changes in attitude, or its antenna's positioning to keep them pointing toward the earth. The orbit of the satellite is changed due to movements in the north and south plane in a figure-eight pattern over a 24-hour period.

11.2 General Satellite Operation

End users are connected to an earth station at their location, whether it is a personal system or a network serving multiple users. This earth station provides two-way communications between it and the satellite, referred to as the space station. The space station relays the information to another earth station that provides the services, whether it is data links; military operations for command, communications, and control; Internet services; telephone; fax; video; music; or other types of information (Figure 11-2).

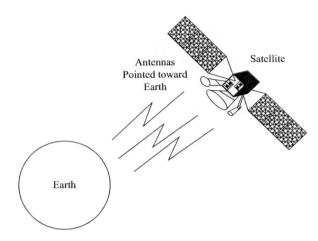

Figure 11-1   Attitude control to ensure antennas pointing direction.

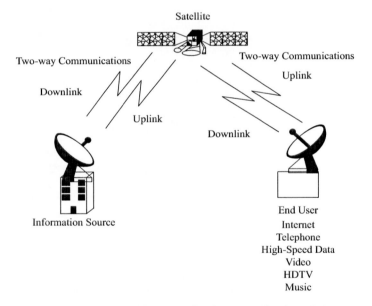

Figure 11-2   Satellite operational communications link.

11.2.1 Operational Frequencies

Five different bands of operation are used for satellite communications. The lowest band is called the L-band, which operates with an uplink at 1.6 GHz and a downlink at 1.5 GHz using a narrow bandwidth. The next band is the C-band, which operates at around 6 GHz for the uplink to the satellite and 4 GHz for the downlink from the satellite to the ground station. The next band, which is generally used by the military, is the X-band, which operates at

Table 11-1   Standard and VSAT satellite frequency bands of operation

VSAT Satellite Frequency Bands		
Frequency Band	**Uplink (GHz)** **Earth Station to Satellite**	**Downlink (GHz)** **Satellite to Earth Station**
C Bank	5.925 to 6.425	3.700 to 4.200
Extended C Band	6.725 to 7.025	4.500 to 4.800
Ku Bank	14.000 to 14.500	10.950 to 11.700

Standard Satellite Frequency Bands				
Band	**Freq GHz**	**Uplink GHz**	**Downlink GHz**	**Other**
L-band	1.6/1.5	1.6	1.5	narrow BW
C-band	6/4	5.850–6.425	3.625–4.200	500 MHz BW
X-band	8/7	7.925–8.425	7.250–7.750	500 MHz BW mostly military
Ku-band	14/12	14–14.5	11.5–12.75	500 MHz BW
Ka-band	30/20	27.5–31	17.7–21.2	3.5 GHz BW

around 8 GHz for the uplink and 7 GHz for the downlink. The next band, which has become popular for telecommunications, is the Ku-band, which operates at around 14 GHz for the uplink and 11–12 GHz for the downlink. The most widely used of these four are the C- and Ku-bands.

A fifth band, the highest band of operation, called the Ka-band, is becoming popular for broadband communications and military applications. It operates at 30 GHz for the uplink and 20 GHz for the downlink and provides a much higher bandwidth for high-speed data and more simultaneous end users. A summary of the different frequency bands including very small aperture terminal (VSAT) frequency bands is shown in Table 11-1.

These are approximate frequencies and represent bands currently in use. In addition to the five standard satellite bands mentioned previously, very high frequency and ultra high frequency (VHF/UHF) satellite bands are also being used for lower data rates and operate across multiple frequencies from 136 to 400 MHz. Bands are continually expanding and changing, so these numbers may not be exact and may vary slightly. Also, along with frequency changes, bandwidths change for different systems and uses. In the future, other bands will eventually open up for satellite communications as the demand increases.

11.2.2 Modulation

The main modulation scheme for the Intelsat/Eutelsat time division multiple access (TDMA) systems is coherent quadrature phase-shift keying (QPSK). Coherent QPSK uses four phase states providing 2 bits of information for each phase state. Since this is coherent QPSK, the absolute phase states 0, 90, 180, and 270 are used to send the information. The data rate is approximately equal to 120 Mbps using a bandwidth of 80 MHz. The Intelsat system uses 6/4 GHz, while the Eutelsat system uses 14/11 GHz. The radio frequency (RF) is down-converted to typical intermediate frequencies (IFs) including 70 MHz, 140 MHz, and 1 GHz.

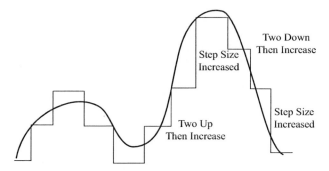

Figure 11-3   ADPCM to convert analog to digital signals.

Techniques have been developed to increase the efficiency of the satellite link. One of these techniques is called digital speech interpolation (DSI), in which data signals are transmitted during the dead times of voice channels or telephone calls. This provides a method of sending data and utilizing times when voice is not being sent. Another technique to increase link efficiency combines DSI with a decrease in voice speed from 64 to 32 kbps; this is called digital circuit multiplication equipment (DCME).

11.2.3 Adaptive Differential Pulse Code Modulation

One of the best and most efficient ways to quantize analog signals is via adaptive differential pulse code modulation (ADPCM). The analog signals are sampled at a rate greater than the Nyquist rate. Each of the samples represents a code value for that sample, similar to an analog-to-digital converter (ADC). The differential part of the ADPCM describes a process by which the sampled value is compared with the previous sampled value and measures the difference. The adaptive part of the modulation scheme means that the step size can be made finer or coarser depending on the difference measured. This helps in tracking large analog voltage excursions. If the samples are continually increasing, then the step value is increased. If the samples are continually decreasing, then the step size is decreased (Figure 11-3).

11.3 Fixed Satellite Service

Satellite communications using geostationary satellites to provide services to the end user are known as fixed-satellite service (FSS). FSS is a radio communications service between fixed points on the earth using one or more geostationary satellites.

11.4 Geosynchronous and Geostationary Orbits

Most satellites are approximately 22,000 miles above the earth's surface. If they follow a geosynchronous orbit, then the period of the orbit is equal to the earth's rotation. If they are only geosynchronous, then they are not in a fixed position as far as the view from the earth. The orbits are inclined with respect to an orbit around the equator (Figure 11-4).

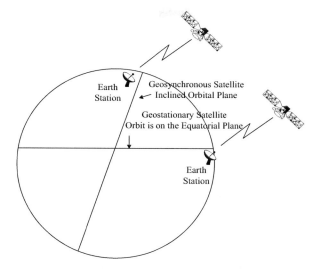

Figure 11-4 Geosynchronous and geostationary satellite orbit.

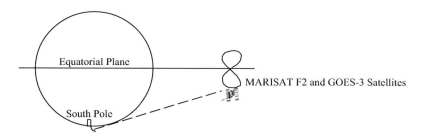

Figure 11-5 Satellite tilt orbit in a figure-eight pattern.

Most communications satellites follow a geostationary orbit. These orbits need to be geosynchronous but should travel in a circular orbit on the equatorial plane. This allows satellites to be in a fixed position in relation to the view from the earth. Therefore, they follow a circular orbit on the equatorial plane, circling the earth once every 24 hours, the same time it takes the earth to rotate once on its axis. By synchronizing to the earth's rotation, so that the satellite follows the earth at approximately the same speed as the angular rotation of the earth, the satellite will look stationary to a fixed point on the earth (Figure 11-4).

This provides a continuous link with a given satellite at any time during the day or night. Therefore, the communications link will be available all the time unless there is a problem with the data link between the earth and the satellite. Another advantage is that the ground equipment becomes less expensive and easier to operate since tracking a reasonably stationary satellite is much easier than tracking a satellite with high angular velocity.

Even though geostationary satellites appear stationary, they actually drift in an orbital figure-eight pattern north and south of the equator (Figure 11-5). Satellites are designed to help mitigate this drift so that the earth stations can be stationary without automatic tracking controls. After they have been in orbit for close to their space lives, this figure-eight pattern becomes much larger. For 30-year-old satellites, the orbital tilt becomes large. Because of this, satellites such as the Marisat F2 and GOES-3 were used for communications to remote places such as the South Pole. There it is impossible to see satellites on the equatorial plane, but because these older satellites follow this figure-eight pattern when they travel south of their orbit they become visible from the South Pole (Figure 11-5).

11.5 Ground Station Antennas

Even though the satellite appears to be stationary for a geostationary orbit, large antennas with very narrow beam widths generally require an automatic tracking device to provide the best performance. Because smaller antennas have small apertures and wide beams, they generally do not require a tracking device. Digital satellite systems for television use a small antenna and do not require any tracking device. The antenna is installed at the site and usually does not have to be adjusted for several years.

The gain of a ground station antenna is largely dependent on the diameter of the antenna and is also frequency related. The equation for the gain of the parabolic antenna that is often used for satellite communications systems is

$$G_t = 10\log[n(\pi(D)/\lambda)^2]$$

where

$n =$ efficiency factor < 1
$D =$ diameter of the parabolic dish
$\lambda =$ wavelength

Low-cost systems operating in the Ku- and Ka-bands using small antennas approximately 1–2 m in diameter are known as very small aperture terminals. VSATs provide two-way communications to a central location called a hub. They are used mainly for businesses, schools, and remote areas. They basically connect remote computers and data equipment to the hub via satellites.

Three types of antenna systems are incorporated in satellite communications:

- Primary focus antenna system. The feed is positioned in front of the primary reflector, and the signal is reflected once from the feed to the intended direction of radiation. The single reflector antenna system is generally less expensive and provides the simplest design (Figure 11-6a).
- Cassegrain antenna system. This antenna system uses a dual-reflector arrangement. The feed comes from the back of the primary reflector and sends the signal to a convex subreflector, which is mounted in front of the primary reflector. Then the signal is fed to the primary reflector where it is reflected in the desired direction (Figure 11-6b). Cassegrain antennas are more efficient than primary focus antennas for several reasons, the main one that the subreflector can be adjusted or formed to optimize the signal reflection and to focus less energy toward the blockage of the subreflector, which falls in the path of the primary reflector. They are also easier to maintain, since the feed horn is located at

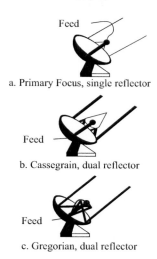

a. Primary Focus, single reflector

b. Cassegrain, dual reflector

c. Gregorian, dual reflector

Figure 11-6   Types of antennas for satellite communications systems.

the base of the reflector, which provides easy access compared with having the feed horn on structures in front of the reflector.

- Gregorian antenna system. This antenna system is also a dual-reflector system, very similar to the Cassegrain antenna. The main difference is that the subreflector is concave instead of convex. The feed is mounted on the rear of the primary reflector and sends the signal to the subreflector. The signal is reflected in the concave subreflector, which causes a crossover of the reflected signal. The reflected signal is sent to the primary reflector, where it is reflected in the desired direction (Figure 11-6c). Because of this crossover, the subreflector must be a greater distance from the primary reflector.

11.6 Noise and the Low-Noise Amplifier

The low-noise amplifier (LNA) is critical to the design of a satellite communications system. The system performance is improved directly for each dB of improvement in the LNA. The LNA is the main element that sets the noise figure for the satellite receiver unless there are large losses after the LNA or the bandwidth becomes larger, which may degrade the receiver performance significantly. A complete noise analysis is addressed in Chapter 1.

Another method of doing link budgets that is commonly used for satellite communications is evaluating the system using equivalent noise temperature. Most link budgets for terrestrial systems use noise power, signal power, and the standard $kTBF$ at the output of the LNA, with the losses from the antenna affecting the signal level in the link on a one-for-one basis. However, for satellite transmission systems, the analysis uses equivalent temperatures and converts the noise factor of the LNA to an equivalent temperature:

$$T_e = (F - 1)T_o$$

where

T_e = equivalent temperature due to the increase in noise of the LNA noise factor
F = noise factor of the LNA (noise factor = 10log(noise factor))
T_o = 290 K

Working in reverse to determine the noise factor:

$$T_e = (F - 1)T_o = T_oF - T_o$$
$$F = (T_e + T_o)/T_o$$

Therefore, the noise factor, which is the increase in noise for the receiver, is equal to the additional temperature (T_e) added in front of an ideal receiver that produces the same amount of increased noise. So using this equation, if the noise figure of the LNA is low, say 0.5 dB, then the noise factor is 1.12 and the equivalent temperature T_e is equal to 34.8 K. To illustrate this, the noise power at the output of the LNA is equal to kT_oF if the bandwidth and gain is ignored. The input noise power of the LNA is equal to kT_o with the same assumptions. Thus, the increase in noise power due to the noise factor of the LNA is equal to

$$\text{Increase in noise power} = N_i = kT_oF - kT_o$$

Since the equivalent noise increase is desired for the link budget, the equivalent noise T_e is solved by eliminating the k:

$$T_e = T_oF - T_o = (F - 1)T_o = (1.12 - 1)290 = 34.8 \text{ K}$$

Since the satellite link budget is in terms of temperature, the losses in the system between the antenna and the LNA need to be converted into temperature (Figure 11-7). This is done by taking the difference in noise power due to the total losses:

$$\text{Noise power difference of attenuator} = N_d = kT_a - kT_a/L_T$$
$$\text{Temperature difference} = T_d = T_a - T_a/L_T = T_a(1 - 1/L_T)$$

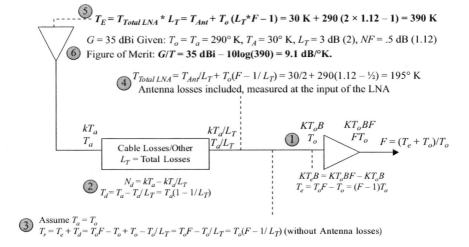

Figure 11-7   Equivalent noise temperature analysis, G/T.

where

> T_d = temperature difference due to losses or attenuation
> L_T = total losses
> T_a = temperature of the losses or attenuation, generally equal to $T_o = 290$ K

Note that if $L_T = 1$, there are no losses and $T_d = 0$. As L_T approaches infinity, then $T_d = T_a = T_o$. Therefore, the total equivalent temperature out of the antenna for the receiver (T_r) is

$$T_r = T_e + T_d = (F - 1)T_o + T_o/L_T = T_oF - T_o/L_T = T_o(F - 1/L_T)$$

assuming T_o is the temperature of the losses.

Using the previous example and 3 dB (2) of cable losses and solving for T_r:

$$T_r = T_o(F - 1/L_T) = 290(1.12 - \frac{1}{2}) = 180 \text{ K}$$

This is the equivalent temperature measured at the input to the LNA.

To find the total equivalent temperature for a system measured at the input to the LNA, the noise due to the antenna (T_{Ant}) needs to be included. The antenna noise is attenuated by the cable loss (L_T), and the total equivalent temperate at the LNA would be

$$T_{Total\ LNA} = T_{Ant}/L_T + T_o(F - 1/L_T)$$

The equivalent temperature at the antenna input is calculated by integrating the gain and noise temperature in the direction of the antenna. This is accomplished by multiplying the previous equation by the cable loss between the LNA and the antenna to provide the temperature equivalent at the location of the antenna. The noise of the antenna is calculated as follows:

$$T_E = L_T(T_{Ant}/L_T) + T_o(F - 1/L_T) = \boldsymbol{T_{Ant} + T_o(L_T \times F - 1)}$$

This is the total equivalent noise temperature at the antenna (Figure 11-7).

Once the temperature of the antenna is calculated, then a figure of merit, which is used to evaluate different satellite systems, can be used to determine the quality of the receiver. The figure of merit is equal to the gain of the antenna divided by the total temperature at the antenna, (G/T), which is usually given as dB/K. Since the gain of the antenna is usually expressed in dBi, G/T is calculated by

$$G/T = G \text{ dBi} - 10\log(T_E)$$

A typical value is around 30 dB/K, but this value will vary tremendously with the size of the antenna, frequency, LNA, losses, and other factors.

11.7 The Link Budget

The link budget uses the noise and LNA along with the signal level to determine the range and quality of the satellite link. To calculate the signal or carrier level at the input to the receiver, an effective isotropic radiated power (EIRP) is determined from the transmitter.

11.7.1 EIRP

The EIRP out of the transmitting antenna is the power output of the power amplifier (PA), losses from the PA through the antenna, and the gain of the antenna. The power out of the PA is usually expressed in dBm (10log(mW)) or dBW (10log(W)).

Several types of PAs are used for transmitting signals to the satellite, including traveling wave tubes (TWTs), Klystrons, and low-cost field-effect transistor (FET) amplifiers. TWTs offer wide bandwidths, 500 MHz and higher, good group delay, and the ability to handle many signal inputs. Klystrons are expensive, are high-power only, and have narrow bandwidths in the 40–80 MHz range. The FET solid-state (SS) amplifiers offer low-cost solutions and include many of the features of the other technologies, though generally not for extremely high-power outputs. A general power requirement for different types of signals includes 1 W per channel for telephone signals and 1 KW per television carrier.

The losses are in dB and are subtracted from the PA output power; the antenna gain is in dB and is added to the final result. The EIRP for the satellite is calculated as follows:

$$EIRP = P_t - L_{ta} + G_{ta}$$

where

P_t = power output of the PA (in dBm)
L_{ta} = total attenuation in the transmitter (in dB)
G_{ta} = gain of the transmitter antenna (in dB)

11.7.2 Propagation Losses

The propagating channel is calculated for the satellite data link according to the losses discussed in Chapter 1, with the major contributor being free-space loss:

$$A_{fs} = 20\log[\lambda/(4\pi R)] = 20\log[c/4\pi Rf]$$

where

λ = wavelength
R = slant range, same units as λ
c = speed of light, 300×10^6 m/sec, R is in meters
f = frequency

The free-space loss for a satellite system operating in the X-band is approximately 200 dB. Depending on the conditions in the atmosphere, other losses such as clouds, rain, and humidity need to be included in the link budget. These values are usually obtained from curves and vary from day to day and from region to region. Each application is dependent on the location, and a nominal loss, generally not worst case, is used for the link analysis.

11.7.3 Received Power at the Receiver

Other losses include multipath loss, which for satellite systems is generally small if the antennas are sited properly. For low-angle satellites, multipath can become a factor.

In addition, the receiver also contains losses, similar to the transmitter, which need to be included in the link budget. Finally, implementation and spreading losses are included. The received power at the LNA is

$$P_s = EIRP - A_{fs} - L_p - L_{ra} + G_r - L_{multi} - L_{imp} - L_{ss}$$

where

$EIRP$ = effective radiated power with respect to an isotropic radiator (in dBW)
A_{fs} = free-space loss
L_p = propagation loss
L_{ra} = receiver losses
G_r = gain of the receiver antenna
L_{multi} = multipath losses
L_{imp} = implementation losses due to hardware
L_{ss} = spreading losses due to spread spectrum

A typical link budget for satellite communications is provided to evaluate the trade-offs of the system and to ensure that the receivers on either end have enough signal-to noise ratio (SNR) for reliable communications (Table 11-2).

11.7.4 Carrier Power/Equivalent Temperature

Often in satellite systems, the power received by the receiver is called the carrier power (C), which is also compared with an equivalent temperature (C/T). A simplified form of combining the losses, dividing both sides by temperature, and converting all values to actual values (not dB) is

$$C/T = EIRP * G_r/(L_T * T) = G_r/T * EIRP/L_T$$

where

C/T = carrier power/equivalent temperature
G_r/T = figure of merit
$EIRP$ = effective isotropic radiated power
L_T = total losses from the EIRP of the transmitter to the receiver

Power flux density (PFD) is used to determine the amount of power radiated by the antenna in a direction at a large distance per unit of surface area. For an isotropic radiator, the equation equals

$$PFD = P/[4(\pi)d^2]$$

where

PFD = power flux density
P = output power
d = distance

Add the gain to this to find the PFD for a gain antenna.

Table 11-2   Link budget analysis for the uplink from the earth station to the satellite

	Slant Range (Km)	Freq. (GHz)	Power (W)	
Enter Constants	35,000	6	10	
Enter Inputs	**Inputs**	**Power Levels**		**Temp**
Transmitter	Gain/Loss (dB)	Sig. (dBm)	Noise (dBm)	Kelvin
Tran. Pwr (dBm) =		40		
Trans. Line Loss =	−0.5	40		
Other (switches)	−0.5	39		
Trans. Ant. Gain =	47	86		
Ant. Losses =	−1	85		
EIRP		85		
Channel				
Free Space Loss =	−198.89	−113.89		
Rain Loss =	−0.20	−114.09		
Cloud Loss =	−0.10	−114.19		
Atm Loss (etc.) =	−0.10	−114.29		
Multipath Loss =	−2.00	−116.29		
Receiver				
Rx Ant. Gain =	47.00	−69.29		
G/T dB/K				**25.20**
Total Noise Temp. at Ant.				152
Antenna Noise T_A				30
Ant. Losses =	−0.02	−69.31		
Other (switches)	−0.25	−69.56		
Rec. Line Loss =	−0.25	−69.81		
Total Losses L_T	−0.52			
RF BW(MHz)	100.00		−94.00	
Total Noise Temp. at Rec.				134
Equiv. Temp. T_e =				75
LNA Noise Fig. =	1.00		−93.00	
LNA Gain =	25.00	−44.81	−68	
LNA levels =		−44.81	−68.00	
Receiver Gain =	60.00	15.19	−8.00	
Imp. Loss	−4.00	11.19		
Detector BW	10.00		−18	
Det. Levels =		11.19	−18	
S/N =		**29.19**		
Req. E_b/N_o				
Req. E_b/N_o		12	$P_e = 10^{-8}$	
Coding Gain =	4.00	8.00		
E_b/N_o **Margin** =		**21.19**		

With satellite's increased power and more sensitive receivers, earth stations are becoming less costly and smaller. For example, the Intelsat V satellite contains 50 transponders on-board and operates with just 5–10 W of power. The total bandwidth for the Intelsat V is 500 MHz to provide high-speed data for multiple users.

11.8 Multiple Channels in the Same Frequency Band

To utilize the band more efficiently and to provide more data or users, two different schemes are employed. The first scheme uses beam pointing at two different points on Earth using the same satellite. Each beam is focused on an area of the earth so that the same antenna can be used for two systems in the same band and frequency. The narrower the beamwidth, the less coverage area that is illuminated on the earth's surface. However, this provides for more EIRP so that the earth stations can be less expensive.

Another way to increase the data capacity for a given band is via polarization. If orthogonal polarizations are used for two different channels, they can use the same antenna and bandwidth with minimal interference between them. In theory, horizontal and vertical polarizations are orthogonal, as are left-hand circular polarization (LHCP) and right-hand circular polarization (RHCP). Cross-polarization can occur, which degrades the separation of two channels that are orthogonally polarized. Often orthogonal polarization is used to provide increased isolation between the transmitter and the receiver for transceiver operation.

For a system to use polarization for frequency reuse in a satellite communications system, the isolation between the channels should be at least 25–30 dB. The main causes that degrade the isolation are the Faraday effect (Earth's magnetic field) and atmospheric effects (rain or ice crystals). Also, multipath can alter the polarization during reflection of the signals on a surface, and propagation through the troposphere or ionosphere can lead to polarization disturbances.

11.9 Multiple Access Schemes

Multiple access is important in satellite communications to allow multiple users in the same bandwidth, satellite, and antenna system. Two basic methods are used to provide for multiple users. The first is a multiplexing scheme called preassigned multiple access (PAMA), which permanently assigns a user to a channel or time. An example of PAMA is using time slots and assigning a given time slot to each end user. Since they are permanently assigned, the users will have that given time slot and will be multiplexed with all of the other users, similar to time division multiplexing (TDM) used in other communications links.

The other method of multiplexing is called demand assigned multiple access (DAMA). This method is a true multiple access scheme similar to TDMA, which is used in other communications links. This system is used on an "as needed" basis, and each user takes any time slot when needed.

Another method of providing for multiple users is frequency division multiplexing (FDM) or frequency division multiple access (FDMA). These methods use frequency division to provide for multiple users on one band. In systems that incorporate FDMA, the channels are separated using different frequency slots for different users. Other techniques to provide for multiple users are beam focusing, so that each of the beams covers a different area of the earth from the satellite, and antenna polarization, which can separate multiple users via the polarization of the antennas, vertical versus horizontal, or LHCP versus RHCP.

11.10 Propagation Delay

One of the problems with real-time communications using satellites is the large distance between the earth station and the satellite. There is an approximate propagation delay of 275 msec. The two-way propagation delay is double that time, or 550 msec. Since this propagation delay is so long, echo cancellation is vital to the quality of the communications link, especially for voice applications. Video applications, especially one-way broadcasting, generally does not need echo cancellation techniques.

11.11 Cost for Use of the Satellites

The cost is determined by the type of transmission, type of signal sent, and the length of the transmission time. The types of transmissions include PAMA, DAMA, and occasional. PAMA is the most costly, since it basically ties up an entire multiple access slot all of the time. Demand systems such as DAMA are less costly, since the access slot is assigned only when it is in use. The least costly is the occasional use, which is billed only for the time that it is used.

The types of signals sent using the satellite link include voice, video, and data. Many systems do not include voice as an option, with the Internet being the most important. For more information on the cost of use of satellites, see the Intelsat tariff handbook, which specifies detailed charges.

11.12 Regulations

The International Telecommunication Union (ITU) radio regulations provide the rules for satellite communications to avoid interference and confusion. The main specifications supplied by the ITU include frequency band allocations, power output limitations from either the earth station or the satellite, minimum angles of elevation for earth station operation, and pointing accuracy of antennas.

11.13 Types of Satellites Used for Communications

The Inmarsat satellite system's main use is for maritime communications for ships and shore earth stations. Currently it is being employed for broadband communications and extended coverage. The satellites cover the main bodies of water, but earth stations do not have to be located on the shore as long as the satellites are visible. The Inmarsat satellites operate on C- and L-bands. The FSS gateway for telephone service is via the Inmarsat satellites.

Another satellite system is the Intelsat system. Intelsat provides broadband communications, including digital data, video, telephones, and other communications. There are several types of Intelsat satellites in use today.

Other satellites that cover various sections of the world are Eutelsat, which are used mainly for European countries, and PanAmSat, which are used for Central and South American countries. Each satellite's coverage needs to be evaluated for optimal performance, satellites in view, usage, and cost.

With the push toward providing broadband communications, including high-speed data and Internet connections, many companies such as Astrolink, CyberStar, SkyBridge, Spaceway, and Teledesic are working to develop both Ku- and Ka-band satellite communications systems. They use geosynchronous earth orbit satellites (GEOS), low earth orbit satellites (LEOS), and hybrid satellite constellations to provide both Ka- and Ku-band operation, with data capacities ranging from 30 Gbps to systems providing much greater data capacities in the future.

LEOS are broken down into two subclasses: little and big. The orbits of little LEOS are from 90 minutes to 2 hours and are less than 2000 km high. They can be used for approximately 20 minutes per orbit. They operate in the following frequency bands listed:

137−138 MHz

400.15−401 MHz

148−149.9 MHz

Examples of little LEOS include OrbComm, Starsys, and VITA.

Some little LEOS follow a polar orbit at an altitude of 850 km and are sun synchronous, which means that the satellite passes over a certain area at the same time every day.

Big LEOS include satellite systems such as Iridium, Teledesic, Globalstar, Odyssey, ICO Global Communications, Project-P, Aries, and Ellipsat.

Another class of satellites is medium earth orbit satellites (MEOS). They orbit at an altitude of approximately 10,000 km, with a 12-hour orbit, which means that these satellites travel around the earth twice a day. They are generally 55° inclined from the equatorial plane. These types of satellites include global positioning systems (GPSs).

Geostationary satellites were discussed at length earlier in this chapter. They orbit at an altitude of approximately 35,800 km.

And finally, there are highly elliptical orbit satellites (HEOS). Since their orbits are elliptical, the perigee is approximately 500 km, and the apogee is approximately 50,000 km. Their orbit time is between 8 and 24 hours for one revolution around the earth.

11.14 System Design for Satellite Communications

The main design criteria for satellites are the transponder bandwidth, EIRP, and G/T. Typical figure of merit values for 4 GHz range from 41 dB/K using a 30 m antenna using a parametric amplifier to 23 dB/K for a 4.5 m antenna using an FET amplifier. For a space station receiver at 6 GHz, typical figure of merit values ranges from 19 dB/K using an FET transistor LNA providing a wide coverage area using a wide beamwidth to −3 dB/K for a pencil beam antenna. For the earth station, the main design criteria are G/T, antenna gain, system noise temperature, and transmitted power. The overall system design criteria are operation frequency bands, modulation methods, multiple access parameters, system costs, channel capacity, and overall system performance.

11.15 Summary

Satellites are now providing extensive coverage for communications and data link operations in remote areas. The satellite connection consists of a remote earth station, a satellite, and another earth station. This triangle forms a two-way communications link to provide the remote earth station with access to all types of communications, including data links, military operations, Internet, video, voice, and data at high data rates.

Satellite and ground systems use five bands: L, C, X, Ku, and Ka. The latter is becoming popular for both commercial and military sectors. A geostationary orbit is used so that the ground station tracks a fairly stationary transceiver and the satellite appears to be stationary.

A link budget is used to determine the power, gains, and losses in a system and also determines the figure of merit. Multiple access schemes are used to allow multiple users on the same band. Costs are associated with the type of system used and the length of use.

References

Bullock, S. R. "Use Geometry to Analyze Multipath Signals." *Microwaves & RF*, July 1993.

Crane, Robert K. *Electromagnetic Wave Propagation through Rain*. Wiley, NJ, 1996.

Haykin, Simon. *Communication Systems*, 5th ed. Wiley, NJ, 2009.

Holmes, Jack K. *Coherent Spread Spectrum Systems*. Wiley & Sons, NJ, 1982.

Inglis, Andrew F. *Electronic Communications Handbook*. McGraw-Hill, NY, 1988.

International Telecommunications Union. *Handbook on Satellite Communications*. Wiley, NJ, 2002.

Schwartz, Mischa. *Information, Transmission, Modulation and Noise*. McGraw-Hill, NY, 1990.

Tsui, James Bao-yen. *Microwave Receivers with Electronic Warfare Applications*. SciTech Publishing, NJ, 2005.

Problems

1. What system has the most ubiquitous infrastructure for communications systems?
2. What is the difference between a geostationary satellite and a geosynchronous satellite?
3. What phenomenon of a satellite orbit allows communication with the South Pole?

CHAPTER 12

Global Navigation Satellite Systems

The last few years have shown an increased interest in the commercialization of the global navigation satellite system (GNSS), which is often referred to as the global positioning system (GPS) in various applications. A GPS system uses spread spectrum signals—binary phase-shift keying (BPSK)—emitted from satellites in space for position and time determinations. Until recently, the use of GPS was essentially reserved for military use. Now there is great interest in using GPS for navigation of commercial aircraft. The U.S. Federal Aviation Administration (FAA) has implemented a wide area augmentation system (WAAS) for air navigation to cover the whole United States with one system. There are also applications in the automotive industry, surveying, and personal and recreational uses. Due to the increase in popularity of GPS, and since it is a spread spectrum communication system using BPSK a brief introduction is included in this text.

12.1 Satellite Transmissions

The NAVSTAR GPS satellite transmits a direct sequence BPSK signal at a rate of 1.023 Mbps using a code length of 1023 bits. The time between code repetition is 1 msec. This is known as the coarse acquisition (C/A) code, which is used by the military for acquisition of a much longer precision code (P-code). Commercial industries use the C/A code for the majority of their applications. There are 36 different C/A codes that are used with GPS, and they are generated by modulo-2 adding the C/A code with a different delayed version of the same C/A code (Figure 12-1). Therefore, there are 36 different time delays to generate the 36 different C/A codes. Approximately 30 codes are used presently for satellite operation; each is assigned a pseudo-random noise (PRN) code number for identification. The space vehicle identification (SVID) number is also used for identification.

Two frequencies are transmitted by the satellite. The C/A code uses only one of these frequencies, L1, for transmission. The frequency of L1 is 1.57542 GHz. Therefore, there are 1540 carrier cycles of L1 for each C/A chip. The frequency of L2 is 1.2276 GHz, and it does not contain the C/A code. The P-code, which is used by the military, is transmitted on both frequencies. The P-code is transmitted in quadrature (90°) with the C/A code on L1. The carrier phase noise for 10 Hz one-sided noise B is 0.1 radian root mean square (RMS) for both carriers, and the in-band spurious signals are less than −40 dBc.

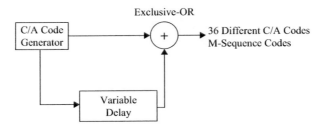

Figure 12-1   Generating 36 different C/A codes.

A summary chart is provided for both the C/A code and the P-code as follows:

- Coarse acquisition (C/A) code:

  BPSK spread spectrum modulation waveform
  1.023 Mbps chipping rate
  Code length = 1023 bits
  Code repetition rate = 1 msec
  C/A code 3 dB higher than P-code
  36 different C/A codes
  Sliding correlator demodulation
  Frequency L1 = 1.57542 GHz, 1540 cycles/chip

- Military uses precision code (P-code):

  BPSK spread spectrum modulation waveform
  10.23 Mbps chipping rate
  Code length approximately 1 week long
  Uses both L1 = 1.57542 GHz and L2 = 1.2276 GHz
  P-code transmitted in quadrature with C/A code on L1

The antenna used in the satellite for transmissions has a 3 dBi gain and is linear, right-hand circularly polarized. The group delay deviation for the transmitter is within ±3 nsec, 2σ.

12.2 Data Signal Structure

The satellite navigation data are sent out at a 50 bps rate. As a general rule, the information bandwidth should be at least 1 kHz. The data are sent out in 25 full frames with 5 subframes in each frame. Each subframe is 300 bits long (10 words at 30 bits each) providing frames that are 1500 (5 × 300) bits long. The total number of bits required to send out the total satellite information is 37.5 kbits (25 × 1500). Therefore, the time it takes to send out the total number of bits at a 50 bps rate is 12.5 minutes:

$$37.5 \text{ kbits}/50 \text{ bps} = 750 \text{ sec} = 12.5 \text{ min}$$

The following five subframes contain the data necessary for GPS operation:

- Subframe 1: SV clock corrections.
- Subframes 2 and 3: Complete SV ephemeris data. Ephemeris data contain such things as velocity, acceleration, and detailed orbit definitions for each satellite.

- Subframes 4 and 5: Subframes 4 and 5 accumulated for all 25 frames provide almanac data for 1 to 32 satellites. The almanac data are all the position data for all the satellites in orbit.

Note that the clock corrections and space vehicle (SV) or satellite ephemeris data are updated every 30 seconds.

Each of the subframes has an identification (ID) code 001,010,011,100,101 and also uses a parity check, which involves simply adding the number of "1"s sent. If an odd number is sent, the parity is a "1," and if an even number is sent, then the parity is a "0."

12.3 GPS Receiver

The standard GPS receiver is designed to receive the C/A coded signal with a signal input of −130 dBm. Since the C/A code amplitude is 3 dB higher than the P-code amplitude and the frequency of the C/A code is 1/10 of the P-code, then using a filter the receiver is able to detect the C/A code messages. Note that the C/A code also lags the P-code by 90° to keep the codes orthogonal and further separated. For most systems, the SV needs to be greater than 5° elevation to keep multipath and jamming signals at a minimum. Since the GPS signal is a continuous BPSK signal, a sliding correlator can be used to strip off the 1.023 MHz chipping signal, leaving the 50 Hz data rate signal.

12.3.1 GPS Process Gain

Although the received signal is very small, approximately −130 dBm, the data rate, bandwidth and the noise is small:

$$\text{Date rate} = 50 \text{ bps}$$
$$\text{BW} < 50 \text{ Hz}$$
$$\text{Noise} < -155 \text{ dBm}$$

The chip rate is equal to 1.023 Mcps for the C/A code and 10.23 Mcps for the P-code so that the process gain for each code is equal to

$$Gp = 10\log(1.023 \text{ mcps}/50 \text{ bps}) = 43.1 \text{ dB for C/A code}$$
$$Gp = 10\log(10.23 \text{ mcps}/50 \text{ bps}) = 53.1 \text{ dB for C/A code}$$

12.3.2 Positioning Calculations

The GPS position is calculated using three satellite signals, each of which is received by the GPS unit, which calculates the time difference corresponding to range, called a pseudo-range. For each time measured from the satellite to the GPS receiver, corresponding pseudo-range solutions generate a sphere around the satellite (Figure 12-2). Using two satellites and measuring the time or pseudo-range solutions for each satellite, these two spheres intersect producing pseudo-range solutions around a circle (Figure 12-2). Using three satellites, the three intersecting spheres produce two points that are the only two possible solutions. Since one of the solutions will not be located on the earth's surface, the other solution determines

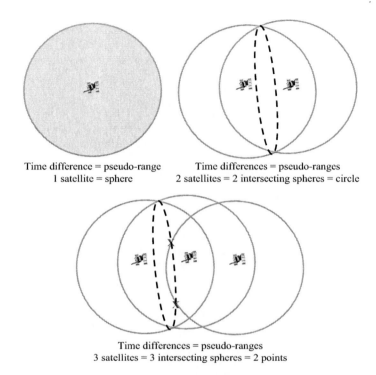

Time difference = pseudo-range
1 satellite = sphere

Time differences = pseudo-ranges
2 satellites = 2 intersecting spheres = circle

Time differences = pseudo-ranges
3 satellites = 3 intersecting spheres = 2 points

Figure 12-2   Intersecting spheres from each satellite produce the position solution.

the position of the user (Figure 12-2). The equation for the position of the user with three satellites is therefore

$$PR1 = \sqrt{(x1 - xp)^2 + (y1 - yp)^2 + (z1 + zp)^2}$$
$$PR2 = \sqrt{(x2 - xp)^2 + (y2 - yp)^2 + (z2 - zp)^2}$$
$$PR3 = \sqrt{(x3 - xp)^2 + (y3 - yp)^2 + (z3 - zp)^2}$$

With three equations and three unknowns, these equations solve the xp, yp, zp position of the user.

To eliminate a clock bias that exists between the satellites and the GPS user, another variable needs to be solved, and it requires one more satellite and one more equation (four equations and four unknowns):

$$PR1 = \sqrt{(x1 - xp)^2 + (y1 - yp)^2 + (z1 + zp)^2} + C_{bias}$$
$$PR2 = \sqrt{(x2 - xp)^2 + (y2 - yp)^2 + (z2 - zp)^2} + C_{bias}$$
$$PR3 = \sqrt{(x3 - xp)^2 + (y3 - yp)^2 + (z3 - zp)^2} + C_{bias}$$
$$PR4 = \sqrt{(x4 - xp)^2 + (y4 - yp)^2 + (z4 - zp)^2} + C_{bias}$$

Since all of the GPS satellites are synchronized with GPS master time and are also corrected as needed, the clock bias is assumed to be equal.

12.4 Atmospheric Errors

The atmospheric path loss for the satellite link is approximately 2 dB. The troposphere and ionosphere cause variable delays, distorting the time of arrival (TOA) and position (errors are dependent on atmosphere, angle, and time of day). These delays are extreme for low-elevation satellites. To compensate for the errors, ionosphere corrections are implemented using both frequencies (f_{L1}, f_{L2}):

$$PR = \frac{f_{L2}}{f_{L2}^2 - f_{L1}^2} (PR(f_{L2}) - PR(f_{L1}))$$

where

PR = compensated pseudo-range
f_{L2} = frequency of $L2$
f_{L1} = frequency of $L1$
$PR(f_{L2})$ = pseudo-range of $L2$
$PR(f_{L1})$ = pseudo-range of $L1$

If the pseudo-ranges of $L1$ and $L2$ are not known, then the measured and predicted graphs over time need to be used. This would be the case for a C/A code-only receiver. A simple C/A code receiver uses the ionospheric correction data sent by the satellites for a coarse ionosphere correction solution.

12.5 Multipath Errors

The angular accuracy for platform pointing is affected by multipath:

$$\sigma_\theta = \sigma_R / L$$

where

σ_θ = angular accuracy of platform pointing
σ_R = range difference caused by multipath
L = baseline

Good antenna design can reduce multipath to approximately 1 m. The receiver antenna is designed to reduce ground multipath by ensuring that the gain of the antenna is low toward the ground and other potential multipath sources compared to the gain toward the satellites of interest.

Using integrated Doppler (or carrier phase) can also improve the effects of multipath on the overall positioning solution to help smooth the code solution. Further discussion on smoothing the code solution with the integrated Doppler of the carrier follows later in this chapter.

Another way to reduce multipath errors is to first determine that the error is caused by multipath and then enter it as a state variable in a Kalman filter. This solution requires that the multipath can be measured accurately and is constant.

It is interesting to note that multipath errors can actually give you a closer range measurement. By definition, multipath arrives later than the direct path; however, the resultant autocorrelation peak can be distorted such that the tracking loop (e.g., early–late gate) produces a solution that is earlier than the direct path.

12.6 Narrow Correlator

Multipath can cause errors in range due to distortion of the autocorrelation peak in the tracking process. One way to reduce errors is to narrow the correlation peak, which is known as the narrow correlator. The early–late and tau-dither loops are generally operated on half-chips (i.e., a half-chip early and a half-chip late), which provides points on the cross-correlation peak halfway down from the top on both sides. By using less than half-chip dithering, the points are closer to the peak, reducing the ambiguity caused by multipath (Figure 12-3). The accuracy of the measurement is also greatly improved for the same reasons.

When using the narrow correlator, it is important to note that the precorrelation bandwidth is larger depending on the step size of the correlator. For example:

2 MHz is required for 1 C/A chip ($\pm$[0.5 chip]) standard correlator.
8 MHz is required for 0.1 C/A chip ($\pm$0.05 chip) narrow correlator.

The narrower the correlator, the wider the precorrelation bandwidth needs to be because the peak needs to be sharper; thus the input code needs to have sharper rise times to create this sharper peak. If the peak is rounded, then the narrow correlator does not improve performance because of the ambiguity of the samples. One of the drawbacks of the narrow

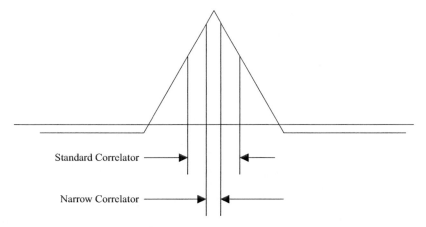

Figure 12-3   Comparison of the standard correlator versus the narrow correlator on the correlation peak.

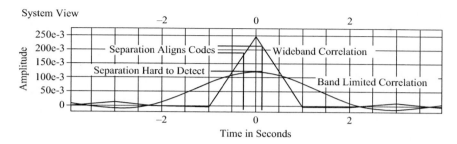

Figure 12-4   Limitations of the narrow correlator.

correlator is a larger precorrelation bandwidth (8 MHz compared with 2 MHz), which is also more vulnerable to jammers and noise, and a faster clock is required. However, in many precision navigation applications, this technology is preferred.

For P-code receivers, the P-code null-to-null bandwidth is 20 MHz. Therefore, for the narrow correlator to work, the bandwidth should be approximately 80 MHz, approximately four times, to ensure that the peak of the correlator is sharp. The problem is that the satellite filters this output signal using a bandwidth of approximately 20 MHz. The increased bandwidth in the receiver does not produce sharper peaks in the autocorrelation peak because the high-frequency components are filtered out at the satellite transmitter and therefore the correlation peak is rounded (Figure 12-4). This makes it hard to detect differences in the values that drive the locked conditions.

The locking mechanism that aligns the correlation peak is called a delay-locked loop (DLL). DLL processing can be accomplished by searching early and late (as shown in Figure 12-4) or by punctual and early-minus-late that generates the error off of the punctual. For low signal-to-noise ratios (SNRs), this may prove advantageous since the punctual signal is higher in amplitude than the early or late gate.

12.7 Selective Availability

Selective availability (SA) was developed by the U.S. Department of Defense (DOD) to make C/A code positioning less accurate. SA is caused by jittering or dithering the code clock and provides an improper description of the satellite orbit, that is, ephemeris data corruption. This clock dithering reduces the accuracy of the C/A code receiver since the method of how SA is generated is not available to the public. With SA turned on, the worst-case accuracy of a C/A code receiver is approximately 100 m, which also includes other error sources. SA was turned off in May 1, 2000, and is currently not a factor in the accuracy of the GPS C/A code receiver. This improves positioning accuracy from approximately 100 m to approximately 20 m for stand-alone GPS systems. The actual accuracy of the GPS C/A code receiver is dependent on several other parameters and conditions.

Note that many of the specifications written for accuracy are done with SA turned off. This provides a good baseline for comparison, but caution needs to be used when using the actual accuracy numbers if SA is ever turned back on again. The probability of SA being turned back on is remote since there are techniques to cancel SA for accurate C/A code receivers; these will be discussed later in this chapter.

12.8 Carrier Smoothed Code

The code solution by itself is noisy, particularly with the effects of multipath, SA, and atmosphere. This makes it difficult to obtain accurate measurements for the code-derived pseudo-ranges and introduces variations in the measurements. To aid in the code solution, the carrier is used to help smooth the variations in the code solution.

The carrier solution is a low-noise solution. It contains a wavelength ambiguity so that it does not know which of the wavelengths it is supposed to be comparing for the phase offset. For example, the difference in phase may be 10°, but it may be off several cycles or wavelengths—for example, 370°, 730°. By using the code solution that contains no wave-length ambiguity and the carrier phase solution, a carrier smoothed code can be produced. The phase change at selected repeatable points of the carrier is used to smooth the noise of the received code (Figure 12-5).

This method of measuring the carrier phase is referred to as integrated Doppler and is performed on each satellite. By integrating the Doppler frequency of the carrier $d\varphi/dt$, φ is obtained. This is the phase shift that changes with time, which is directly proportional to the change of range with time. Therefore, this phase plotted with time has the same slope as range plotted with time. The absolute range is not required, since the phase measurements are used only to smooth out the code solution. The low-noise carrier integrated Doppler plot (corresponding to range change with time) is better than the noisy code solution of the range, which is also changing with time. The Doppler is a changing frequency, and integrating a frequency change produces a phase change, which is plotted on the same graph as the range change with time (Figure 12-5).

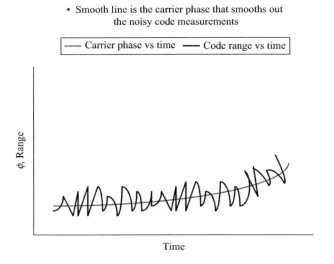

- Smooth line is the carrier phase that smooths out the noisy code measurements

| —— Carrier phase vs time   —— Code range vs time |

ϕ, Range

Time

Figure 12-5   Carrier smoothed code using integrated Doppler.

Another way to evaluate the integrated Doppler is to simply use the frequency times the period of integration:

$$\text{Integrated Doppler} = \int fdt = f * t$$

For example, if the Doppler frequency is 1 kHz and the time of integration is from 0 to 0.5 sec, then the equation becomes

$$\text{Integrated Doppler} = \int_0^{.5} 1000 dt = 1000 * t \Big|_0^{.5} = 1000 * 0.5 = 500 \text{ sec}$$

This is used to smooth out the code variations. For example, if the satellite is moving away from the receiver, the range (R) is positively increasing as the phase (φ) is negatively increasing. If the satellite is moving toward the receiver, the range is positively decreasing while the phase is positively increasing. Note that the Doppler is negative for a satellite going away, so the phase (φ) is negative yet the range is positive. A Kalman filter is often used to incorporate the integrated Doppler in the code solution. The Kalman filter is designed to achieve minimum variance and zero bias error. This means that the filter follows the mean of the time-varying signal with minimum variation to the mean. The Kalman filter uses the phase change to smooth the estimated range value of the solution as follows:

$$R_a = R_e - ID$$

where

R_a = actual range
R_e = measured range
ID = integrated Doppler

A filter could be used to smooth the noise of the code, but it would be difficult to know where to set the cutoff frequency to filter out the changes since this would vary with each of the satellites being tracked.

12.9 Differential GPS

Differential GPS (DGPS) uses a differencing scheme to reduce the common error in two different receivers. For example, if the GPS receivers are receiving approximately the same errors due to ionosphere, troposphere, and SA, these errors can be subtracted out in the solution. This would increase the accuracy tremendously, since these errors are the main contributors to reduced accuracy. As noted earlier, this technique was used to virtually eliminate the problems with SA, which has now been turned off. The main contributor to errors presently is the ionosphere.

DGPS uses a ground-based station with a known position. The ground station calculates the differential pseudo-range corrections, that is, how the pseudo-ranges are different from the surveyed position. These corrections are sent to other nonsurveyed GPS receivers, such as an aircraft or ground vehicle. These pseudo-range corrections are then applied to the aircraft's or

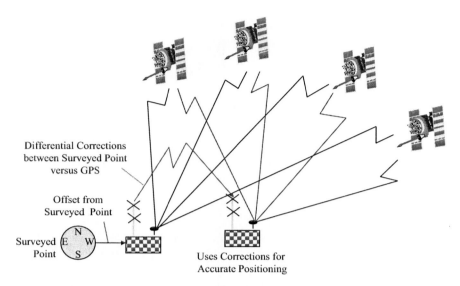

Figure 12-6   Differential GPS system for accurate positioning.

ground vehicle's pseudo-ranges for the same satellites (Figure 12-6). The accuracy obtained by differential means can produce a position solution that is approximately 100 times more accurate than an autonomous C/A code receiver: approximately 1 to 3 m.

12.10 DGPS Time Synchronization

The GPS time provided in the satellite navigation data is used to synchronize the receiver's time reference. Selective availability can affect the time reference. DGPS helps to reduce SA effects on both time and orbital information. It is assumed that the time stamping of the aircraft and ground station occur at the same time even though absolute time is varied by SA. Currently, with SA turned off, the time synchronization is not affected by SA.

12.11 Relative GPS

Relative GPS is the determination of a relative vector by one receiver given its own satellite measurements and also the raw pseudo-range measurements from the second receiver. The objective is to produce an accurate positioning solution between two systems relative to each other and not absolute positioning of each system.

For example, if an aircraft desires to land on an aircraft carrier using GPS, the aircraft carrier pseudo-ranges are uplinked to the aircraft. The aircraft calculates its relative position using its own measured pseudo-ranges with the pseudo-ranges that were sent by the aircraft carrier. The positioning errors that are in the pseudo-ranges from the aircraft carrier will be basically the same as the pseudo-ranges in the aircraft, so when the relative position solution is calculated, the errors are subtracted out. To enhance accuracy and integrity, the same set of satellites must be used so that the errors are common.

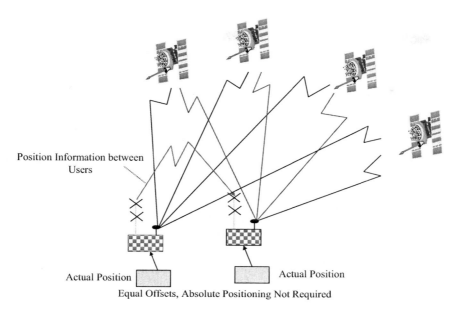

Figure 12-7 Relative GPS providing accurate relative positioning.

In a relative solution, atmospheric effects are assumed to be identical for both the ground, or aircraft carrier, and airborne units. Therefore, no tropospheric or ionospheric corrections are performed on the independent measurements. In a relative GPS system, the absolute accuracy depends on the accuracy of both the ground and airborne receivers. However, the absolute position does not need to be known, only the relative position with respect to each other.

Suppose that the GPS positioning solution for one GPS is off by 20 m toward the north. If the same satellites are used in the calculation of the GPS positioning solution for the other GPS, the assumption is that the other GPS would be off by 20 m toward the north. Therefore, both of their absolute positions would be off by 20 m toward the north. Since they are off by the same amount, their relative position would be accurate since the errors that caused them to be off are close to the same and can be subtracted out (Figure 12-7).

The accuracy for a relative GPS is equivalent to the accuracy for a DGPS, or approximately 100 times more accurate than standard GPS—approximately 1 to 3 m. Therefore, using relative GPS, what matters is not the absolute accuracy but only the relative accuracy between the aircraft and the aircraft carrier (Figure 12-8). Both relative GPS and DGPS use two different GPS receivers for accuracy. However, DGPS uses the difference from a surveyed point to calculate the errors needed to update the other GPS receiver, and the relative GPS uses the difference in the two GPS receivers to cancel the common errors.

12.12 Doppler

Doppler or range rate caused by the satellite moving toward or away from the GPS receiver needs to be determined so that it does not cause a problem in carrier phase tracking systems

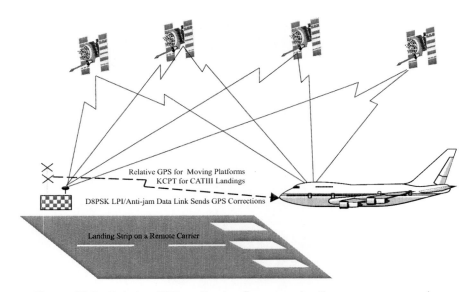

Figure 12-8   Relative GPS application for remote landing on a remote carrier.

or when the carrier is used to smooth the code. Note that in a DGPS or relative GPS system, this Doppler is corrected in the process. The Doppler frequency range is about ±5 kHz for the worst case. Doppler causes the number of cycles to be off when using the carrier phase approach. If Doppler is known, then the number of cycles is known but not λ ambiguities inherent with carrier phase tracking.

12.13 Kinematic Carrier Phase Tracking

Kinematic carrier phase tracking (KCPT) is used for very high accuracy positioning, less than 0.1 m. KCPT is most commonly used in survey applications but has recently been applied to high-accuracy commercial aircraft approach and landing systems. The KCPT solution uses the difference in phase of the carrier to determine range. The problem with the KCPT solution is the cycle ambiguity or wavelength ambiguity. The phase difference may be known, but the number of cycles for the range is not. But there are techniques to resolve the ambiguity of the number of cycles. If the receiver is moving, both the Doppler caused by the satellite and the Doppler due to receiver movement need to be considered.

One of the concerns with the KCPT solution after acquisition is what is known as cycle slip, especially when it is undetected. This is where the number of calculated cycles changes. If the cycle count slips, the range error is off by the number of cycles that have slipped. For L1, one cycle slip affects the pseudo-range by approximately 0.6 ft or 19 cm. The system needs to detect cycle slip to adjust the cycle number, maintain accuracy and continuity of function, and ensure the integrity and safety of the application.

There are several ways to detect cycle slip and correct the number of cycles, one of which is using a differential autonomous integrity monitor, where the ground observables are compared with the airborne observables for each satellite. Another way involves performing multiple solutions in parallel using the same raw measurement data and then comparing them. And other real-time tests considering bias and noise can be performed. Regardless of the method used to detect cycle slips, the ambiguities must be resolved again using some a priori data from the previous resolution.

12.14 Double Difference

The double difference is used to solve for the number of wavelengths to prevent cycle slip. The process involves taking the phase difference between satellites and subtracting the result from the baseline difference between antennas on the ground or air. Any two receivers can be used for the double difference. The main objective is to solve for N, the number of wavelengths:

$$\nabla\Delta = \Delta\varphi_1 - \Delta\varphi_2 = b(e_1 - e_2) + N\lambda$$

where

> b = baseline vector
> e_1, e_2 = unit vectors to satellites

Given the following absolute phase measurements

$$\varphi = \rho = d\rho = c(dt - dT) = \lambda V - d_{ion} = d_{trop} = \varepsilon\varphi$$

where

> ρ = geometric range
> $d\rho$ = orbital errors
> dt = satellite clock offset
> dT = receiver clock offset
> d_{ion} = ionosphere delay
> d_{trop} = troposphere delay
> ε = noise

There are two ways to achieve the double difference. One is to take the difference for one receiver and two satellites for first the carrier phase and then the code:

$$\varphi = \delta\rho = \delta d\rho + c\delta(dt - dT) + \lambda\delta V - \delta d_{ion} + \delta d_{trop} + \varepsilon\delta\varphi$$
$$\delta p = \delta\rho + \delta d\rho + c\delta(dt - dT) + \delta d_{ion} + \delta d_{trop} + \varepsilon\delta\varphi$$

The d_{ion} in the top equation is very small or zero in many cases if the ionospheric effects are the same for both the aircraft and the ground station. The second receiver does the same, and then the resultants are subtracted to achieve the double difference (Figure 12-9).

Another approach is to take the difference between the two receivers and one satellite and then between the two receivers and the next satellite. Finally, take the difference of the two resultants.

Note that the atmospheric losses are reduced or eliminated by receiver differences. For short antenna baselines, they are eliminated. For large baselines, they are reduced inversely proportional to the separation of the antennas.

$$\Delta^\nabla = (\phi_1 - \phi_2)_{Rec1} - (\phi_1 - \phi_2)_{Rec2}$$
$$\Delta^\nabla = \Delta\phi_1 - \Delta\phi_2 = b(e_1 - e_2) + N\lambda$$
$$\Delta\phi = \rho + \delta\rho + c(\delta t - \delta T) + \lambda N - d_{ion} + d_{trop} + \varepsilon\phi$$

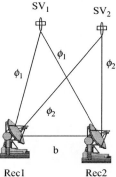

Figure 12-9   Double difference for kinematic carrier phase ambiguity resolution.

12.15 Wide Lane/Narrow Lane

Wide lane is used to reduce the cycle ambiguity inherent in the carrier phase tracking process. It uses the difference in the received frequencies L1 and L2 (L1 − L2), which produces a lower frequency (348 MHz) containing a larger wavelength, approximately 3.5 to 4.5 times greater (Figure 12-10). Therefore, the ambiguity search time is 3.5 to 4.5 times smaller. The lower frequency contains less cycle ambiguities because there are fewer cycles over the same distance. However, the disadvantage is that the accuracy is reduced.

Narrow lane uses the combination of L1 and L2 (L1 + L2), which provides more accuracy at the expense of more ambiguities to search over (Figure 12-10). A combination can be used: wide lane for ambiguity search and narrow lane for accuracy.

The frequencies and wavelengths are shown as follows:

$$L1 = 1.57542 \text{ GHz} = 19 \text{ cm}$$
$$L2 = 1.2276 \text{ GHz} = 24 \text{ cm}$$
$$L1 - L2 = 348 \text{ MHz} = 86 \text{ cm}$$
$$L1 + L2 = 2.8 \text{ GHz} = 11 \text{ cm}$$

12.16 Other Satellite Positioning Systems

Two other satellite positioning systems use their own satellites in space to provide positioning. Future satellite receivers are going to be capable of receiving these additional satellite positioning systems for better coverage worldwide.

The Global Navigation Satellite System developed by the Soviet Union, which is known as GLObal'naya NAvigatsionnaya Sputnikovaya Sistema (GLONASS), has been in operation since the mid-1990s. The system consists of 24 medium earth orbit satellites (MEOS): 21 are used for positioning, with 3 spares. However, the total number of operational satellites

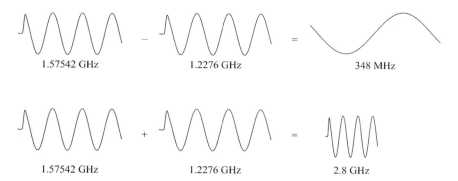

| 1.57542 GHz | | 1.2276 GHz | | 348 MHz |

Figure 12-10   Wide lane and narrow lane for GPS waveforms.

has varied over the years. The satellites are split into three inclined orbital planes separated by 120°, with an orbital time of 11 hours and 15 minutes. The GLONASS consists of a low-accuracy C/A code and a high-accuracy P-code for military use. The low-accuracy code is similar to GPS stand-alone C/A code with SA turned on, less than 100 m. The high-accuracy P-code provides accuracy of about 10 m. The GLONASS operates in two frequency bands— 1602.5625–1615.5 MHz and 1240–1260 MHz—using 25 channels separated by 0.5625 MHz and frequency division multiplexing for separation of users.

Galileo is a European satellite positioning system that is being developed and implemented. The first satellites are in orbit, and the plan is to have a total of 30 satellites in orbit. This satellite system shares the frequency bands with the United States and provides separation of users by code division multiplexing using different pseudo-noise codes for each satellite.

12.17 Summary

Global positioning system technology is being used for many applications including surveying, air traffic control and landing, position location for hikers, and mapping and location functions for automobiles. Originally developed for military applications, GPS has been adapted for commercial use using mainly C/A code receivers. Industry is now focusing more effort on using GPS signals in innovative ways that enhance the accuracy of measurements and therefore improve the availability and integrity of user services.

References

NavtechGPS. "Differential GPS." ION GPS-93 Tutorial. NavtechGPS, 1993.

NavtechGPS. "Dynamic Real Time Precise Positioning." ION GPS-93 Tutorial. NavtechGPS, 1993.

NavtechGPS. "Fundamentals of GPS." ION GPS-93 Tutorial. NavtechGPS, 1993.

Institute of Navigation. *Global Positioning System*. Institute of Navigation, 1980.

Problems

1. What is the null-to-null bandwidth of C/A code and P-code GPS signals?
2. What is the theoretical process gain of C/A code and P-code signals using a 50 Hz data rate?
3. What are the advantages of using C/A code over P-code?
4. What are the advantages of using P-code over C/A code?
5. What is the advantage of using the narrow correlator detection process?
6. Name at least two disadvantages of using the narrow correlator detection process.
7. Name the two harmful effects of SA on a GPS receiver.
8. What is carrier smoothing? Why is the code noisier than the carrier?
9. What is the main reason that DGPS is more accurate than standard GPS?
10. What is the main obstacle in providing a KCPT solution?
11. Why is the wide lane technique better for solving wavelength ambiguities?
12. How are multiple pseudo-noise codes produced for GPS systems?
13. How are common errors reduced in GPS systems to obtain more accurate results?

CHAPTER 13

Direction Finding and Interferometer Analysis

Direction finding is a method to determine the direction of a transmitted signal by using two antennas and measuring the phase difference between the antennas, as shown in Figure 13-1. This process is called interferometry. In addition to using a static interferometer, further analysis needs to be done to calculate the direction when the interferometer baseline is dynamic; that is, the interferometer is moving and rotating in a three-dimensional plane. Thus coordinate conversion processes need to be applied to the nonstabilized antenna baseline to provide accurate measurement of the direction in a three-dimensional plane.

13.1 Interferometer Analysis

For a nonstabilized antenna baseline, the elevation, roll, and pitch of the antenna baseline need to be included as part of the interferometer process for accurate azimuth determination using interferometer techniques. Therefore, the azimuth angle calculation involves a three-dimensional solution using coordinate conversions and direction cosines. The overall concept is that the coordinates are moved due to the movement of the structure. The direction phasor is calculated on the moved coordinates, and then a coordinate conversion is done to put the phasor on the absolute coordinates. This will become apparent in the following discussions. A brief discussion on direction cosines follows since they are the basis for the final solution.

13.2 Direction Cosines

Direction cosines provide a means of defining the direction for a given phasor in a three-dimensional plane. They are the key in defining and calculating the interferometer equations.

A phasor A is represented as

$$A = |A|(\cos\alpha i + \cos\beta j + \cos\gamma k)$$

where i, j, and k are unit vectors. For example, $i = (1,0,0), j = (0,1,0)$, and $k = (0,0,1)$ related to the magnitudes in x, y, z, respectively. For example, i is just for the x direction. Therefore, α

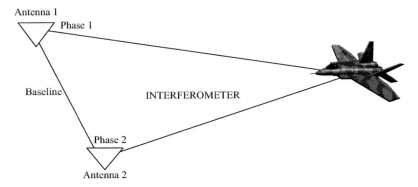

Figure 13-1   Basic interferometer used for direction finding.

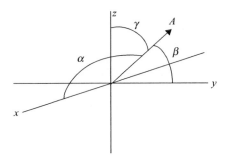

Figure 13-2   Angles for computing the direction cosines.

is the angle between phasor A and the x axis, β is the angle between phasor A and the y axis, and γ is the angle between phasor A and the z axis, as shown in Figure 13-2.

The direction cosines are the cosines of each of the angles specified: $\cos\alpha$, $\cos\beta$, and $\cos\gamma$. The cosines are equal to the adjacent side, which is the projection on the specified axis divided by the magnitude of phasor A. The projection on the x axis, for example, is simply the x component of A. It is sometimes written as $\mathrm{comp}_i A$. The vector A can be described by (x, y, z). The direction cosines are defined in the same manner, with x, y, and z replacing i, j, and k.

Another phasor is defined and is the horizontal baseline in the earth's plane. This is the baseline vector between two interferometer antennas. Therefore

$$B = |B|(\cos\alpha x + \cos\beta y + \cos\gamma z)$$

If $\gamma = 90°$, which means that the phasor is horizontal with no vertical component, then

$$A = |A|(\cos\alpha x + \cos\beta y + 0)$$

If the baseline is on the y-axis, then

$$B = |B|(\cos90x + \cos0y + \cos90z) = |B|(0x + 1y + 0z)$$

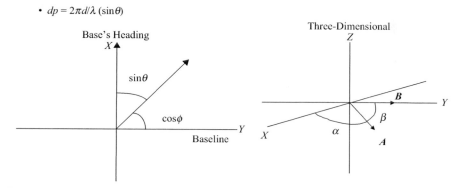

Figure 13-3 Standard Interferometer two-dimensional equation with baseline on the y-axis.

Therefore

$$A \cdot B = |A||B|\cos\beta$$

For this example, if the baseline interferometer is mounted on the y-axis, then α is the angle from the boresight, or the x-axis, to the baseline, which is the desired angle. This leads to the standard interferometer two-dimensional equation. For a two-dimensional case, this dot product equals $|A||B|\sin\alpha$, as shown in Figure 13-3.

13.3 Basic Interferometer Equation

The basic interferometer equation, used by many for approximate calculations, is only a two-dimensional solution:

$$dp = (2\pi d)/\lambda \sin\theta$$

where

dp = measured electrical phase difference (in radians)
d = separation of the interferometer antennas
θ = true azimuth angle
λ = wavelength

Note: θ is equal to α in the previous example.

This is a familiar equation used in most textbooks and in the direction finding literature. It is derived by simply taking the dot product of the direction phasor and the interferometer baseline, which produces the cosine of the angle from the baseline. Using simple geometry, the relationships between the additional distance traveled for one interferometer element (d_2) with respect to the baseline difference (d_1) and the measured electrical phase difference (d_p) are easily calculated:

$$d_2 = d_1\cos(\varphi)$$

where

φ = angle of the phasor from the baseline

and

$$d_p = (d_2/\lambda)2\pi$$

where

d_2 = extra distance traveled from the target

Figure 13-3 shows the analysis with the baseline on the y-axis.

Since the angle is usually specified from the boresight, which is perpendicular to the interferometer baseline, the equation uses the sine of the angle between the boresight and the phasor (θ), which results in the standard interferometer equation (Figure 13-3):

$$d_p = (2\pi d_1)/\lambda \sin\theta$$

This still assumes that the target is a long distance from the interferometer with respect to the distance between the interferometer elements, which is a good assumption in most cases.

This works fine for a two-dimensional analysis and gives an accurate azimuth angle with slight error for close-in targets, since $d\sin(\theta)$ is a geometrical estimate and A1 and A2 are assumed parallel, as shown in Figure 13-4.

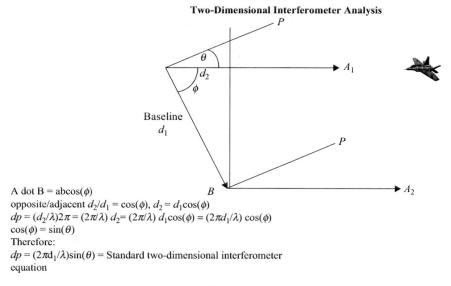

Two-Dimensional Interferometer Analysis

A dot B = abcos(ϕ)
opposite/adjacent d_2/d_1 = cos(ϕ), $d_2 = d_1$cos(ϕ)
$dp = (d_2/\lambda)2\pi = (2\pi/\lambda) d_2 = (2\pi/\lambda) d_1$cos($\phi$) = $(2\pi d_1/\lambda)$ cos(ϕ)
cos(ϕ) = sin(θ)
Therefore:
$dp = (2\pi d_1/\lambda)$sin(θ) = Standard two-dimensional interferometer equation

Figure 13-4   Two-dimensional interferometry.

13.4 Three-Dimensional Approach

If γ is not equal to 90°, then

$$A = |A|(\cos\alpha x + \cos\beta y + \cos\gamma z)$$
$$B = |B|(0x + 1y + 0z)$$

The dot product remains the same, $A \cdot B = |A||B| \cos\beta$; however, both the angles α and β change and the dot product no longer equals $|A||B| \sin\alpha$. For example, if γ is 0, α and β would be 90°. Therefore, for a given A with constant amplitude, the α and β change with γ.

If the alignment is off or there is pitch and roll, these angles change. Since the angle offsets are from the mounted baseline, then the B vector is defined with the angles offset from the B baseline, as shown in Figure 13-5. Therefore, the resultant equation is

$$B = |B|(\sin\alpha_1 x + \cos\beta_1 y + \sin\gamma_1 z)$$

where

$\alpha_1, \beta_1, \gamma_1$ = the angles from the B desired baseline caused by misalignment or movement.

Therefore, if the interferometer baseline is rotated, pitched, and rolled, then the results are

$$A = |A|(\cos\alpha x + \cos\beta y + \cos\gamma z)$$
$$B = |B|(\sin\alpha_1 x + \cos\beta_1 y + \cos\gamma_1 z)$$

This B is the new phasor, offset from the earth's coordinate system:

$$AB = |A||B|(\cos\alpha\sin\alpha_1 x + \cos\beta\cos\beta_1 y + \cos\gamma\sin\gamma_1 z)$$

The main solution to this problem is to find the angle from the boresight, which is a three-dimensional problem when considering the elevation angle and the dynamics of the baseline. Approaching the problem in three dimensions using direction cosines produces a very straightforward solution, and the results are shown in the following analysis. The standard interferometer equation is not used, and $\sin(\theta)$ is meaningless for this analysis since a two-dimensional dot product to resolve azimuth is not used. The effects that the elevation angle has on the azimuth solution are shown in Appendix C.

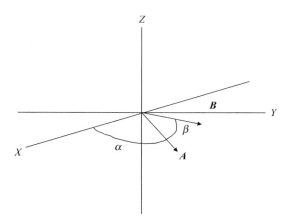

Figure 13-5   Interferometer analysis with the baseline not on the y-axis.

13.5 Antenna Position Matrix

The first part of the analysis is to define the three-dimensional interferometer coordinate system relative to the dynamic coordinate system. The definition of this axis is imperative for any analysis. For example, mounting an interferometer on a ship produces the following analysis using the y-coordinate from port to starboard (starboard is positive y) with the interferometer baseline along the y-axis, the x-coordinate from bow to stern (bow is positive x), and the z-coordinate up and down (up is positive z) (Figure 13-6). The position matrix for the interferometer on the ship's coordinate system is defined as $\{\sin(\alpha), \cos(\beta), \sin(\gamma)\}$.

If there is a misalignment in the x–y plane and not in the z-plane, the position matrix is $\{\sin(\alpha), \cos(\beta), 0\}$, as shown in Figure 13-7. If there are no offsets, then the position matrix

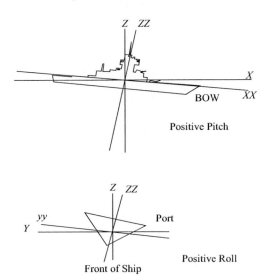

Figure 13-6   Roll and pitch definitions.

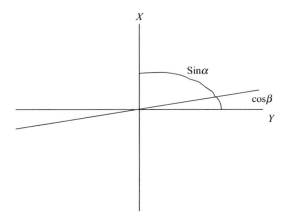

Figure 13-7   Position matrix offset.

is (0,1,0), which means that the interferometer is mounted along the y-axis. The antenna position matrix compensates for the misalignment of the interferometer baseline with respect to the ship's coordinate system.

13.6 Coordinate Conversion Due to Pitch and Roll

The coordinate conversion transformations modify the antenna position matrix to obtain the earth coordinate antenna position x,y,z. This is done for yaw*, pitch, and roll. Heave is generally insignificant but can be included for specific cases. Yaw is the movement in the horizontal plane that affects bearing, with positive yaw rotating in the clockwise direction. Pitch is the bow of the ship moving up and down, with down being in the positive direction. Roll is port and starboard movement of the ship rotating up and down, with port side down being positive. Heave is the movement of the ship as an entire unit in the vertical direction, with positive heave being up. The conversion matrices are shown in Appendix A. This example is done for 15° roll, 5° pitch, and 3° yaw. The order of the analysis is yaw, pitch, and roll. The order of the transformation needs to be specified, and different orders will alter the final solution. Further information on coordinate transformations is included in Appendix A.

The interferometer measurement is done with the ship's antenna position at the location caused by the yaw, pitch, and roll. To bring the coordinates back to earth or level coordinates, the process is done in the reverse order and opposite direction, that is, −roll, −pitch, −yaw*. The movement defined as positive pitch is bow down, and positive roll is port down. The negative numbers are put in the original coordinate transformations as shown in Figure 13-8.

Also, yaw can be solved after taking the roll and pitch azimuth calculation and simply offsetting the angle by the amount of the yaw. This may prove simpler for implementation.

$$
\begin{vmatrix}
\textbf{Roll Matrix} & & \\
1 & 0 & 0 \\
0 & \cos(r) & \sin(r) \\
0 & -\sin(r) & \cos(r)
\end{vmatrix}
=
\begin{vmatrix}
\textbf{Roll Matrix} & & \\
1 & 0 & 0 \\
0 & 0.966 & -0.259 \\
0 & 0.259 & 0.966
\end{vmatrix}
$$

$$
\begin{vmatrix}
\textbf{Pitch Matrix} & & \\
\cos(p) & 0 & -\sin(p) \\
0 & 1 & 0 \\
\sin(p) & 0 & \cos(p)
\end{vmatrix}
=
\begin{vmatrix}
\textbf{Pitch Matrix} & & \\
0.996 & 0 & 0.087 \\
0 & 1 & 0 \\
-0.087 & 0 & 0.996
\end{vmatrix}
$$

$$
\begin{vmatrix}
\textbf{Yaw Matrix} & & \\
\cos(y) & \sin(y) & 0 \\
-\sin(y) & \cos(y) & 0 \\
0 & 0 & 1
\end{vmatrix}
=
\begin{vmatrix}
\textbf{Yaw Matrix} & & \\
0.999 & -0.052 & 0 \\
0.052 & 0.999 & 0 \\
0 & 0 & 1
\end{vmatrix}
$$

Figure 13-8   Coordinate transformation.

13.7 Using Direction Cosines

Now that the coordinate conversion and position matrix have been solved, the x,y, and z values as a result of the these processes are used with the direction cosines to achieve the solution. The basic direction cosine equation, as mentioned earlier, is

$$a = |a|(\cos\alpha x + \cos\beta y + \cos\gamma z)$$

For a unit vector, the scalar components are the direction cosines:

$$u = \cos\alpha x + \cos\beta y + \cos\gamma z = Xx + Yy + Zz$$

The x,y, and z values are the coordinate converted and offset compensated x,y, and z. The X,Y, and Z values are the direction cosines of the Poynting vector for the horizontal baseline coordinate system. The unit vector is equal to the phase interferometer difference divided by the phase gain.

Since x,y, and z are known from previous analysis and u is measured and calculated, the direction cosines are the only unknowns and are solved by the following identities and equations:

$$(X^2 + Y^2 + Z^2)^{1/2} = 1, \text{ so } X^2 + Y^2 + Z^2 = 1$$

Note that $Z = \cos\gamma = \sin\Psi$, since the direction cosine is defined from the top down and Ψ is the angle from the horizontal up. The elevation angle (Ψ) is calculated using altitude and range. Also, the effects of the earth's surface can cause the elevation angle to have some slight amount of error. This can be compensated for in the calculation of the elevation angle and is shown in Appendix D. Therefore

$$X^2 + Y^2 + \sin^2\Psi = 1$$

and from above

$$u = Xx + Yy + \sin\Psi z$$

Solving for Y

$$Y = (u - Xx - \sin\Psi z)/y$$

Solving for X^2

$$X^2 = 1 - \sin^2\Psi - [(u - Xx - \sin\Psi z)/y]^2$$

Solving simultaneous equations produces a quadratic equation for X in the form $AX^2 + BX + C$, where

$$[1 + (x/y)^2]X^2 - [2x/y(u - \sin\Psi z)/y]X + \sin^2\Psi - 1 + [(u - \sin\Psi z)/y]^2 = 0$$

where

$A = [1 + (x/y)^2]$
$B = [2x/y(u - \sin\Psi z)/y] = [-2x/y^2(u - z\sin\Psi)]$
$u = $ electrical phase difference (in radians) (phase interferometer difference divided by the phase gain)
$C = \sin^2\Psi - 1 + [(u - z\sin\Psi)/y]^2$

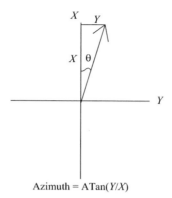

$$\text{Azimuth} = \text{ATan}(Y/X)$$

Figure 13-9   Baseline located on the y-axis.

Therefore, solving for X using the quadratic equation:

$$X = \frac{-B + \sqrt{B^2 - 4AC}}{2A}$$

and solving for Y

$$Y = \frac{u - xX - z\sin\psi}{y}$$

then the azimuth angle θ is

$$\theta = a\tan(Y/X)$$

The azimuth angle using the previous analysis is shown in Figure 13-9.

The direction cosines are the values in the x- and y-directions in the final converted coordinate system, $Xx + Yy + Zz$. The x, y, and z coordinates include the coordinate conversion constants.

An example showing the steps in producing the interferometer solution is found in Table 13-1. This is programmed into a spreadsheet for ease in changing parameters for different interferometer configurations. Other factors considered when determining azimuth are true north calculations and phase ambiguities, which are included in Appendix B.

13.8 Alternate Method

The previous analysis was done with the X-axis on the bow–stern axis and the Y-axis on the port–starboard axis. If another axis is used, then the equations need to be modified to reflect the correct axis. If the X-axis is on the port–starboard axis and the Y-axis on the bow–stern, then a right-hand coordinate system is defined. This analysis can be used with the following changes:

1. The quadratic equation for solving x needs to be

$$X = \frac{-B - \sqrt{B^2 - 4AC}}{2A}$$

Table 13-1 Interferometer analysis example

EXAMPLE: 15,5,3 DEGREES ROLL,PITCH,YAW, 45 DEGREES EL AT 24 DEGREES AZ.

x-axis on the bow-stern axis
y-axis on port-starboard axis

Start with Baseline

Roll = -15

Roll and Pitch Only:

	A	B	C
	1.0005454	-0.012517	-0.42818
	X	Y	
	0.6604613	0.2525685	

Az. Angle **20.927463** ATAN(Y/X)

Roll Matrix | Pos. Mat.

	Roll Matrix			Pos. Mat.	
x' =	0	0	0	0	x''
y' =	0.9659258	0.9659258	-0.258819	1	y''
z' =	0.258819	0.258819	0.9659258	0	z''

Pitch = -5

Pitch Matrix

	Pitch Matrix				
x =	0.0225576	0.9961947	0	0.0871557	x'
y =	0.9659258	0	1	0	y'
z =	0.2578342	-0.087156	0	0.9961947	z'

Roll, Pitch, Yaw:

	A	B	C
	1.0008421	0.015556	-0.428159
	X	Y	
	0.6463378	0.2867882	

Az. Angle **23.927463**

Note that the Yaw is simply an offset of 3 degrees.

Yaw = -3

Yaw Matrix

	Yaw Matrix				
xyaw =	-0.028026	0.9986295	-0.052336	0	x
yyaw =	0.9657826	0.052336	0.9986295	0	y
zyaw =	0.2578342	0	0	1	z

Error: 0.0725368

		Radians		Azimuth:	**Answer:**
Elevation Angle (deg.) =	45	0.7853982		24	**23.927463**
Electrical Phase Diff. (deg) =	635.295	25.277585	Elect. Phase/Phase Gain		
Electrical Phase Diff. (rad) =	11.087989	38.602861			
Phase Gain =	25.132741	0.4411771	Elec. Phase Diff. (rad)/Phase Gain		

For an error in alignment, so that the interferometer is not lined up with the axis, the following example is shown:

Alignment Error:	Inputs:	Calculations:	
Y =	0	Angle =	0
Dist.	1.312	x'' =	1
		y'' =	0

2. The azimuth angle is calculated as

$$\varphi = a\tan(X/Y)$$

Note: The formula for converting the differential phase in electrical degrees to angular degrees is

$$\text{Angular\_degrees} = (1/\text{phase\_gain}) * \text{differential\_phase}$$

where

angular_degrees = number of degrees in azimuth that the target is from the boresight and right of boresight is defined as positive
phase_gain = $(2\pi d)$/wavelength
differential_phase = difference in phase between the two antennas

This parameter is measured in electrical degrees.

13.9 Quaternions

Quaternions are used in place of coordinate conversions as discussed in this chapter. The advantages of using quaternions are to prevent ambiguities in the coordinate conversion process and to make it more efficient to implement. The coordinate conversion process compensates for the movement, roll, pitch, and yaw in three matrix multiplication around three axis. Quaternions compensate the roll, pitch, and yaw by rotating the position around one axis that is calculated in the process.

13.10 Summary

Interferometers use phase differencing to calculate the direction of the source of transmission. The basic mathematical tool used in interferometer calculations is the direction cosine. Direction cosines define the direction of the phasor in a three-dimensional plane. The basic interferometer equation deals with only two-dimensional analyses. The third dimension alters the two-dimensional solution significantly. This three-dimensional approach using coordinate conversion techniques to compensate for baseline rotations provides an accurate solution for the interferometer.

References

Halliday, David, Robert Resnick, and Jearl Walker. *Fundamental of Physics*, 9th ed. John Wiley & Sons, NJ, 2010.

Salas, Saturnino L., and Einar Hille. *Calculus: One of Several Variables*, 10th ed. John Wiley & Sons, NJ, 2007.

Shea, Don. "Notes on Interferometer Analysis." 1993.

Problems

1. What is the true azimuth angle for a two-dimensional interferometer given that the operational frequency is 1 GHz, the phase difference between the two antennas is 10 radians, and the antennas are separated by 3 m?
2. Why is a three-dimensional approach required for a typical interferometer calculation?
3. Why is the $\sin\theta$ used for the elevation angle calculation?
4. Is the order of the coordinate conversion process critical? Why?

Coordinate Conversions

The following are diagrams showing the coordinate conversions for motion including heading, roll, pitch, and yaw. These conversions are for the alternate method using the x-axis on the port–starboard plane and the y-axis on the bow–stern plane.

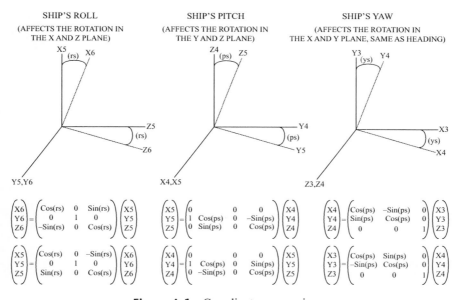

Figure A-1 Coordinate conversions.

True North Calculations

The formula to be used for correcting from true north is

bearing_true(n) = Relative_bearing(n) + ships_heading(n)

where

bearing_true(n) = The target bearing from the ship with respect to true north

ships_heading = The ship's heading as measured by the ship's inertial navigation system or a magnetic compass

Relative_bearing(n) = The current target bearing relative to ship's longitudinal axis starting at the bow for zero degrees relative and rotating clockwise.

The formula to be used for correcting from true north to magnetic north is

bearing_mag(n) = bearing_true(n) + magnetic_variation

where

bearing_mag(n) = The target bearing from the ship with respect to magnetic north

bearing_true(n) = The target bearing from the ship with respect to true north

magnetic_variation = The variation between magnitic north and true north. Magnetic variation may be plus or minus.

Phase Ambiguities:

For an interferometer to have no phase ambiguities, the spacing between the antennas should be less than $\lambda/(2\pi)$ wavelength apart. This provides a phase gain of less than 1. Note that a phase gain of exactly 1 gives a one-to-one conversion from azimuth angle to electrical angle. Therefore, there are no phase ambiguities. If the separation is greater than 1 wavelength, giving a greater than 1 phase gain, then ambiguities exist. For example, if the separation is 2λ, then half the circle covers 360° and then repeats for the second half of the circle. Therefore, a phase measurement of 40° could be two spatial positions.

Elevation Effects on Azimuth Error

The elevation effects on the azimuth error are geometric in nature and were evaluated to determine if they are needed in the azimuth determination and to calculate the magnitude and root mean square azimuth error. A simulation was done using three different angles 10°, 25°, and 45°. The simulation sweeps from 0° to 24° in azimuth angle, and the error is plotted in degrees. The results of the simulations are shown in Figure C-1.

The azimuth error is directly proportional to the elevation angle; the higher the angle the greater the error. Also, the azimuth error is directly proportional to the azimuth angle off interferometer boresight. This analysis was performed with a horizontal interferometer baseline with no pitch and roll. The azimuth error at an azimuth angle of 22.5° and an elevation angle of 45° was equal to 6.8°, which is the worst case error.

The azimuth error caused by elevation angle can be calculated by

$$
\begin{aligned}
\text{Az Error} &= \text{True Az} \\
&\quad - a\sin[\cos(\text{true el})\sin(\text{true az})/\cos(\text{assumed El angle})] \\
&= 22.5 - a\sin[\cos(45)\sin(22.5)/\cos(0)] = 6.8°
\end{aligned}
$$

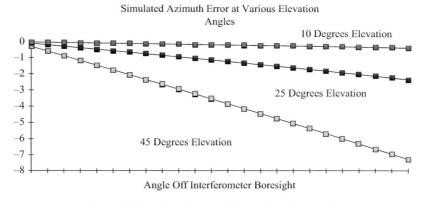

Figure C-1   Azimuth error due to elevation angle.

For elevation compensation only, an approximate solution can be used:

$$dp = (2\pi d)/\lambda \; \sin\theta \, \cos\Psi$$

where

$$\Psi = \text{elevation angle}$$

However, including roll and pitch produces a three-dimensional analysis.

Using the standard interferometer, two-dimensional equation, and trying to compensate for the elevation, roll, and pitch results in a very complex transcendental equation, without making too many assumptions, and is generally solved by iterative methods.

Earth's Radius Compensation for Elevation Angle Calculation

The angle (θ), which is the desired elevation angle, is calculated for the curved earth as shown below.

Solve for the angle α using the law of cosines:

$$c^2 = a^2 + b^2 - 2ab\cos(\alpha)$$

where

   $a = $ slant range
   $b = $ altitude of the ship plus the earth's radius
   $c = $ altitude of the aircraft plus the earth's radius

Therefore

$$\theta = \alpha - 90° = a\cos((a^2 + b^2 - c^2)/2ab) - 90°$$

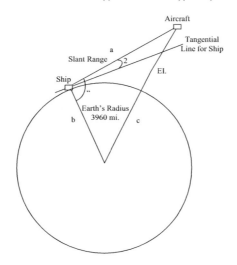

Figure D-1   Earth's radius compensation for elevation angle calculation.

Answers

Chapter 1:

1. Answer: 0 dBm is a power level of 1 milliwatt. 2 dBm is a power level of 1.58 milliwatts. 0 dBm $\pm$ 2 dBm = 1 mW $\pm$ 1.58 mW = 1 mW + 1.58 mW = 2.58 mW = 4.1 dBm and 1 mW $-$ 1.58 mW = $-$.58 mW can't have negative power.

   The correct expression is: 0 dBm $\pm$ 2 dB = $\pm$2 dBm: +2 dBm = 1.58 mW and -2 dBm = .63 mW.

2. Answer: 1 nW/3.16 * 316 / 6.3 * 31.6 /2 = 0.00025 mW. 10log(.00025) = $-$36 dBm. Or: -60 dBm $-$ 5 dB + 25 dB $-$ 8 dB + 15 dB $-$ 3 dB = $-$36 dBm

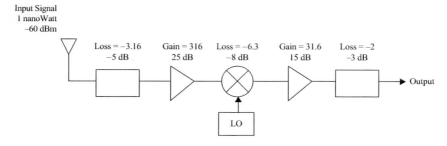

Figure P-2 Receiver diagram.

3. Answer: 10log1 mW = 0 dBm. 10log(.01 mW) = -20 dBm. Therefore, the spurious response is -20 dBm $-$ 0 dBm = -20 dBc. For -40 dBc, the spur level is -40 dBc + 0 dBm = -40 dBm = .1 μW.

4. Answer: The MDS will be reduced by 1.5 dB. Increase the transmitter power by 1.5 dB, reduce the losses after the power amplifier of the transmitter and before the LNA of the receiver by 1.5 dB, or reduce the noise figure of the LNA by 1.5 dB.

5. Answer: $G_t = 30$ dBi = $10\log[n(\pi D/\lambda)^2]$. $1000 = n(\pi D/\lambda)^2$. Therefore, $D = (1000/n)^{1/2} * \lambda/\pi = (1000/.5)^{1/2} * .06\text{m}/\pi = .854$ meters.

6. Answer: -131.8 dB of attenuation. $A_{fs} = 20\log[4\pi R/\lambda]$, 1 nautical mile = 1852 meters. 10 nmile = 18520 meters. Afs = $20\log(4\pi(18520)/.06) = 131.8$ dB attenuation.

7. Answer: $KT + 10\log(BW) + NF = -174$ dBm $+ 10\log(10$ MHz$) + 3$ dB $= -101$ dBm.
8. Answer: You must convert to actual power and use the noise factor equation and then convert the answer to dB. The results are as follows:

$F_t = F_1 + [(F_2*\text{Losses}) + -1]/G_1 = 1.995 + [(3.98 * 3.16 - 1)/100] = 2.11$:
Noise Figure $= 10\log(2.11) = 3.24$ dB.

Chapter 2:

1. The desired signal is $\cos(0.1$ MHz$)t$, and the LO is $\cos(10$ MHz$)t$. The transmitted signal is therefore

$$\cos(.1 \text{ MHz})t\cos(10 \text{ MHz})t = \frac{1}{2}\cos(10.1 \text{ MHz})t + \frac{1}{2}\cos(-9.9 \text{ MHz})t$$

The received signal using the worst case by multiplying by $\sin(10$ MHz$)t$ is

$$\left[\frac{1}{2}\cos(10.1 \text{ MHz})t + \frac{1}{2}\cos(-9.9 \text{ MHz})t\right][\sin(10 \text{ MHz})t]$$

$$= \frac{1}{4}[\sin(20.1 \text{ MHz})t + \sin(.1 \text{ MHz})t] - \frac{1}{4}[\sin(.1 \text{ MHz})t + \sin(-19.9 \text{ MHz})t]$$

$$= \frac{1}{4}[\sin(20.1 \text{ MHz})t - \sin(-19.9 \text{ MHz})t]$$

The desired signal frequency $\cos(.1$ MHz$)t$ is canceled out.

2. The transmitted filtered waveform is as follows:

$$\cos(.1 \text{ MHz})t\cos(10 \text{ MHz})t = \frac{1}{2}\cos(10.1 \text{ MHz})t + \frac{1}{2}\cos(-9.9 \text{ MHz})t$$

$$= \frac{1}{2}\cos(-9.9 \text{ MHz})t \quad \textit{for filtered output.}$$

The received signal using the worst case by multiplying by $\sin(10$ MHz$)t$ is

$$\frac{1}{2}[\cos(-9.9 \text{ MHz})t][\sin(10 \text{ MHz})t] = \frac{1}{4}\sin(19.9 \text{ MHz})t - \frac{1}{4}\sin(.1 \text{ MHz})t$$

By filtering the signal waveform, the desired signal $\sin(.1$ MHz$)$ is retrieved.

3. The phasors are shown for the summation of two quadrature phase-shift keying (QPSK) modulators to produce 8-PSK generator. The possible phases if the phasor diagram is rotated by 22.5° are 0°, 45°, 90°, 135°, 180°, 225°, 270°, and 315°. The phase between the two QPSK modulators is 45°.

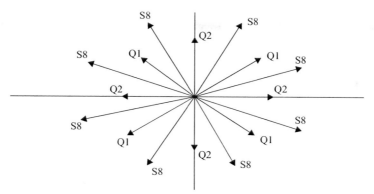

Q1 = the phasors of the first QPSK Generator
Q2 = the phasors of the second QPSK Generator offset in phase from the first by 45 degrees
S8 = the 8 resultant phasors to make up the 8-PSK modulator

4. By building two QPSK modulators, and then off-setting the modulators in time so that only one QPSK modulator switches phase at one given time. This prevents the 180° phase shift. The 180° phase shift is a problem because it produces unwanted AM on the waveform since the resultant phasor travels through 0 amplitude on the transition. Since a practical system cannot process an instantaneous signal, (infinite bandwidth), then AM is present.

5. The "1" is necessary since the output of the modulo-2 adder will be "0" and never change the pattern in the shift registers, all zeros are resultant and never change.

6. The process gain is equal to

$$10\log(1/\text{duty cycle}) = 10\log(1/2) = 6.99 \text{ dB}$$

7. Since the jammer can eliminate two frequencies at a time, only half of the frequency cells can be used for the process gain:

$$10\log(10) = 10 \text{ dB}$$

8. The only difference is that the 16-OQAM has two amplitude states for each phase shift producing four times as many resultant phasors.

9. The only difference is that the π/4DQPSK only uses four phase states ($\pm45°$, $\pm135°$), and D8PSK uses all eight phase states ($\pm45°$, $\pm135°$, $\pm90°$, $0°$, $180°$) on the phasor diagram.

10. Process gain is equal to the ratio of the spread bandwidth with respect to the data bandwidth needed to process the data. Process gain in dB needs to be 10 * log spread bandwidth/data bandwidth.

11. Implementation loss and required SJR for the desired probability of error.

12. CDMA has limited jamming margin. For multiple uses, when one user is close to the base unit and another user is far away, the jamming margin is not sufficient to handle that much difference in power, the close user would jam the far user. Power control is designed to adjust the powers of the users so that they are received with equal powers.

13. Digital data links can achieve perfect reconstruction of the digital waveform assuming no bit errors.

14. Differential systems are less complex and lower cost; however, the disadvantage of the differential system is that they experience more errors because they depend on the previous bit as a reference. If the previous bit is in error, then the next bit is in error.
15. MSK or CP-PSK experience minimum AM and also minimum spectral regrowth due to nonlinearities of the hardware.
16. MSK can be generated by sinusoidally weighting an OQPSK or can be generated by FSK with a minimum spacing for a given data rate.
17. 16-QAM can send more data for a given bandwidth; however, it needs more SNR than BPSK since the spacing of the phasors is less.
18. Some filters that are used to approach the ideal match filter are Gaussian, raised cosine, raised cosine squared, and root raised cosines. An ideal matched filter in the time domain produces a perfect square wave in the frequency domain.
19. The reasons to use spread spectrum are anti-jam, multiple users, multipath, and low probability of intercept. A reason to not use spread spectrum is it takes more SNR if there are no jammers present due to implementation and spreading losses. Also it is more complex.
20. Hedy Lamarr.
21. Phase shift keying or PSK of a continuous wave oscillator, generally the LO.
22. A shift register and an exclusive-OR function.
23. Code division multiple access (CDMA), time division multiple access (TDMA), and frequency division multiple access (FDMA).

Chapter 3:

1. A circulator or a T/R switch.
2. $G_r = 20\log[V_o/2^n] - MDSI = 20\log[1/2^8] - [-114 - 10\log10 + 3 + 4] = 68.8$ dB.
3. Third-order SFDR $= 2/3(IP_3 + 174 - NF - 10\log B) = 2/3(20 + 174 - 3 - 10\log(10$ MHz$)) = 80.66$ dB.
4. $80.66 - 10 = 70.66$ dB.
5. 3 dB + 10 dB = 13 dB.
6. 1X0 = 10 MHz
   0X1 = 12 MHz
   2X0 = 20 MHz
   0X2 = 24 MHz
   3X0 = 30 MHz
   0X3 = 36 MHz
   1X1 = 2 MHz, 22 MHz
   1X2 = 14 MHz, 34 MHz
   2X1 = 8 MHz, 32 MHz
7. 120 MHz/100 MHz = 1.2.
8. According to the Nyquist criteria, Sample rate = 2 * 1 MHz = 2 Msps.
9. Maximum phase error $= 45 - \tan^{-1}(1/2) = 18.43°$
   Maximum amplitude error = 6 dB
10. The advantages for oversampling the received signal are better resolution for determining the signal and better accuracy for determining the time of arrival. The disadvantages are generally more expensive and require more processing for the increased number of samples taken.

11. So that the quality of the time domain square pulse signal is preserved. The result of nonconstant group delay is distortion of the pulse which causes intersymbol interference (ISI).
12. LNA stands for low-noise amplifier. It is important because it is the main contributor to the noise figure of the receiver and is a direct correlation to the link budget. 1 dB higher noise figure results in 1dB lower link margin.
13. Aliasing occurs where the high frequencies can alias or appear as lower frequencies and distort the design signal.
14. Each ADC bit divides the voltage range by ½, and ½ the voltage is equivalent to 6 dB.

Chapter 4:

1. Voltage-controlled attenuator and a voltage-controlled amplifier.
2. The RC time constant should be much larger than the period of the carrier and much smaller than the period of any desired modulating signal. The period of the carrier is .1 μs, 1/10 MHz, and the desired modulating signal period is 1 μs, 1/1 MHz. Therefore, the period should be about .5 μs, 2 MHz. This will depend on how much distortion of the desired signal there is compared to how much of the carrier is allowed through the processing.
3. The integrator provides a zero steady state error for a step response.
4. As the error approaches steady state (very slow changing error), then the gain of the integrator approaches infinity. Therefore, any small change in error will be amplified by a very large gain and drives the error to zero.
5. An integrator.
6. If the slope is nonlinear, using a linear approximation causes the loop gain in some operational periods to be less or more than the approximation, therefore, the response time will be slower or faster, respectively.
7. The diode is nonlinear and is the point where the piecewise linear connections are. It provides a smoothing function due to the nonlinearity.
8. The integrator for the PLL is built in and does not need to be added, whereas the AGC needs to have an integrator added to the circuit.
9. They both are feedback systems. They just have different parameters that are in the feedback loop.
10. The PLL is in the lock state. This analysis does not include the capture state where the PLL has to search across a wider bandwidth to bring it into the lock state.
11. In all feedback systems, careful design needs to be done to prevent oscillations or instability.

Chapter 5:

1. Weights need to be time reversed and are $\pm$ 1 not 0. $\times 1 = 1, \times 2 = 1, \times 3 = -1$, $\times 4 = -1, \times 5 = 1, \times 6 = -1, \times 7 = 1$.
2. $[\cos(\omega t + (0,\pi))]^2 = \cos^2(\omega t + (0,\pi)) = 1/2[1 + \cos(2\omega t + 2(0,\pi))] = 1/2[1 + \cos(2\omega t + (0,2\pi))] = 1/2[1 + \cos(2\omega t + 0)]$. Therefore, the phase ambiguity is eliminated. However, the frequency needs to be divided in half to obtain the correct frequency.

3. Since the chipping rate is 50 Mchips/sec(Mcps), the minimum pulse width would be 1/50 Mcps. The null-to-null bandwidth $= 2/PW = 2/(1/50 \text{ Mcps}) = 2(50 \text{ Mcps}) = 100$ MHz.
4. (a)   The bandwidth is unchanged.
   (b)   The bandwidth is reduced by the process gain.
5. Intersymbol interference is the distortion caused by a chip or bit that interferes with the other chips or bits, usually with the adjacent chips or bits.
6. The results of the overlaying traces on an oscilloscope of a PN code.
7. The best place is the center of the eye. The worst place is transition point of the eye.
8. (a)   Signal would need to be squared three times.
   (b)   Signal would need to be squared four times.
9. (a)   45° for 8 PSK.
   (b)   22.5° for 16 PSK.
10. A two-frequency discriminator could detect the MSK waveform.
11. The MSK waveform sidelobe levels are significantly reduced. Also, sidelobes do not regenerate due to nonlinearities in the system.
12. A sliding correlator and a matched filter.
13. An early–late gate.
14. A squaring loop and a Costas loop.
15. SNR and bandwidth.

Chapter 6:

1. Since the integral of the probability density function is equal to $1 = 100\% = P_{wrong} + 37\%$; $P_{wrong} = 63\%$.
2. $E[x] = \Sigma x f_x(x) = 1(.4) + 2(.6) = 1.6$. The mean is equal to $E[x] = 1.6$.
3. Closer to 2 because there is a higher probability that the answer is going to be 2 than 1.
4. $E[x^2] = \Sigma x^2 f_x(x) = 1(.4) + 4(.6) = 2.8$.
5. $\text{Var} = E[x^2] - \text{mean}^2 = 2.8 - 2.56 = .24$.
6. $\text{Std Dev} = (\text{var})^{1/2} = (.24)^{1/2} = .49$.
7. $1 - .954 = .046 = 4.6\%$.
8. Increase the number of ADC bits.
9. The probability receiving 1 pulse is .98. The probability of receiving all 20 pulses is

$$.98^{20} = 66.8\%$$

10. Using the binomial distribution function the probability of only one error is

$$p(19) = \binom{20}{19} p^{19} (1-p)^{(20-19)} = \binom{20}{19} (.98)^{19} (.02)^{1}$$

$$= \frac{20!}{(20-19)!19!} (.98)^{19} (.02)^{1} = 26.7\%$$

Therefore, the percentage of the errors that result in only 1 pulse lost out of 20 is

$$\text{Percent (1 pulse lost)} = 66.8\%/(100\% - 26.7\%) = 48.9\%$$

11. The probability of error is a predicted value of what you should get given a Eb/No value, Bit error rate is the actual measurement of the system, number of errors of the total number of bits sent.
12. Parity, check sum, cyclic redundancy check CRC. CRC is the best.
13. Block codes and convolutional codes.
14. Since interference causes burst errors, interleaving disperses these errors over multiple messages so that fewer errors need to be corrected for a given message.

Chapter 7:

1. Glint errors are angle of arrival errors and scintillation errors are amplitude fluctuations.
2. Specular multipath affects the solution the most since it is more coherent with the signal and directly changes the amplitude and phase. Diffuse multipath is a constantly changing more random signal which looks more like noise and is usually smaller in amplitude.
3. The effect of multipath at the pseudo-Brewster angle for vertically polarized signal is reduced.
4. There is basically no effect of multipath at the pseudo-Brewster angle for horizontally polarized signal. Therefore, the multipath still highly affects the incoming signal.
5. Rayleigh criterion.
6. The Rayleigh criterion is as follows:

$$h_d \sin d < \frac{\lambda}{8}$$

where

h_d = peak variation in the height of the surface
d = grazing angle

If the Rayleigh criterion is met, then the multipath is a specular reflection on a rough surface. Therefore,

$$10\sin(10) = \frac{\lambda}{8}$$

$$\lambda = 80\sin(10) = 13.89 \text{ meters}$$

$$f = \frac{c}{\lambda} = \frac{3 \bullet 10^8}{13.89} = 21.6 \text{ MHz}$$

7. The divergence factor is the spreading factor caused by the curvature of the earth. This factor spreads out the reflecting surface. This is generally assumed to be unity since the effects are negligible for most applications, except for possibly satellites.

8. Leading edge tracking. The multipath returns are delayed from the desired return, so the radar detects the leading edge of the pulse and disregards the rest of the returns.

9. The vector addition is shown in the figure:

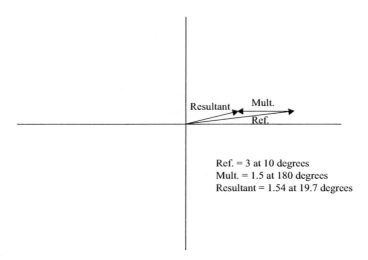

Ref. = 3 at 10 degrees
Mult. = 1.5 at 180 degrees
Resultant = 1.54 at 19.7 degrees

Figure P7-9 Vector addition showing multipath effects.

10. The power summation uses power instead of voltage. The reflection coefficient is squared to represent a power reflection coefficient.

11. Since multipath is spatial dependent, antenna diversity uses multiple antennas at different spatial positions. If one antenna is in a multipath null, there is a high probability that the other antennas are not and the signal is reliably received.

Chapter 8:

1. A pulse jammer duty cycle is equal to 1/response time of the AGC. Therefore, with a jammer present, the AGC adjusts the gain to minimum, then the jammer turns off and it takes the response time of the AGC to recover, and then the jammer turns on again. This captures the AGC and optimizes the jammer.

2. Approximately 1 MHz.

3. The FIR filter has fixed weight values for a given filter response, and the adaptive filter uses feedback to update or adjust the weight values.

4. The unwanted sidebands during the mix down and up need to be eliminated.

5. The μ value is the gain of the feedback process.

6. Increasing the μ value does the following:
   (a) Increases convergence time.
   (b) Generally decreases stability.
   (c) The steady state error is larger.

7. The assumption is good when the jammer is much larger than the signal, and the directional antenna has significant gain toward the desired signal and provides a reduction in J/S.
8. The assumption is bad when the jammer is not larger than the signal and the directional antenna does not provide a reduction in J/S.
9. Channelized receiver. This receiver covers the largest instantaneous bandwidth with the best sensitivity.
10. Adaptive filter.

Chapter 9:

1. Hostile jammers, friendly or cosite jammers, noise.
2. DSA, power control, modulation, adaptive filters, null steering, beam spoiling, multi-hop network.
3. DSA, switch to different frequency to avoid the jamming signal.
4. Base switches to a random frequency and the remote searches. Or they both know a random sequence and upon loss of signal switch to the next frequency in the sequence.
5. Using RSS received with feedback. Navigational data using range.
6. Use basic control theory.
7. Only one, generally the base.
8. Hysteresis.
9. An integrator in the feedback loop.
10. Adaptive filter.
11. Utilize beam steering to steer the beam away from the jammer or narrow the beam so that the jammer is not in the beamwidth.
12. Higher data rates or provide a more robust signal.
13. Uses two antennas for the receiver, separated by a distance such that a multipath null affects only one of the antenna and not the other.
14. Point the beam in the direction of the multipath and away from the direct path, and use the multipath as the desired signal path.
15. Multi-hops provide extended range and also provides a different path for communication if the direct path is being jammed.
16. Takes into account all of the capabilities of the system, not just the radio, and can provide trade-offs and optimal system solutions.

Chapter 10:

1. Power line, phone line, and RF.
2. Orthogonal frequency division multiplexing (OFDM).
3. The inner product is equal to zero.
4. The inner product of the signal with itself is equal to one. For orthogonal signals the inner product is zero. Therefore, taking the inner product with a duplicate of the known signal will produce only the desired signal.

5. A phone line needs to be present and not all homes have phone lines going to every room in the house.
6. Interoperability, anti-jam, and security.
7. SDRs can be programmed on the fly to produce different waveforms and modulation schemes so they can be programmed for different users.
8. Star, bus, ring, and mesh.

Chapter 11:

1. Satellite system is the most ubiquitous infrastructure since it covers the globe.
2. Geosynchronous satellites are synchronous with the earth's rotation, and geostationary satellites are geosynchronous satellites with their orbits on the equatorial plane. Therefore, they appear as if they are stationary with respect to the earth. Geosynchronous satellites usually have inclined orbits so they move around with respect to the earth.
3. The figure-eight pattern around the orbital plane that increases with age of the satellites. This phenomenon allowed a point on the South Pole to see the satellites for a period of time at the bottom of the figure-eight pattern for communications.

Chapter 12:

1. C/A code = 2 × 1.023 MHz = 2.046 MHz
   P-code = 2 × 10.23 MHz = 20.46 MHz.
2. Theoretical process gain for C/A code using the data rate of 50 Hz = 10 log 1.023 MHz/ 50 = 43 dB.
   Theoretical process gain for P code using the data rate of 50 Hz = 10 log 10.23 MHz/ 50 = 53 dB.
3. Short length code for faster acquisition times. Slower rate code for a higher signal to noise in a smaller bandwidth.
4. Long length code providing a more covert signal for detection. Faster rate code for higher process gain against unwanted signals.
5. The narrow correlator is better accuracy of the GPS solution.
6. The narrow correlator is less stable, is easier to jam due to the wider bandwidth required to process the signal, and provides no benefit using a P code receiver due to the fact that the bandwidth is already limited in the transmitter.
7. Jitters the clock timing and distorts ephemeris data regarding the orbits of the SVs.
8. Carrier smoothing is using the carrier change of phase with time to filter the code measured range data with time.
   Carrier data is not affected by the SA, multipath, and ionospheric effects as much as the code is affected.
9. Common errors, for example ionospheric errors, exist in both receivers and are subtracted out in the solution.
10. Wavelength ambiguity and cycle slips.

11. The wavelength of the difference frequency is larger; therefore, there are fewer wavelength ambiguities to search over.
12. Use one code and exclusive-OR with different delayed versions of that same code. This generates multiple codes dependent on the delay.
13. Differential and relative GPS techniques.

Chapter 13:

1. $\lambda = 3 \times 10^8/1 \times 10^9 = .3$ meters
   $\theta = a\sin[dp/((2\pi d)/\lambda)] = a\sin[10/((2\pi 3)/.3)] = 9.16$ degrees.
2. The elevation angle affects the azimuth interferometer calculation.
3. The direction cosines are defined for the $\cos(\alpha)$ from the top down. Therefore, the angle up is the $\sin\theta$, where θ is the elevation angle from the horizontal baseline.
4. The order is critical. There is a difference in the solution for the order of the coordinate conversion. For example, if the platform is pitched and then rolled, this gives a different solution from if the platform is rolled and then pitched.

Index